TOWARD A HISTORY OF GEOLOGY

TOWARD A HISTORY

THE M.I.T. PRESS

Proceedings of the
New Hampshire Inter-Disciplinary Conference
on the History of Geology
September 7–12, 1967

OF GEOLOGY

Cecil J. Schneer, Editor

Cambridge, Massachusetts, and London, England

Designed by Dwight E. Agner. Set in Linotype Janson, and printed and bound in the United States of America by The Riverside Press, Inc.

SBN 262 19058 3 (hardcover)

Library of Congress catalog card number: 76-87306

Contents

Introduction

Cecil J. Schneer

With the publication of Darwin's *Origin of Species* in 1859, the triumph of the geological world view was assured. For the next half century, this science and the historical and evolutionary frame of reference to which it had given rise were firmly enthroned in the world mind. Not since Newton's union of mechanics and astronomy in the seventeenth century had there been a comparable conquest of the minds of men. The parallel strikes deep. For a century after Newton, the brilliance of the *Principia* was the despair of physics. For a century after Darwin, major geological concepts were to win acceptance on the strength of the inspiration that was the theory of evolution.

For three thousand years or more, the question of the origin of things, the search for a history, has shared our attention with the Judaic problem of the purpose of existence. As in the crust of the earth itself, so in the history of our ideas about the nature, causes, origins, of this present configuration of the earth, we can find "no trace of a beginning, no prospect of an end." For the ancients, Cosmos was an organism — intelligent, conscious, orderly; the earth was a living animal growing minerals and fossils within, just as it bred grass and tree without. The seeds of our present concepts, the kinds of solutions to the problems of the origins of the crust, the very limits to which pure reason might penetrate, were explored by the ancients; fire and ice, flood and catastrophic earthquake — the earth begun by a gathering of the waters, the great cycles of the years, the central fire.

Anaximander, in the seventh century B.C., seems proto-Darwinian in brilliant observations of parallels between phylogeny and ontogeny. The great questions were raised: What is nature? What is the source of the confusion and change that we see around us, and how are we to reconcile this with the concept of eternal and absolute order? Collingwood[1] says that there was a curious turn to the Ionian mind (and therefore to our Graeco-Hebraic civilization) that turned this to the question, "What is the world made of?" The Alexandrians (Strato, Strabo) proposed the equally significant questions of quantity: How large is the earth?

These ideas matured slowly in the centuries that saw the decline of classical civilization and the flowering of Islamic culture. Geological and mineralogical ideas of the tenth century Tadzhik, Avicenna, and the Uzbek, al Biruni, reflect the blending of Eastern and Western cultures in the oasis cities of the old silk route, the historic line of the migrations of peoples. A brief discussion of this Central Asian science in V. V. Tikhomirov's paper alerts us to the exhaustive scholarly studies under way at the Oriental Institute of Tashkent.

In Islamic civilization, we find the seeds of classical thought carefully nurtured. Aristotle's concept of *minima,* or the smallest parts into which a heterogeneous substance such as flesh may be divided and still remain flesh, was rendered quantitative by Averroes (1126–1198), the Cadi of Seville, who observed that the increase or decrease of a substance must involve a change in the number of these *minima,*[2] in this way bringing Aristotelian synthesis and the atomistic concepts of Democritus closer — a stage on the path to atomism and crystallography. The Arabic element of practicality, the manufactures, the trade, brought about the fusion with empiricism that differentiates modern science from ancient thought.

Cyril Smith's paper treats of the much later attempts in the eighteenth century to imitate the superb porcelain of the Chinese (first introduced about the tenth century in the Thang dynasty) and the experimentation that was to demonstrate vulcanist ideas. The earliest use of minerals and metals was in jewelry and ornament. Drives that can only be described as aesthetic led Paleolithic men to collect brightly colored, or lustrous, or geometrically figured gems. In the Neolithic bronzes of the Shang or the even earlier figurines of Urartu in the mountains of Armenia, we see practical, religious, and acquisitive impulses given common catharsis by art.

The union, even identity, of science and aesthetics (philosophy) in the Renaissance is to be seen in the notebooks of Leonardo da Vinci (1452–1519). However, the tradition of the notebooks of the artist-

[1] R. G. Collingwood, *The Idea Of Nature* (New York: Oxford, 1960).

[2] A. G. Van Melsen, *From Atomos to Atom,* trans. H. J. Koren (Pittsburgh: Duquesne, 1952).

architect-engineer is more ancient than Leonardo; we have those of Villard d'Honecourt (c. 1250), for example. Leonardo as artist apprehended the world visually. As artisan and craftsman, as well as anatomist and dissectionist, his perceptions were kinesthetic.[3] The artist who had himself spent long hours supervising the imperfect casting of the bronze statue of the Sforza must have understood at once the true origin of the fossils that the Sforza's workmen uncovered in canal excavations. Leonardo made oblique aerial sketches — bird's-eye views, he called them — as military maps. I submit that these were geomorphic studies and that only the combination of visual and kinesthetic perception, developed to the highest degree in this remarkable man, would have led to this understanding of topography and geological structure. The man whose sketches and paintings reveal profound anatomical insight gleaned from multitudinous dissections left not only anatomical sketches, but sketches of the structure of the folded rocks of Tuscany and Lombardy, which could only have been gleaned from the closest attention to the dissections of the earth carried out by the Italian canal builders and engineers. When Leonardo applied his incisive yet curiously neo-Platonic mind to these observations, the result was an actualism not to be surpassed until the appearance of Lyell.

The artist who found models for the Virgin and Child among the peasant girls of Tuscany and their infants had no difficulty in seeing in the living fauna of the Italian coast the models for the fossils in the rocks. The same intelligence found laws of perspective, looked for naturalism in all phenomena, saw the origins of valleys in the action of their present streams, connected the plains of Lombardy with the delta of the Po, and grasped the connection between rainfall and stream flow. At the same time, Leonardo, as mystic and neo-Platonist, could see in the flow of ground water analogy to an imperfectly conceived circulation of the blood.

In Professor La Rocque's study of the potter-heretic Bernard Palissy, we see again the fusion of the aesthetic impulse with the kinesthetic apprehension of the craftsman. Palissy, a man who understood processes of petrifaction by virtue of his own manufacturing of stoneware, is the prototype for the experimental geologist of today. The ceramic arts date back to the Neolithic. The relationship of these to

[3] Professor Heymann distinguishes two approaches to the Franco-Cantabrigian cave art, one based on primarily visual perception, the other with a "haptic-kinesthetically" perceived concept of reality, with parallels to modern engineering drawing. See R. E. Heymann "An Approach to Early Art from Technical Drawing," *Proceedings of the Tenth International Congress of the History of Science, Ithaca, 1962* (Paris: Hermann, 1964), p. 373.

geology and crystallography required the neo-Platonic preconceptions of sixteenth-century humanism. By the eighteenth century, the high level of activity and achievement described in Cyril Smith's study warrants the application of the term *experimental science* to ceramics, to metallurgy, and to crystallography.

The technology of silicate fusion and the science (one hesitates to employ the term *ideology*) of crystallinity are ready for the experimentation of Sir James Hall and a new level of sophistication in the basalt controversy.

Dr. Debus' account of the English physician Edward Jorden makes clear the fusion that was achieved by the Renaissance iatrochemists. As man, the microcosm, mirrored in his person the great world, the macrocosm, so this generation explained geological processes in biological-chemical terms. The earth, Johann Kepler wrote in explaining the forms of crystals, is learned in the whole of geometry.

> Kepler supposed, or pretended to suppose, that the earth contained a circulating vital fluid, and was possessed of living powers . . . and that a process of assimilation goes on in it as well as in other animals. Every particle of matter, according to him, is alive, and possesses volition and instinct. Each kind of mineral substance is capable of converting masses of other matter into its own peculiar kind, as animals convert their aliment into blood. The burning mountains are the respiratory organs of the globe; and the slates are the organs of secretion, as the glands are those of the animal. The slates decompose the waters of the ocean, in order to prepare its elements to produce earthquakes and volcanic eruptions. The metallic veins in the strata of the earth are caries or abscesses of the mineral kingdom, and the metals themselves are the products of decay and disease, and hence the offensive odour of some of these products.[4]

These views, which Comstock in 1834 found so "void of common sense" as to suspect Kepler of mocking his contemporaries, place the great cosmologist squarely in the tradition of Paracelsus, who placed a living spirit, an *archeus*, in every particle of matter, and van Helmont, whose exploding furnaces testified to the *spiritus sylvestre*, the chaos, the *gas*, and the beginnings of pneumatic chemistry. In the seventeenth century, the ancient Milesian concept of a single unifying substrate to existence stood in opposition to Descartes' separation of the organic from the inorganic world.

Just as Comstock had in 1834, Adams, writing in 1936, as quoted in the first note to Professor Kelly's study, could find little but amuse-

[4] J. L. Comstock, *Outline of Geology* (Hartford: Robinson, 1834), p. 14. Kepler, J., *Strena Seu de Nive Sexangula* (Frankfort: Tampach, 1611), p. 19.

ment in the speculations of Renaissance theorists. But the unprejudiced reader impartially examining modern accounts of the origins of mountains, the drift of continents, the segregation of metals and the origins of veins, will see that we are not closer to the solutions of certain fundamental problems.

Professor Kelly's paper deals with the geological content of Renaissance cosmology antecedent to the great theoretical systems of the scientific revolution. Dr. Debus' iatrochemists are concerned with the detailed explanation in Paracelsan terms of the processes of the earth, of the differentiation and segregation of minerals and veins. The protogeological speculations of Sr. Kelly's Platonists are subordinate to their concern with cosmology, seen as a branch of theology and an outgrowth of scholastic philosophy. We note with particular attention the division between static or eternal and continuous concepts of existence, such as those of Cuse and Bruno, and temporal, evolutionary concepts, such as those of Genesis, Paracelsus, and Carpenter. Genesis begins with the spirit of God moving over universal waters. A water-encompassed earth is also the starting point for Paracelsus and Carpenter. Were these simply the fruits of a priori reasoning, or is it possible that the unconscious apprehension of sedimentary processes and marine deposits prepared the way for Neptunist geology?

In another paper,[5] I discussed the rise of historical geology in the seventeenth century. The reconstruction of the detailed chronology-process resulting in the present configuration of the earth was proposed by Robert Hooke in lectures at Gresham College and at meetings of the Royal Society of London during the last quarter of the seventeenth century. Although the fossil controversy (organic versus either fortuitous or mechanical-chemical origin of fossils) and the diluvialist controversy (attribution of numerous geological phenomena to the biblical deluge) continued to rage for at least another century, the idea of earth history through Baconian methods, that is, by inference from observation and biological, physical, and chemical analysis, was established. Hooke clearly saw fossils as evidences of the time and conditions of deposition of the enclosing strata, referring to rock strata as analogous to the pages of "civil history" and the fossils as hieroglyphics to be deciphered. The concept of the geological map (Lister), the description and classification of fossils (Lister, Steno, Lhuwyd), the laws of the constancy of the interfacial angles of crystals (Hooke), and

5 C. J. Schneer, "The Rise of Historical Geology in the Seventeenth Century," *Isis*, *45* (1954), 266–267.

sedimentation and geological structure (Steno) may all be traced to this period.

Hooke wrote in his *Lectures on Earthquakes* (published posthumously in 1705):

The doctrine aimed at, is, the Cause and Reason of the present Figure, Shape and Constitution of the Surface of this Body of the Earth. . . . Now, because when we look into Natural Histories of past Times, we find very few, if any, Footsteps of what alterations or transactions of this Nature have been performed, we must be fain to make use of other helps than what Natural Historians will furnish us with, to make out an account of the History thereof: Nor are there any Monuments or Medals with Literal, Graphical, or Hieroglyphical Inscriptions that will help us out in this our Inquiry, by which the writers of Civil Histories have of late Years been much assisted from the great curiosity of modern Travellers and Collectors of such curiosities. . . .

If in digging a Mine, or the like, an artificial Coin or Urne, or the like Substance be found, no one scruples to affirm it to be of this or that Metal or Earth he finds them by trial to be of: Nor that they are Roman, Saxon, Norman, or the like, according to the Relievo, Impression, Characters, or Form they find them of. Now these Shells and other Bodies are the Medals, Urnes, or Monuments of Nature whose Relievoes, Impressions, Characters, Forms, Substances, &c. are much more plain and discoverable. . . . to correct natural Chronology . . . nor will there be wanting *Media* or *Criteria* of Chronology. . . .

Hooke himself considered that Steno's *Prodromus* was derived from his *Lectures on Earthquakes*, charging the English translator of the *Prodromus*, Henry Oldenburg, with plagiarism.[6] A full century later, Buffon wrote in the introduction to *Epochs of Nature* (1779):

As in Civil history, one consults titles, searches for medals, deciphers antique inscriptions, to determine the epochs of human revolutions and to fix the date of moral events: so in natural history it is necessary to plunder the archives of the world, to draw old monuments from the entrails of the earth, to bring together their debris and to reassemble in a body of proofs, all the indices of the physical changes which can enable us to attain the different ages of nature . . .[7]

It is the phraseology of Bergman in 1766, quoted in the paper by Dr. Hedberg. By 1812, the simile of Hooke's had become geological commonplace, and Cuvier could begin his *Discourse* by describing himself as "Antiquarian of a new species, I had at the same time to learn to restore the monuments of past revolutions and to decipher their meaning. . . ."[8] In 1844, Gideon Mantell titled his popular work on geology *Medals of Creation*, a title that he attributed to Knorr.

[6] C. J. Schneer, "Kepler's New Year's Gift of a Snowflake," *Isis, 51* (1960), 540.
[7] G. Buffon, *Epoques de la Nature* (1779). In G. Laurent, *Les Grands Ecrivains Scientifiques* (Paris: Colin, 1919), p. 102.
[8] G. Cuvier, *Discours sur les Révolutions du Globe, 1812* (Paris: Berche et Tralin, 1881).

Credit for the first attempt at a secular account of the origins of the earth since classical times should perhaps go to Descartes and his cosmological hypothesis of vortices. Descartes accounted for the origins of the solar system by a series of "vortices," with the earth as a star "differing from the sun only in being smaller," collecting dense and dark matter by what we would designate gravitational attraction and losing (energy) as it falls into place by a process of gaseous condensation. Descartes' full theory was to be published under the title *Le Monde*, but on his learning of the condemnation of Galileo, it was withdrawn. We have only a fragmentary knowledge from portions included in the *Principles* (1644) and in other works published posthumously.

"It is, perhaps, universally admitted, that the *fluid agent* employed in the formation of minerals, must have been either water, or caloric,"[9] Parker Cleaveland wrote in 1817 from Bowdoin College in Maine. Leibniz — whose cosmology rested on a base of percipient monads and therefore fundamentally challenged the Cartesian dualism of spiritual and organic versus lifeless and inorganic entities — proposed a theoretical history to the earth in the *Protogaea*, written about 1680 but not published until 1749.

According to Lyell (1830), Leibniz proposed that an originally fiery earth cooled from the outside in forming a stony and irregular crust. The condensation of vapors on this crust produced a vast primordial sea over the highest mountains, in which the sedimentary strata were deposited. In the next sequence, the sea subsided, breaking into vast subterranean caverns, producing the disarrangement of strata that Leibniz said Steno had described. But it was Georges Louis Leclerc, Marquis de Buffon, whose *Epochs of Nature* was the most fully developed of those theories of the earth that began with the earth as a cooling star.

Buffon, too, began with a molten earth (first epoch), cooling from the outside in (second epoch), and included in a third epoch, the condensation of vapors into the primordial sea.[10] Descartes, Leibniz, and even Buffon cast their thought in the form of systems of analysis, deriving the details of matter from vortices for Descartes, from activity of monads for Leibniz, or, for Buffon, from more actualistic generalizations of information on the earth's figure, heat balance, and fossil lithology.

9 P. Cleaveland, *Mineralogy and Geology* (2nd ed.; Boston: Cummings and Hilliard, 1822), p. 423.
10 K. V. Zittel, *History of Geology*, trans. M. Ogilvie-Gordon (London: W. Scott, 1901; reprint, New York: Hafner, 1962).

> Imagine for a moment the present surface of the globe despoiled of all its seas, of the calcareous downs, as well as all of its horizontal beds of gravel, of chalk, of tufa, of vegetative soils, of clay, — in a word of all the materials, liquid or solid, which have been formed or deposited by water . . . there would only remain the skeleton of the earth, that is to say the fusible rock which constitutes the interior mass . . . the base which serves to sustain all the earthy materials brought there next (after congelation) by water.[11]

The seven epochs into which Buffon divides earth history are a prototype geological time scale, and the importance of Buffon is in providing this model for the geological chronology that Hooke had proposed a century before. The association of particular kinds of rocks with specified periods of time is properly designated *metastoria* to distinguish it from history proper, the true concern of which is the mind of men.[12] The metastorical subdivisions of rocks (chiefly stratified) are the province of *stratigraphy*, the early history of which is the subject of Dr. Hedberg's paper.

The grand theories of the earth's origin as a fiery star (later to evolve into the vulcanist-plutonist school of geological thought) can be traced in this tenuous fashion through Descartes and also through Athanasius Kircher to Paracelsus. But a metastory based on the alternative fluid agent, water, was also developed.

Bishop Burnet's *Sacred Theory of the Earth* was an early quasi-naturalistic attempt to ascribe the origins of the present earth's configuration to the action of water. Like John Woodward, he described the earth's original configuration as a vast interior cavity filled with water and bursting through the earth's crust on the command of God. In Woodward's *Essay Towards a Natural History of the Earth*, 1695, particularly, the arguments were from observations of fossils and sedimentary processes. Melvin Jahn's paper on Johann Scheuchzer and his remarkable skeletal reconstruction of *Homo diluvii testis* is an admirable illustration of the strength of the conviction that the discoveries of natural science would only render more convincing the truths of revealed religion. The diluvial hypothesis, as these attempts to place natural history and the Bible in a one-to-one correspondence were designated, was revived and refurbished with striking proofs by William Buckland in his *Vindiciae Geologicae* of 1820 — a last attempt to find, this time in the widespread glacial drift, direct evidences of the Mosaic flood.

11 Buffon, *Epoques de la Nature*, p. 104. Buffon had previously assigned a major role to action of the sea.

12 B. Croce, *History, Its Theory and Practice* (1916), trans. D. Ainslie (New York: Russell & Russell, 1960).

At the beginning of the eighteenth century, de Maillet's *Telliamed* was written. The origins and the relationships of this work to eighteenth-century geological thought are carefully studied in Professor Carozzi's paper.

The survey of the extent and diffusion of geological knowledge in the eighteenth century by Dr. V. A. Eyles complements Dr. Hedberg's meticulous study of Bergman's influence on stratigraphy. Professor Tikhomirov's paper treating developments in Russia, the paper by Dr. Wegmann on developments in Scandinavia, and the paper by Dr. White on the American Midwest illustrate the fundamental unity of science in the age of reason. In the midst of civil and intellectual conflicts, the common problems of eighteenth-century culture produced a cross-fertilization of ideas and a common effort at the development of science. This was rooted, in part, in the search for minerals, a search prosecuted vigorously by Peter the Great. An Academy of Sciences at Petersburg in 1725, the Prussian Academy in 1700, a Munich Academy in 1759, Benjamin Franklin's Junto, which grew into the American Philosophical Society (1743), and the Swedish Academy of Science (1741) testify to the eighteenth-century attitudes toward science and the climate in which geology developed. In England in the first quarter of the century, John Strachey began to describe the strata associated with the coals of Somerset. In Russia, major and successful expeditions to locate mineral resources were mounted throughout the seventeenth century, leading to the establishment of mining schools in the eighteenth century. In Germany, the mining tradition extended to Roman times. The vast silver deposits of Saxony were opened in the eighth century and were the focus of the medieval mining and metallurgy described by Agricola in 1546. The Bergakademie at Freiberg technically began in 1765, but Freiberg was a site of vigorous mining and metallurgical activity for eight centuries.[13]

The distinction between the so-called crystalline rocks — coarse, relatively massive granites and gneisses — at the cores of mountainous regions, on the one hand, and the slightly inclined to flat, frequently fossiliferous strata of the plains and foothills of the mountains, particularly the Appenines, begins in the eighteenth century with the Abbé Moro's recognition of Primary mountains of unstratified rock and Secondary mountains of sedimentary strata. The primary rocks were recognized as primitive — the original crust for neo-Cartesians such as Buffon, the first submarine irregularities for neptunist followers of

13 H. Baumgärtel, "Die 200-Jahrfeier der Bergakademie—Probleme und Aufgaben," *Freiberger Forschungshefte, D 42* (1962), 77.

de Maillet — the secondary were the younger products of later sedimentary stages. G. Arduino (1714–1795), a mining expert for the Venetian Republic, distinguished four separate stages, or "orders," which he said were very large strata arranged one above the other.[14] The four orders — Primary, Secondary, Tertiary, and Quaternary — comprised the Atesine Alps, the Alpine foothills, the sub-Alpine hills, and the plains of the Po, respectively. J. G. Lehmann, Professor at Berlin, both generalized the Italian classification — Gangebirge, Flötzgebirge, and postdiluvial — and particularized the description of the rocks of central Germany, a process of subdivision continued and extended by Georg Lasius (1752–1833), surveyor for the state of Oldenburg. These classifications identified strata by the common and mining terms for beds of rock whose distinctive appearance or situation with respect to mineral deposits (coal, salt) made them significant. Such were the chalk (Cretaceous) underlying the rolling downs cutting across England from the Channel to the North Sea; the Old Red Sandstone beneath the coal (Carboniferous); the New Red Sandstone above; the Kupferschiefer, or copper slates, and Zechstein above the coal in Germany. G. C. Füchsel (1722–1773) in 1762 divided the rocks of central Germany, from a basal vein series in the Harz and Thuringian mountains to an uppermost limestone, into nine *series montana*, or successive rock units, each lithologically consistent, representing successive epochs of time and conditions of formation.

Lehmann was called to Petersburg by Catherine the Great and named professor of chemistry and director of the imperial museum, "It is thus," Cuvier wrote, "that Bernouilli, Bayer, Euler, Gmelin, Muller, Amman, Lowitz, Duvernoy . . . have made known to us the immense territory of Russia. . . ."[15] In mounting an astronomical expedition to Siberia in 1769 to observe a transit of Venus, Pierre Simon Pallas (1741–1811) was appointed as naturalist. At age 33, on his return to Petersburg six years after Count Orlof, Catherine's favorite, had dispatched the expedition, Pallas' hair was completely white, dysentery had wracked him, and his eyesight was failing. His observations on the two vast mountain ranges of Siberia, Cuvier wrote, had changed the theory of the earth. No eighteenth-century writer, from Bergman and Buffon on, would fail to accept his "law of mountains," extending observations such as Arduino's on the sequence from the primitive granite

[14] F. Rodolico, "Arduino, G.," *Dictionary of Scientific Biography* (New York: Scribners, 1969).

[15] G. Cuvier, *Recueil des Éloges Historiques* (Strasbourg: Levrault, 1819), Vol. II, p. 119.

core to younger calcareous flanks, to cover all mountains throughout the earth. Cuvier would model his life's work (described in Dr. Bourdier's paper) on Pallas' reports of the bones and even frozen carcasses of elephant, rhinoceros, buffalo, and other large quadrupeds, which he had unearthed.

Expeditions of this scope and enlightened purpose do not fail to inspire the imagination of patriotic and enlightened men. Jefferson's Lewis and Clarke expedition was surely in the mold of Pallas' *Travels Through the Russian Empire*, the English edition of which made its appearance in 1802,[16] just before the Louisiana Purchase. The Lewis and Clarke expedition was the forerunner of the Powell, the Hayden, the Wheeler, and the King surveys of the mid-nineteenth century, by which the extraordinary geological phenomena of the American West were first brought into geological cognizance. Cook's voyage to the Pacific was also in pursuit of a transit of Venus. Surely the greatest of these surveys was the voyage of H.M.S. *Beagle*, from 1831 to 1835, carrying Charles Darwin through the Atlantic and South America to the islands of the Pacific, Australia, and New Zealand.

In this context, the exploration of the Alps by Horace Benedict de Saussure, the explanation of their topography by erosion, and the study of glaciers began that preoccupation with mountains which became so characteristic an attraction of geology. The most dramatic of the great geological explorers was von Humboldt, the subject of the admirable essay by Director Baumgärtel. Alexander von Humboldt (1769–1859), beside whose journeys those of all other explorers are dwarfed, was a man who took for his province nothing less than the sum of all things. His works were "gigantic publications in almost every branch of science." At the age of 74 he published *Cosmos: A Sketch of a Physical Description of the Universe*.[17] Such men saw the earth in the largest sense. Themselves larger than life — they crossed continents on foot and tried to see the whole of the world with their own eyes.

Von Humboldt, like von Buch, Jameson, Kirwan, and a score of others, equally distinguished, was a pupil of the remarkable Abraham Gottlob Werner (1750–1817), a man who so stamped the science with his ideas and his personality that during his lifetime, wrote Cuvier, ". . . whole academies were formed taking his name (Wernerian societies) as if they wished to invoke his genius."[18] The Wernerian philoso-

16 P. S. Pallas, *Travels Through the Southern Provinces of the Russian Empire in the Years 1793 and 1794* (London: Strahan, 1802).

17 A. von Humboldt, *Cosmos: A Sketch of a Physical Description of the Universe*, trans. E. C. Otté (London: Bohn, 1849).

18 Cuvier, *Recueil des Éloges Historiques*, Vol. II, p. 309.

phy was dominant for a time in geological thought. In Werner, we see the marriage of the marine mechanism (but a restrained sedimentological approach) and the concept of a great historical sequence of strata deposited sequentially and simultaneously all over the globe from a vast primordial ocean. The orders of Arduino and the epochs of Buffon are combined with the *series montana* of Füchsel, and the result is the concept of the geological time scale. Werner and his pupils and followers, including Cuvier, and Brongniart for France, Maclure and Eaton for America, Jameson and Greenough for Britain, made up the school of neptunists. Their opponents, followers of Hutton, were the vulcanists or, later, plutonists. Their great battle, a kind of intellectual world war, divided geognosy and geologists then as it divides historians of geology today. It divided historians at the New Hampshire Conference, and their views are expressed in the several studies in this volume. Conflicting views of the geological roles played by the two fluid agents, water and caloric, persist even today in the granite controversy. Professor Ospovat's essay on Werner's major publication, the *Kurze Klassifikation*, should be read together with Professor Taylor's paper on the vulcanist Desmarest and the papers of Professor Rappaport and Dr. Eyles on the development of geological thought in the eighteenth century.

The efforts of Jean Guettard (1715–1786) in carrying out the proposal made originally by Lister in the previous century, to prepare a map showing the kind of rocks to be found and their boundaries within a limited territory, are to be contrasted with the broad generalizations and the search for universal geognostic laws of men such as Pallas or the theorists Buffon and, later, von Humboldt. In 1751, Guettard identified basalt at Volvic in the Auvergne as volcanic in origin. This dark, compact rock, hard as granite and ubiquitous in occurrence, was to prove the focus of the neptunist-vulcanist controversy for the next century. Werner and his followers were rigidly metastorical in their approach. Basalt appears as the uppermost of the formations of the Secondary class, or Flötzgebirge, in Jameson's *Mineralogy*. "All basalt originally belonged to one widely extended and very thick layer," Werner wrote.

The principal weakness of the Wernerian System was precisely in its most important contribution, that of historicity. As Dr. Ospovat and Dr. Page point out, neptunism per se did not invoke catastrophes as the agents of geological change and topographic disorder. But it lent itself to this interpretation, and the latter-day diluvialists described in Professor Page's paper found in neptunism scientific verification of the Mosaic deluge, which was, in turn, the ideal example of catastrophe. In Werner's system geology was over, the earth's configuration finished, metastory complete. A system precisely opposed to Werner's, stressing the dynamics and continuity of earth processes, was promulgated by Werner's rival, James Hutton (1726–1797), whose ideas form the subject of Dr. Dott's contribution. In place of the epochs of Buffon, the worldwide layers of strata beginning with granite settling sequentially from the universal ocean, Hutton was to write, "I can find no traces of a beginning, no prospect of an end."

At this point in our history, at the close of the eighteenth century, geology underwent a profound transformation, destined to produce major changes in the geological world view. A new dimension was added to geology by the work of William Smith, and Cuvier and Brongniart.[19] The life and work of the remarkable English surveyor-observer William Smith, form the subject of Mrs. Eyles's contribution to this volume. It was Smith who observed the similarities of the sequence of the strata of southern England persisting from place to place through changes in the inclination and thickness of the beds, making his identifications positive by cataloguing their fossil contents. "It may be positively asserted," Lardner Vanuxem wrote in 1823, "that . . . formations may be at all times unequivocally identified by their fossil remains."[20] At almost the same time, the classification, sequential ordering, and paleontological determination of the Tertiary strata of the Paris Basin were being carried out by Cuvier (a principal figure in Dr. Bourdier's paper) and Brongniart. The "Napoleon of Science," the Baron Georges Dagobert Cuvier, dreamed of a "physical history of the globe," saying that as astronomers had unraveled the celestial motions, so finding the history of the globe before man was a task awaiting its Newton. The reader of his great work, *Discourse on the Revolutions of the Globe*, was left to make the obvious identification to Newton. Nor was the similarity of the title to Copernicus' *De Revolutionibus* unconscious.

Smith's influence was felt directly by Greenough, Conybeare, and Phillips, and most significantly by Sedgwick and Murchison, who in the fourth decade of the century undertook the subdivision of Werner's transition series, directly beneath the Secondary, or Flötzgebirge. The influence of Werner and Cuvier extended to America. Professor White's paper is an account of early geological observations in the central states before the publication of Maclure's geological map.

19 G. Cuvier and A. Brongniart, *Description Géologique des Environs de Paris* (Paris: Dufour et D'Ocagne, 1822).

20 S. G. Morton, "Geological Observations on the Secondary, Tertiary and Alluvial Formations of the Atlantic Coast of the United States of America. Arranged from the notes of Lardner Vanuxem," *J. Acad. Nat. Sci. Phila.*, *1* (1829), 59.

William Maclure (1763–1840), who undertook the extension of the Wernerian system to the United States, published a geological map in 1809 and again in 1817. It was Maclure who set out down the Ohio River with the famous "boat-load of knowledge" carrying Lesueur, Troost, and Thomas Say, to the "communist" settlement at New Harmony, Indiana, which he founded with Robert Owen of Lanark. Maclure's map was the basis of Cleaveland's (1819) and also of the extensive efforts of Amos Eaton, who proclaimed the task of American geology to be one of filling in the details of Werner's system and whose classification of New York strata was based on their degree of inclination and their lithology.[21]

Lardner Vanuxem brought the method of Cuvier and Brongniart back from the École des Mines, where he was the first American student (by nomination of Lafayette). Vanuxem correctly separated older coastal plains sediments into Tertiary and Secondary and identified the strata of the interior with the Transition period on the basis of their fossil contents. These Transition rocks, in turn, were subdivided through efforts beginning with those of Adam Sedgwick, a parson turned geologist on his nomination to the Woodwardian Chair at Cambridge. His friend Sir Roderick Impey Murchison abandoned fox hunting to join with him in unraveling the details of the contorted strata of the Welsh borderlands passing beneath the Secondary strata of William Smith. Murchison named the sequence of Transition strata that he was describing, the Silurian system, after the tribe of ancient Britons of the area, but when Sedgwick endeavored to emulate him by separating out a lower sequence of Transition rocks as a "Cambrian" system, Murchison turned upon him, and the two men quarreled. At almost the same time, Vanuxem, Ebenezer Emmons, W. W. Mather, T. A. Conrad, and James Hall began the survey of the strata of New York, which was to result in a similar and extensive subdivision of the New York Transition strata and a curiously parallel quarrel over the lowest of the Transition strata, Emmons' Taconic series.

Dr. Greene and Dr. Sherwood discuss the development of mineralogical and theoretical activity in American geology — urban, civilized geology — in contrast to the kind of geology that led Cuvier to say of Pallas that he had spent two-thirds of his life in the desert.

Hooke and Steno in the seventeenth century had grasped the fundamental principle of crystallography, the invariance of the angular relationships between the plane faces of crystals of the same species.

[21] Amos Eaton, *A Geological and Agricultural Survey of the District Adjoining the Erie Canal* (Albany: Packard and Van Benthuysen, 1824), p. 18.

Haüy made the Pythagorean discovery that the intercepts of these plane faces when referred to their intersections as axes, stood in simple integral ratios.[22] Suddenly, from the extraordinary complexity of this natural science, with its fossils, its subterreannean caverns of flood waters and volcanic fires, its porphyries and graywackes, there emerged a rigidly mathematical and exact discipline — crystallography.[23] The description of the further development of crystallographic and chemical mineralogy as a branch of geology in the first half of the nineteenth century was the contribution of Dr. Burke to the New Hampshire Conference.

It is the purpose of this Conference and this volume to reconstruct the intellectual climate within which our science evolved — to recreate the world view that was available to young Charles Darwin, out of which the theory of evolution was to emerge. By 1800, the early battles of the eighteenth century over diluvialism and neptunism were receding. As Professor Sherwood demonstrates, some geologists had become natural theologians, while theologians found themselves protecting the Bible from physical and geological defenses. The basalt controversy had ended in a complete rout for the neptunist position, and the experiments of the Scottish James Hall and the work of G. Poulett Scrope had begun the modern science of vulcanology. But if the neptunist origin for basalt and granite was now discredited, the Wernerian system of stratigraphic chronology, of earth history and therefore progression and development, had been developed out of all bounds and had become the major portion of the science. The dominant figure in the culmination of this multiplication of systems, which was the geological time scale and the idealized geological column of successive strata, was Georges Cuvier. For a time, Louis Agassiz was Cuvier's student. The master won fame through his reconstructions of the large vertebrata — saurians, flying reptiles, elephant, rhinoceros — many from the quarries of the Paris Basin, mapped with Brongniart at the outset of his career. The pupil's fame, his exhaustive studies of fossil fish aside, rested on the introduction of the glacial hypothesis of Venetz and Charpentier, according to which great blankets of ice had covered northern Europe and northern North America after the Tertiary and accounted for the widespread diluvium. The deluge, the glacial hypothesis, vulcanism, the attempts to explain the still-frozen carcasses of elephant-like mammals in the gravels of Russia, and the succession of remains of vertebrata in the strata of the Paris Basin brought men like Buckland, Cuvier, Sedgwick, and Mur-

22 R. J. Haüy, *Traité de Minéralogie* (Paris: Chez Louis, 1807).
23 A. Bravais, "Mémoire sur les Systèmes Formés par des Points Distribués, Régulièrement sur un Plan ou dans l'Espace," *J. École Polytech.*, *20*, (1851), 34.

chison to a cataclysmic view of earth history — catastrophism. Hutton's intellectual heir was Charles Lyell, the man who stood in relation to Cuvier as Hutton had to Werner. He opposed catastrophism with an extreme version of actualism called uniformitarianism. It was not sufficient to explain the phenomena of the earth's crust by actual physical and chemical causes, or even by "causes now in operation." Uniformitarianism would admit no significant change in the intensity of those forces and saw, in the greatest effects, only the cumulative action through geological time of the relatively trifling forces of the present time. With what triumph, then, the uniformitarians were to explain the falls of Niagara by the gradual eating back of the gorge in the Niagaran escarpment, the Grand Canyon by the erosion of the Colorado River, a mile below the canyon rim! The polarization of Lyell's thought on his visit to Sicily, the intellectual processes that culminated in the publication of his *Principles*, and his extraordinary influence on subsequent thought were discussed at the Conference by Dr. Rudwick, Dr. Wilson, and Dr. Cannon.

As Hutton could see nothing but cycles of repetition, his oldest rocks made from the debris of ruined worlds, the uniformitarians would admit no progression and reacted uneasily to the paleontological-stratigraphic development of an earth history beneath the new terminology of which the Wernerian-neptunist origins were clearly discernible. Cuvier, the catastrophist, was primarily a zoologist, crediting Brongniart with the geological part in their joint enterprises. The relationships of Cuvier, Lamarck, and Geoffroy Saint-Hilaire; the growing sciences of embryology and Cuvier's comparative zoology; the concept of a *plan*, a great chain of nature, relating all organic beings are treated in the paper sent to the Conference by Dr. Bourdier. When Lamarck proposed the theory of evolution, of the transformation through geological time of all species of living things, neither Cuvier or Lyell would admit the possibility of change in a species with time.

In the Bridgewater treatise "on the power, wisdom, and goodness of God as manifested in the creation," *Geology and Mineralogy considered with reference to natural theology* (1836),[24] Rev. William Buckland, no evolutionist, published an ideal cross-section through the earth's crust in twelve panels and sixteen colors. The section was by Webster and included eight subdivisions of the stratified rocks from the Primary to the Alluvium. For each of the periods, beginning with the first fossiliferous strata (the Transition), Buckland selected illustrations of the representative fossil fauna and flora. The fern trees, the flying

[24] W. Buckland, *Geology and Mineralogy Considered with Reference to Natural Theology* (Philadelphia: Carey, Lea and Blanchard, 1837), Vol. II.

reptiles, and the armored fish are associated with the lower strata, the Transition, and the Secondary. The large mammals and the birds are associated with the uppermost period of the Tertiary. No modern geologist would ask for a more succinct graphic demonstration of evolution. The evidences and the ideas for a historical science, a metastory, had accumulated by methods that had, themselves, evolved since classical times. The stage was set. Erasmus Darwin, Goethe, Lamarck, Geoffroy Saint-Hilaire, even von Buch, Rafinesque, Robert Chambers, Omalius d'Halloy, Herbert Spencer, von Keyserling, Rouillier, von Baer, and a host of others, were in the cast.

On December 27, 1831, H.M.S. *Beagle* stood out from Devonport for Teneriffe.[25] On board was Charles Darwin, and among his books, the first volume of the newly published *Principles* of Lyell.

The New Hampshire Inter-Disciplinary Conference on the History of Geology before Darwin met September 7 through 12, 1967, at the Farragut House at Rye Beach, a sprawling, old-fashioned New England hotel facing the sea and the Isles of Shoals. It was two months after the Constituent Assembly for the Committee on the History of the Geological Sciences of the International Union, at Yerevan, Armenia, and a year before the first meeting of the Committee scheduled for the ill-fated XXIII International Congress at Prague, Czechoslovakia in August, 1968.

The New England weather was seldom more cooperative. In the small hours of the morning, the sands and rocks were washed clean. The meetings took place with the muted roar of the surf for background. A clambake was followed by Mozart and Brahms in the Victorian parlors of the Farragut. A field trip to the Ossipee ring-dike complex and a picnic on the shores of Lake Chocorua were followed by a cocktail-buffet in a country house on the banks of the tidal Squamscott in Newfields. The conferees formed a uniquely qualified assembly of experts on a common theme. First versions of the papers in this volume had been duplicated and distributed in advance. The meetings were confined to discussions — discussions that spilled over the Conference rooms onto the sands and rocks along the coast road.

The Conference had been conceived in an effort to bring together scholars working in the two fields of geology and the history of science, with a common interest in the history of the geological sciences. The History of Science Society and the Council on Education in the

25 C. Darwin, *Journal of Researches into the Geology and Natural History of the Various Countries Visited by H.M.S. Beagle, Under the Command of Captain Fitzroy, R.N., from 1832 to 1836* (London: Colburn, 1839).

Geological Sciences of the American Geological Institute were cosponsors. The Division of Social Sciences of the National Science Foundation, through a grant to the University of New Hampshire, made the Conference possible. The New England Center for Continuing Education assisted. Besides these authors, Andrew von Benschoten of Harvard, and Harold Wehlau of Hafner Publishers participated in the Conference. Dr. Bourdier and Dr. Tikhomirov participated in absentia.

On behalf of the Conferees and myself, I wish to acknowledge the assistance of Charles Gillispie, then President of the History of Science Society; O. T. Hayward, then Chairman of the Council on Education in the Geological Sciences of the American Geological Institute; Peter Horne and Stella Whitehouse of the New England Center for Continuing Education; Herbert Tischler, Henry Gaudette, Wallace Bothner, and Franz Anderson of the Geology Department of the University of New Hampshire; my wife, Mary Schneer; and William Cotter and the staff of the Farragut House. George White and John Greene served with me as a Steering Committee.

This Conference, having given rise to twenty-five papers on the history of geology totaling nearly 800 typescript pages, represents a new level, if only quantitatively, of activity in the history of a single science. In bringing together scholars of disciplines so disparate as history and geology, it may well have been unique. For myself, and indeed a number of conferees have written to the same effect, the Conference resulted in a major expansion and reorganization of my ideas on the history of the science that is my profession. If history is what is written by historians — if it is true that each generation views its past in a different light, reorganizing its sense of its own roots in the light of an ever-growing richness of detail — we can conclude that the New Hampshire Inter-Disciplinary Conference on the History of Geology has called into being a new composite view of the development of geology up to the time of Darwin. This volume should identify, both by what is included and by what is omitted, the significant problems in the history of our science. At the same time, the standards for scholarship in the field, as they have been modified by the interactions of historians and scientists, should be clarified by the twenty-five papers in this book. In these Conferences, in the new groups forming, in the publications, the letters, the exchange of ideas, there is the sense of a new order of achievement, a new level, qualitatively as well as quantitatively, of research into the origins of geology.

Alexander von Humboldt: Remarks on the Meaning of Hypothesis in His Geological Researches

Hans Baumgärtel

In 1769, when Alexander von Humboldt, whom an admiring contemporary called the "scientific rediscoverer of America,"[1] was born, the natural sciences were only beginning their development; only mechanics had already attained a certain degree of perfection. Geology did not even have a name. The geological knowledge of that time was fragmentary and consisted in part of more or less one-sided and uncohesive observations and descriptions by naturalists, miners, and mineralogists (Agricola, Gesner, Steno, Lister, Woodward, Lomonosov, Hutton, and Füchsel, to name only a few) and in part of cosmogonic and geogonic systems, which, to be sure, contained many ingenious ideas, but which all in all revealed more about the imagined worlds of their authors than about geological reality (*Mundus Subterraneous* by A. Kircher, 1664; *Protogaea* by G. W. Leibniz, 1693; *Telliamed* by de Maillet, published 1749; and so forth). For these works, even when they are to be recognized as noteworthy theoretical achievements, the quotation from Francis Bacon still holds true, that each of them "builds a ship with materials insufficient for the rowing pins of a boat."[2]

When Humboldt died in 1859 at the age of 90, after a full life, rich in work and honors, geology was a very extensive, recognized natural science, which already had its "classical period" behind it. In 1769, there was no systematic theory of this science, and the researches of

Translated from the German by Deborah Watson and Alexander Ospovat.

1 Carl Ritter, *Geschichte der Erdkunde und der Entdeckungen* (Berlin: 1861), p. 264.
2 Frank Dawson Adams, *The Birth and Development of the Geological Sciences* (New York: Dover Publications, 1954), p. 209.

the few men who already merited the name "geologist" were confined almost exclusively to Europe; by 1859, schools of mining and geological chairs at almost all the large universities had already produced thousands of scholars and teachers of geology, whose influence extended over the whole world.

Alexander von Humboldt helped shape the first golden age of geology, even though he in no way gave it its changing character. Scholars like Georges Cuvier, William Smith (who were both born in 1769, the same year as von Humboldt), Leopold von Buch, and Charles Lyell had far more effect on the special development of geology than Humboldt, for whose all-encompassing mind this science remained only a branch — though one of the most important — of his interest. Humboldt was perhaps the last true universal scholar in the field of the physical sciences; in the first half of the nineteenth century, these sciences were already divided into autonomous special fields.

That was doubtless a necessary occurrence. The concrete knowledge of nature must first be vastly enlarged in many separate areas before a complete and scientifically based understanding of it can be formed. Was Humboldt, therefore, with his universal method of thinking, already an anachronism in his own time? His much younger friend, Emil Du Bois-Reymond, said about him later: "But he lost the desire and later even the intellectual habit of analyzing the phenomena past a certain point and tracing them back to the last discernible causes. He allowed himself to be satisfied with the establishment and observation of facts."[3]

Naturally, a man who pursued geology, petrology, geography, meteorology, oceanography, ethnography, biology, zoology, botany, physics, physiology, and several other sciences at the same time could not simultaneously assume leadership in any one particular science, since each of them expanded more and more. Through his researches and discoveries, however, Humboldt contributed abundant material to all of these sciences. He made important contributions to geography, climatology, plant geography, and ethnography, as well as to geology, particularly its special disciplines of vulcanology and the study of earthquakes; but his importance is not in this alone. Because Humboldt was not a specialist in any single area, he could, in opposition to all the tendencies toward specialization, affirm the inner relationship of the sciences, a relationship that is indicated in the material unity of the

[3] Emil DuBois-Reymond, "Die Humboldt-Denkmäler von der Berliner Universität," Rector's address, August 3, 1883.

world and the connection of all the laws of nature, and he could clarify this relationship even for the laymen of his time.[4]

Beyond that, Humboldt, through his many-sided observations and comparative studies, opened the way for the development of several branches of science; this is true especially for comparative climatology and plant geography. So we still today regard him as a universal scholar who cherished the idea of the unity of all the sciences, without which all isolated research would remain only fruitless and imperfect. We see him also, however, as a man who, in the study of nature, never forgot the destiny of human society, and who, in his mental attitude, was always closely associated with the liberal middle-class thinkers and revolutionaries of his time, whether in France, North and South America, or Germany.

When Humboldt was scarcely 24 years old, his older brother William, who later earned fame as a philologist and liberal and political scientist, wrote about him: "It is his nature to connect ideas, to perceive relationships among things which for generations would have remained undiscovered without him."[5] Here the concept "idea" is to be understood in its original meaning — namely, an image, a view of reality. Mere speculations were foreign to Humboldt. He formulated very early the fundamental principle underlying his scientific work: to separate the facts from his own suppositions. "This manner of handling phenomena of nature," he wrote, "seems to me the most fruitful and the most basic. Facts remain even long after the hastily produced theoretical scientific framework has collapsed."[6] Several times in his later works he repeated this basic sentence in similar words, and he always remained true to it.

His manner of working was predominantly empirical and inductive; however, he was no empiricist in the narrow sense of the word, but always strove, next to the "exposition of the details of phenomena, to generalize ideas, and, to approach [more closely the solution of] some of the great problems in the philosophy of nature."[7]

4 See his "Kosmos-Vorlesungen," Berlin, 1827–1828, and a large part of his writings, especially his *Ansichten der Natur* (Stuttgart and Tübingen, 1808), and *Kosmos* (5 vols.; Stuttgart, 1845–1862).

5 Letter by Wilhelm von Humboldt to K. G. von Brinkman, dated March 18, 1793, cited in Hanno Beck, *Alexander von Humboldt* (Wiesbaden: 1959), Vol. I, p. 45.

6 Alexander von Humboldt, "Über die gereizte Muskelfaser," *Neues Journal der Physik, II* (1795), 127.

7 Humboldt, *Essai géognostique sur le gisement des roches dans les deux hémisphères* (Paris, 1823), p. vi.

A similar recognition of the unity of technical research and theoretical observation lies in his words: "The philosophical study of nature exceeds the requirements of a mere discription of nature. It does not consist of a sterile accumulation of isolated observations."[8] In this sense he also approached *Kosmos*, the monumental work of his old age: "Every great and important idea which glimmers anywhere must be recorded here next to the facts."[9]

It is evident that Humboldt was concerned not with a separation of ideas from facts, but with the relationship of the facts to a scientific world view. He makes this clear in the following sentence, taken from the unfinished fifth volume of *Kosmos*, which was first published after Humboldt's death: "My purpose was to seek laws and undeniable proofs for a causal relationship among large groups of actual natural processes."[10] Without this unflinching conviction of the conformity of all natural occurrences, Humboldt would have been only one of the most industrious scientific fact collectors of all time; instead, he was one of the greatest scientists, still the model even for the scholars of our day, in that he combined practical experience with theoretical thinking, thus striking a happy medium between sterile empiricism and unscientific speculation, and pointing the way to the only proper scientific method.

The thought of a comprehensive presentation of nature stirred Humboldt even in his youth; "Je conçus l'idée d'une physique du monde," he wrote at the age of 26.[11] It is often said that he realized this idea first in his *Kosmos*, but basically his whole scientific endeavor was always guided by this universal intention. Although in the following discussion we will limit ourselves to a few aspects of his geological researches, it must never be forgotten that at no time was he solely a geologist, nor did he ever wish to be that.

It is not surprising that geology[12] in particular stimulated Humboldt's thinking again and again. The still youthful, little developed character of this science inspired his creative urge; its relationship with other sciences, such as mineralogy, chemistry, physics, geography, zoology, and botany, suited his all-encompassing way of thinking; and, finally, he was deeply influenced by the peculiar union in geology of

8 Humboldt, *Über den Bau und die Wirkungsart der Vulkane in verschiedenen Erdstrichen* (Berlin, 1823), p. 36.
9 Letter to Varnhagen von Ense, dated October 27, 1834, cited in *Briefe von A. v. Humboldt an Varnhagen von Ense aus den Jahren 1827–1858*, (Leipzig, 1860), p. 20.
10 Humboldt, *Kosmos*, Vol. V, p. 9.
11 Letter to Pictet, dated January 24, 1796, cited in Karl Bruhns, *Alexander von Humboldt. Eine wissenschaftliche Biographie* (Leipzig, 1872), Vol. I, p. 221.
12 The word "geologie" was coined by A. De Luc in 1778.

the present with the seemingly boundless past, and the sober discription of what is, inseparably linked to the eternal question of what will be and what was.

In the few decades before and after Humboldt's birth in 1769, the formulation of geology as a science entered into its decisive stage. In 1751, Guettard discovered the volcanoes of Auvergne; his description was published in 1756.[13] Also in 1756, a member of the Prussian mining board, Johann Gottlob Lehmann, published his *Versuch einer Geschichte von Flötz-Gebirgen*, in which he questioned certain points of the hypothesis of the deluge and expounded "partially inundating" earthquakes and volcanoes as vital causes of the present form of the earth's surface.[14] A few years later, the Thuringian geologist Füchsel, in his *Geschichte des Landes und des Meeres*, grouped rocks according to formations, and thus may be considered as the founder of stratigraphy.[15] And in his last work, which appeared in 1773, the year in which he died, Füchsel defined the concept "formation."[16] Applying the uniformitarian interpretation of geological phenomena, he wrote: "The manner, however, in which nature has worked up to the present day and has brought forth substances is to be taken as the norm; we do not know any other."[17] Lomonosov's book *Erste Grundlagen der Metallurgie oder des Hüttenwesens*,[18] which was first published in 1763, two years after his death, contains the well-known discussion "Über die Erdschichten," wherein Lomonosov also followed the principle of actuality, rejecting the theory of the flood even more sharply than any of his predecessors and contemporaries. He also postulated a theory of the creation of mountains, which in many respects constitutes almost an anticipation of the volcanic elevation theory, as advanced by Leopold von Buch half a century later. He assumed the formation of the earth's surface to be a result of very slow changes brought about by endogenic forces. However, Lomonosov's work remained for a long time almost unknown outside the Russian Empire.

In 1778, Buffon published his much read work *Les Époques de la*

13 Carl Christoph Beringer, *Geschichte der Geologie und des geologischen Weltbildes* (Stuttgart, 1954), p. 44.

14 Johann Gottlob Lehmann, *Versuch einer Geschichte von Flötz-Gebirgen* (Berlin, 1756).

15 Georg Christian Füchsel, "Historia terrae et maris." In *Actorum Acd. elect. Moguntinae* (Erfordiae, 1761) Tom. II.

16 Füchsel, *Entwurf zur ältesten Erd- und Menschheitsgeschichte*, 1773.

17 Füchsel, "Historia terrae et maris," p. 43, as cited by R. Möller, "Mitteilungen zur Biographie Georg Christian Füchsels," *Freiberger Forschungshefte D 43*, (1963), p. 23.

18 M. V. Lomonosov, *Pervyje osnovanija metallurgii ili rudnych djel* (St. Petersburg, 1763).

nature,[19] probably the last really meaningful book of the series of speculative geognostic writings, which more and more had to yield to the sober exploration of facts. In 1795, James Hutton published his life's work, *Theory of the Earth with Proofs and Illustrations*,[20] in which he advanced a comprehensive plutonistic hypothesis of the earth's history. It was the revised and much enlarged draft of a lecture published in 1788 in the *Transactions of the Royal Society of Edinburgh*, which had given rise to strong criticisms of it on the part of the Scottish neptunists.[21] It was these men, and a few others, altogether only a small number, who through their works formed the first geological knowledge of the young Alexander von Humboldt. In particular, however, it was his teacher, Abraham Gottlob Werner. From Werner, Humboldt received not only a great many ideas that we find in his geological hypotheses, but also the decisive training necessary for his comprehensive and fruitful geological exploratory work.

In the geological literature and histories of geology, Werner is often presented primarily as the leader of the neptunian school. Doubtless he was that for decades, but such an evaluation leads to a distortion of the picture of this great scientist, because Werner's importance lies not primarily in his neptunism. If he had been nothing more than a leading neptunist, then with the final collapse of this school — and this resulted practically with his death — his reputation would have vanished, and his name would have been forgotten. But Werner's reputation remained. His greatest students, among them Humboldt, praised him for decades after his death, although almost all of them had long since broken with neptunism.[22] And Werner today is still considered the prototype of a researcher and teacher of geology.

It is not our purpose here to go more deeply into Werner's achievements and particularly into the famous quarrel between the neptunists and the plutonists. But neither Humboldt's geological research methods

[19] Leclerc de Buffon, *Les époques de la nature* (Paris, 1778).

[20] James Hutton, *Theory of the Earth with Proofs and Illustrations* (Edinburgh, 1795) 2 Vols.

[21] Adams, *Birth and Development*, p. 240.

[22] On the occasion of the 1850 Werner-Feier, the 81-year-old Humboldt wrote to the Bergakademie Freiberg: "As probably the oldest among the still living students of the great unforgettable Werner . . . I must express my deep regret, that my personal circumstances, my ancient age . . . deny me the long hoped for joy to express orally how I owe an important part of my education and the direction of my efforts to the vastly comprehensive and orderly mind of our Werner. . . ." *Jahrbruch für den sächsischen Berg- und Hüttenmann* (1851), 61. And D'Aubuisson wrote: "One can almost say of Werner what has been said of Linné: the earth is covered with his students, and from one pole to the other nature has been interrogated in his name." "Sur les travaux de M. Werner en mineralogie," *Annales de Chemie* (31 March 1809), 239.

nor his attitude toward geological hypotheses would be conceivable without Werner's influence. In 1775, Werner, who had studied first at the Bergakademie at Freiberg and then at the University of Leipzig, was named inspector and professor of mining at the Bergakademie at Freiberg. The previous year, whle he was still a student in Leipzig, he had published his epochal work on the external characters of minerals.[23] The system developed therein for the description and recognition of minerals soon became one of the most important tools of all practical mineralogists, geologists, and mining engineers. In his teaching, Werner laid great stress on the further systematic development of mineralogy, but he soon realized that geology, as well as mineralogy and mining, had its own subject matter and demanded a method of its own. He therefore decided to make geology an independent field of study. In 1779, for the first time, he gave his famous lectures on geognosy. With that, geology began its career as an academic discipline and soon was incorporated into the curricula of many institutions of higher learning.

Werner purposely chose the name "geognosy" because the term "geology," coined shortly before, seemed to him too heavily burdened with the speculative character of geogonic hypotheses. Geognosy was not to include unproved ideas, but only the factual "knowledge of the earth." About geological speculation and his own basis for geognosy, Werner said: "On the other hand, geognostic systems formulated in the study, that is, without any or with insufficient observations and field experiences, can contain nothing but sterile, wrong ideas, nothing but fancy, of these we have unfortunately all too many examples. Such systems, often beautifully written, should be considered only as novels about nature. A system of geognosy will be the more valid the more it is based on correct mineralogical-geographical observations and is verified through such."[24]

Three things are particularly striking in this quotation: First, the matter-of-fact way in which a "system" of geognosy, in other words, a comprehensive theory, is advanced; a solely empirical-descriptive, and therefore unsystematic, geognosy was inconceivable to Werner. Second, the use of the word "verified" in the same sense in which we use it today: the theory must be substantiated by application. Third, and finally, the emphasis on "mineralogical geography": the most comprehensive establishment and comparison of minerals and rocks from the most varied localities and countries should be the incontestable basis of

23 Abraham Gottlob Werner, *Von den äuszerlichen Kennzeichen der Fossilien* (Leipzig, 1774).
24 Werner manuscript, Vol. I, p. 267.

the geological system. One would probably not be wrong in the assumption that this demand in particular made a strong impression on the young Humboldt, because it offered him the opportunity to contribute decisively, through his own researches on trips to areas that until then had scarcely been investigated, to the "verification" of a comprehensive system.

It is not surprising, therefore, that Humboldt, a few years after his study at Freiberg, wrote to Werner: "I am working without pause on a great geognostic work, which is to be published under the title 'Über die Construction des Erdkörpers im mittleren Europa, besonders über Schichtung und Lagerung der Gebirgsmassen.'"[25] As far as is known, such a work by Humboldt was never published. At that time he was working on several different projects. In the years 1796 and 1797 alone there appeared more than twenty of his publications,[26] of which the two-volume work *Über die gereizte Muskel- und Nervenfaser*[27] was the most extensive. Moreover, in 1792 he had already completed a geological treatise, entitled *Über die auf Salzsohle abzutreibenden Bohrlöcher*, and a map, *Karte über den Zusammenhang aller Salzquellen in Deutschland*. The treatise and the map were based on field research of various salt deposits in Germany and Austria that Humboldt had undertaken under the auspices of the Prussian government. However, neither was published, and both were presumably lost later.[28]

Humboldt's intellectual acquaintance with Werner had begun before he began his studies at Freiberg. As a result of a trip to the Rhine in 1789, Humboldt published a booklet, *Mineralogische Beobachtungen über einige Basalte am Rhein*.[29] At that time (1789–1790) he was a student at the progressive (for that time) German university at Göttingen. There, in contrast to almost all the other German universities, which were still strongly dominated by the theological faculties, the teaching of science and research in science was in full bloom. Mineralogy, though, was not taught, and Humboldt remarks about this in letters, writing that most of his mineralogical knowledge at that time was self-taught. He had probably studied mineralogy on his own ever since he

[25] Letter to Werner. Cited in Bruhns, Vol. I, p. 271.

[26] Bruhns, Vol. II, pp. 490–492.

[27] Humboldt, *Versuche über die gereizte Muskel- und Nervenfaser, nebst Vermuthungen über den chemischen Process des Lebens in der Thier- und Pflanzenwelt* (Berlin and Posen: 1797) 2 Vols.

[28] Bruhns, Vol. I, p. 149.

[29] Humboldt, *Mineralogische Beobachtungen über einige Basalte am Rhein. Mit vorangeschickten, zerstreuten Bemerkungen über den Basalt der ältern und neuern Schriftsteller* (Braunschweig, 1790). This work was published anonymously.

began studies at Göttingen in the spring of 1788,[30] but he said about the treatise on basalt: "I wrote it before ever having had any instruction in mineralogy."[31]

The considerable amount of knowledge and scientific accuracy expressed in this booklet by the 20-year-old Humboldt is therefore even more remarkable. But even though he was concerned with giving a sober, factual rendering of his observations, he could not avoid taking a stand on a scientific hypothesis. It was no accident that he devoted himself to an area that at the time was the scene of an agitated controversy among many scholars. The quarrel over neptunism-vulcanism, which revolved primarily around the question of the origin of basalt, was at its peak in those years, and it was just this battle of opinions and hypotheses, which was then still undecided, that attracted the young Humboldt. This attraction to the controversial is doubtless an important trait in his character as a researcher, but not less important, it would seem, is the fact that he conducted his own researches not in order to arrive at a decision in favor of one of these hypotheses, nor to formulate a hypothesis of his own, but rather to make facts known. He strove to observe without any preconceived notions, and to present objectively that which he had observed. These observations (on the stratification of basalt, its cleavage, its lithological associations, and so forth) should not concern us here, nor should the extensive philological-petrographical debate on the content of the word "basalt" in classical and modern authors (in which he reveals a brilliant knowledge of literature); but rather we should concern ourselves with his opinion of the hypotheses on basalt.

"It is far from me to formulate new hypotheses in this treatise; I cite only single facts, I describe only that which I have seen, or rather, what I believe to have seen. Other mineralogists, who by their wider perspective and universality of ideas are qualified to make geogenic conclusions, will certainly not find this method of treatment unfruitful. None of them, I hope, will reproach me if I did not always find the stratigraphic relations of rocks to be in accordance with his own theory . . . ."[32] In this way Humboldt begins the exposition of his observations. He expressly divorces himself from the "reports of enthusiastic travelers, who so often find what they wish to find, who so

30 In a letter to Werner, dated December 13, 1790, Humboldt wrote: "I have been studying mineralogy for almost two years now." Cited in Bruhns, Vol. I, p. 115.
31 Letter to Freiesleben, *Ibid.*, p. 92.
32 Humboldt, *Mineralogische Beobachtungen*, p. 77.

willingly transfer the idealistic world into the real."[33] Humboldt shows himself here — like every serious scientific researcher — to be a spontaneous materialist, who strives simply to take nature "as it is, without foreign additions" (Friedrich Engels). Yet one realizes in reading his booklet that he leans toward one hypothesis on the formation of basalt, and, moreover, the one that proved to be false — namely, the neptunian. To be sure, he carefully presents the various hypotheses as equals, next to one another,[34] but the preference for the neptunian is unmistakable. Later, after his study with Werner, Humboldt declared himself even more openly as an adherent of neptunism, but never with the finality of his teacher.[35]

Humboldt's recognition of factual, unprejudiced observation and his support of the neptunian hypothesis seem to constitute a contradiction, but it is only an apparent contradiction. Fundamentally, no scientist (in the present as well as in the past) can avoid hypotheses. We know a great deal, but not everything by far, and before we can attain new understanding, our knowledge must be revived and enhanced qualitatively. The simple accumulation of empirical facts is sterile (as Humboldt noted, see above) and, in the end, senseless, because in that way we cannot attain any higher level of thinking. With empiricism alone, we would probably still today use Ptolemy's concept of the universe.

The fact that Humboldt at that time leaned more toward neptunism than vulcanism should not be judged from the standpoint that vulcanism — or, better, plutonism — proved in many respects, especially with regard to basalt, to be "more correct" than neptunism. The then available evidence spoke just as much for the neptunian explanation as for the vulcanic and plutonistic, and in some cases offered even better arguments for neptunism. Many extreme exponents of vulcanism arrived at unscientific conclusions, and thereby repulsed the soberminded Humboldt. For example, Professor Witte of Rostock explained that the Egyptian pyramids were formed by volcanoes. Besides that, Humboldt's own observations, such as those of the water-filled vesicles in the basalt near Unkel, were compatible with the neptunian hypothesis, while, on the other hand, he found nothing which would have made a vulcanistic explanation necessary. Humboldt's inclination to-

33 *Ibid.*, p. 115.

34 "Regardless of which hypothesis one approves, whether one places himself in the crater of a volcano with Mr. von Veltheim, whether one considers basalt as an ejected lava as Mr. von Dolomieu does, or as a dried out mass as Mr. Werner does (since everything is equally large and incomprehensible!) . . ."

35 Hans Baumgärtel, "Alexander von Humboldt und der Bergbau," *Alexander von Humboldt. Gedenkschrift zur 100. Wiederkehr seines Todestages* (Berlin: Akademie-Verlag, 1959), pp. 20–21.

ward neptunism should therefore be regarded not as a "mistake," as a contradiction of his factual research work, but rather as of consequence of it. Naturally, it strengthened his desire to go to Freiberg and become Werner's student. It speaks as well for Werner as for Humboldt that the latter did not through his study under Werner become an extreme neptunist but rather a great geologist, who contributed much that was meaningful, especially in the investigation of volcanoes, and who helped greatly to correct the mistakes in the geological system of his teacher. Although this system, seen from today's standpoint, rests on a false basis hypothesis, it nevertheless produced its share of observed, classified, and explainable details of knowledge, which, in proportion to the speculative, was greater than that offered by any other comprehensive geological system up to that time; in that respect, Werner's system was also historically justified and fruitful.[36] While Humboldt, in the quarrel over the formation of basalt, could only refer to the already extant hypotheses, he later developed his own geological hypotheses. One of these, which played a large part in his research activity, was that of a uniform law of the strike and dip of the rock strata in all mountains.[37]

As early as November 2, 1792, in a letter to Freiesleben, Humboldt wrote about an essay, written as a report, on the German salt springs: "The main idea is this, that the Franconian, Swabian and Thuringian Floetz rocks have in general the same kind of relative position. . . ."[38] By 1796, when he was working on the above-mentioned great geological work, which was never published,[39] he had expanded this idea considerably. The observations he had made on trips seemed to prove that the bedding of the rocks in Middle Europe followed a certain general law and that the average strike was between 45° and 60° (toward the north). "I will prove that the strike and dip of rocks[40] is based not on the direction and slope of the mountains,[41] but rather on something far greater, and that all primitive rocks in the great European mountain chain, which I have traversed on foot from Savoy to the Tyrols, have

36 Otfried Wagenbreth, "Abraham Gottlob Werners System der Geologie, Petrographie, und Lagerstättenlehre," *Abraham Gottlob Werner. Gedenkschrift aus Anlasz der Wiederkehr seines Todestages nach 150 Jahren am 30. Juni 1967* (Leipzig: 1967), p. 84.

37 This is not to be confused with the direction of the mountains themselves.

38 See footnote 28.

39 See footnote 25.

40 "Gebirge" is an old miner's expression that has the same meaning as rock, in the sense not of a single rock or hand specimen, but of a rock mass; hence the expressions "Urgebirge," "Flötzgebirge," etc.

41 Here the word "Gebirge" is used as in everyday language, meaning an elevation of the earth's surface, a mountain range.

the general strike between three and four o'clock and dip toward the north and west; all the floetz rocks have the same strike and dip toward the south and east. I have been collecting observations for this work for three years. . . ."[42] In the following years, also, he could not dismiss the thought of this conformity. He called the phenomenon, which he had studied since 1792, "parallelism or, rather, loxodromism of the strata" and hoped to have discovered a universal law of nature, which he thought must have been operative during the formation of the earth's crust. As he himself wrote, this hypothesis was one of the main reasons for his famous American trip. "This investigation, which I believed would offer inducement to physicists to discover a great law of nature, impressed me so powerfully at that time, that it became one of the strongest reasons for my decision to make a trip to the equator."[43]

From the observation made first in Middle Europe of an often conforming strike and dip of rock strata, Humboldt developed the idea that such a regularity could be assumed throughout the world. He believed, then, that further observations in this direction would give science a picture of the "construction of the earth," of the "Erdgezimmer" (two of his frequently used expressions).

This hypothesis was to become very important not only for Humboldt himself, but also for geology and geography in general. It is not important that he never succeeded in proving the existence of such a universal law, even though his idea seemed in many ways to be substantiated through further observations in America and Asia. Humboldt also strove to find a possible physical explanation. Before his American trip, he already conjectured as a cause "magnetic powers" in the earth's interior.[44] In America, also, he pursued this thought further and wrote in 1799, in one of his first comprehensive travel reports, that "the strike and dip of the primitive rock layers . . . are according to laws, that they follow a general parallelism, which can only be caused by gravitation and the rotation of the earth . . ."[45] Thus gravity and the rotation of the earth appeared to him to be the final causal forces. However, in all such remarks, he tried very carefully to separate his observations from the hypothetical explanations.

42 That is, E.S.E. Letter to Werner (1796). Cited in Bruhns, Vol. I, p. 171.

43 "This investigation, which I believed would lead physicists to the discovery of one great law of nature, has always had much attraction for me, which became one of the most powerful motives for my voyage to the equator." See Humboldt, *Essai géognostique*, p. 54.

44 Humboldt, *Über die gereizte Muskel- und Nervenfaser*, Vol. I, p. 9.

45 As cited in Hanno Beck, Vol. I, p. 170.

The attempt to go from this hypothesis to a verified theory was unsuccessful. In his evaluation of the American trip, Humbolt already expressed himself more cautiously with regard to the universality of the conjectured law: "I deduced . . . from these very exact observations on the great uniformity in the strike of the strata . . . general conclusions, which I later regarded as vague and less exact."[46] But he was further convinced of the importance of his idea and had only amended his "opinion of an occurrence which is in many ways more meaningful than it had been formerly believed."[47] Now he reached more specific conclusions: "In no hemisphere does a universal and unqualified uniformity of the strike among rocks exist; but in areas of very considerable extent . . . one recognizes that the strike, less often the dip, has been determined *through a special system of forces*."[48] Here Humboldt assumes not global, but territorially widespread processes as cause.

On his Asian trip in 1829, he pursued the idea of the "loxodromism law" further, referring it now, however, more strongly to the structure of the earth's surface. "The SW–NE direction expresses itself not only in the contours and the continental shape of all of Europe; it repeats itself also very often in the mountain systems and in the strata of the rock layers of Europe. The interior of Asia has apparently been subjected to the same influence and in an even greater measure. . . . This direction (three to four o'clock on the mine surveyor's compass) seems to be connected with one of the oldest revolutions which the raised continental masses have suffered."[49] The last sentence clearly shows the influence of the vulcanic elevation hypothesis which was primarily developed by Humboldt's friend and fellow student under Werner, Leopold von Buch, and to which Humboldt himself had made substantial contributions.

The basic importance of Humboldt's hypothesis of the law of loxodromism as a guide for future research lies, in my opinion, in the following elements: first, in the *high degree of universality* of the hypothesis. Based on exact observations, it proceeded from the supposition that everywhere on earth, like or similar forces formed the earth's crust. Second, it was not bound to previous acknowledgement of a particular geogonic hypothesis, for example, vulcanism or neptunism, but directed the research to advance without prejudice from the

[46] Humboldt, *Essai géognostique*, p. 55.
[47] *Ibid.*
[48] *Ibid.* Italics by the author.
[49] Humboldt, *Central-Asien*. Trans. Wilhelm Mahlmann (Berlin, 1844), Vol. I, p. 60.

observations of natural phenomena to an explanation of them. Furthermore, some aspects of the hypothesis are distinguished by a *higher degree of probability* in relation to later, more broadly based hypotheses and theories. Thus, the idea of gravity as a major cause was later developed in the well-known contraction hypothesis. The idea of assuming a system of forces operative over large areas of space that have shaped the structure of certain areas of the earth corresponds to our present ideas of orogeny and tectonic geology.

Finally, the close *interaction* that Humboldt's hypothesis established between geology and geography should not go unnoticed. The whole work on the American trip (as well as all later writings of Humboldt) is permeated with it. Humboldt's over-all influence even in this area of comparative science alone cannot be discussed here. His basic thought was that only universally conducted geographic-geognostic research on the present condition of the earth could lead to the understanding of past geological processes. "Since for the history of the earth no other monuments than the present state of the earth exist, the earth's history can be handled geognostically only when the description of the earth is available."[50]

Another hypothesis was closely connected with that of the "loxodromism of the strata": the hypothesis of the "identity of the strata," which, in turn, was inseparably connected with Humboldt's ideas about the relative age of rocks. At the time that he first concerned himself with the hypothesis that there is a law that governs the strike, he began this work also. "I had already started it in 1792, a long time before my departure for the Cordilleras, when after leaving the Freiberg School, I was charged (as Chief of Mines) with the direction of mines in the Fichtelgebirge."[51]

The often-encountered phenomenon, that the composition of a rock can show certain differences, led some miners and geologists as late as the close of the eighteenth century to the erroneous conclusion "that each section of the globe has an entirely different geological constitution."[52] This was a typical interpretation for narrow-minded empiricists. It was, from the start, incompatible with Humboldt's universal, "idea-connecting" thoughts. A series of observations by dif-

[50] Hanno Beck, "Das literarische Testament Alexander von Humboldts 1799," *Forschungen und Fortschritte, 31* (1957), 70. In this citation of Humboldt one finds not only a strict separation of the descriptive element from the hypothetical, but also the actualistic characteristic and, of course, by implication, the rejection of all other sources except nature's own.

[51] Humboldt, *Essai géognostique*, p. 7.

[52] *Ibid.*, p. 2.

ferent geologists made it plain, even at that time, that, at least in Europe, in addition to the "types of Primitive rocks," such as gneiss and granite, whose universal distribution was generally recognized, the younger "Floetz rocks," such as limestone and sandstone, occur in completely different places in the same, or at least very similar, stratigraphic relationships. And Humboldt, from his observations, worked out the hypothesis that even the characteristic sedimentary rocks must be distributed throughout the whole of the earth, and that the same rocks must have been deposited everywhere at the same time.[53]

This hypothesis also had a high degree of universal applicability, which proceeded from the assumption of a uniform history of the earth. It must not go unnoted here that Humboldt, at the time he conceived these ideas, was strongly under the influence of Werner's geological system. In this system, the formation of the earth's crust was explained as a result of all-encompassing processes, which everywhere on earth had to lead to structures that are basically the same, although with certain differences in regard to place and time. The assumption that these events almost exclusively bore the character of sedimentation (neptunism) was, to be sure, incorrect in its exclusion of other explanations; but because of the universality of the hypothesis, it raised many questions that had to be answered, and thus it stimulated further geological research.

On the basis of his hypothesis, Humboldt carried out extensive investigations of the rocks and their geological succession in America. These researches contributed more than all his previous researches to the knowledge of geology and petrology of South and Central America. And there Humboldt found his hypotheses substantiated. "There is an identity of rocks, where the organic beings are the most diversely modified."[54] Also, the approximately contemporaneous formation of identical or similar rocks proved to be a fact. However, this problem proved to be certainly more complicated than Humboldt himself had at first assumed. It became apparent that the correspondence among the rock strata in various regions of the earth manifested itself not so much in a correspondence of all the individual strata as in the correspondence of their "geologic succession" (alternation of beds). The "identity of strata" occurs primarily with regard to "formations."

This realization, which was derived from an ever-increasing number of observations, meant not a refutation but an expansion of the hypothesis, thereby proving its adaptability. Humboldt could certainly

[53] Bruhns, Vol. III, p. 126.
[54] Humboldt, *Essai géognostique*, p. 4.

have carried the results of research further had he made full use of the index fossils, a method developed primarily by Smith, rather than determining the rocks almost exclusively on a mineralogical-petrographical basis. His records and collections were so exact and extensive, however, that he and other scholars could subsequently evaluate the geological material collected on his trip from paleontological points of view. In the general part of his *Essai géognostique sur le gisement des roches dans les deux hémisphères* (1823), Humboldt discussed in detail the importance of fossils for geology,[55] and he did not neglect to give Werner credit for his own interest in the study of fossils.[56]

Of course, Humboldt also advanced numerous hypotheses in a number of other sciences, of which several were quite important to later research. The hypotheses I have discussed here, however, seemed to me to be especially suitable as examples of the role of hypothesis in Humboldt's research. They were part of his geological researches from his student days. They changed along with his personal development but remained valid elements in his attempts to arrive at general concepts from his countless observations. Between the speculative dogmatism of an idealistic explanation of nature and what is, after all, an agnostic empiricism in a simple description of nature lies the area in which the scientist deduces the general from isolated and particular observations and formulates the law that governs the constant changes of matter, in order to test it over and over again in nature. Hypothesis is a legitimate and necessary part of theoretical thought, and the hypotheses that I have discussed here show very clearly this relationship between practice and theory and are a fitting proof of how fruitful work with substantiated hypotheses is for the researcher of nature.

Humboldt's relations with his teacher, Werner, have been repeatedly alluded to. In closing, it should be mentioned once more that even in regard to the progress that geological knowledge had made in his day in working out the universal from the many isolated observations, Humboldt considered Werner as a guiding example. In no way did he see in Werner the obstinate dogmatist that several later historians of geology have attempted to make of him; he saw the man who gave his students a foundation and method with which to carry on truly scientific research, in which theory and practice were united.

55 *Ibid.*, 36–57 *passim.*

56 *Ibid.*, p. 51. Humboldt mentions here Werner's great satisfaction when his student von Schlotheim began to devote himself to the study of paleontology. Later, von Schlotheim became one of the outstanding cofounders of the science of paleontology.

Because the thirtieth of June, 1967, was the one hundred fiftieth anniversary of Werner's death, I would like to close this paper with a few sentences in which Humboldt honored his teacher:

> [Werner] was conscious that his real glory was rather founded on the discovery of the principles of the science, and on the means of research, rather than on the results obtained at a particular epoch. Werner showed no less regard for such of his pupils as differed from him on the subject of the relative age, and the origin of some formations.[57]
>
> The physical sciences are in large part based on inductions; the more complete these inductions are, the more local circumstances which accompany each phenomenon are excluded from the statements of the general laws. The history of geognosy itself justifies this assertion. Werner, in creating the science of geognosy, recognized with admirable perception all the relationships necessary to envisage the independence of the primitive formations, of the transitional and secondary. He pointed out what it is necessary to observe, and what it is important to know. In countries which he was not able to visit he prepared, in advance so to speak, a part of the discoveries with which geognosy was enriched after him.[58]

57 Humboldt, *Essai géognostique*, p. 63.
58 *Ibid.*, p. 61.

Geoffroy Saint-Hilaire Versus Cuvier: The Campaign for Paleontological Evolution (1825–1838)

Franck Bourdier

Amlinsky asserted in 1956 that Geoffroy Saint-Hilaire was second only to Lamarck among the precursors of Darwin.[1] In 1959, independently of Amlinsky, I assigned to Geoffroy a prominent place among the originators of evolutionary paleontology,[2] and I have since assembled some new documents on the subject that form the basis for this paper.

In order to understand the differences that arose between Geoffroy and Cuvier, we must go back to their youth, attempt to comprehend the conflict between their temperaments, and follow the development of their very different careers.

Etienne Geoffroy, surnamed Geoffroy Saint-Hilaire, was born near Paris in 1772. He was destined for the priesthood, but he preferred the natural sciences, and, at the age of eighteen, he formed a close friendship with the celebrated Abbé Haüy, the founder of modern crystallography, who was to have a profound influence on him. Haüy lived in a rather strange milieu, where Christian mysticism, classical ideas of friendship, and interest in the natural sciences existed together

Translated from the French by Elaine Economides and Cynthia Ware.

[1] I. E. Amlinsky, "Révision de la valeur historique de l'œuvre d'Etienne Geoffroy Saint-Hilaire," *Actes VIIIème Congrès internat. d'hist. des sci. Florence-Milan, 1956* (1958), II, 743–746. Also, I. E. Amlinsky, *Geoffroy Saint-Hilaire et sa lutte contre Cuvier* (Moscow: Inst. d'hist. des sci. et des Tech. de l'URSS, 1955), 424 p., 26 fig., 391 ref. and bibliog. of Geoffroy (327 titles). Théophile Cahan, *La vie et l'œuvre d'Etienne Geoffroy Saint-Hilaire* (Paris: Presses Universitaires de France, 1962), 318 p., 112 ref., and bibliog. of Geoffroy (259 titles).

[2] F. Bourdier, "Quelques aperçus sur la paléontologie évolutive en France avant Darwin," *Bull. Soc. Géologique de France*, ser. 7, t. I (1959), 882–896, and F. Bourdier, "Trois siècles d'hypothèses sur l'origine et la transformation des êtres vivants (1550–1859)," *Revue d'hist. des sci., 13* (1960), 1–44.

in harmony. Like Maupertuis and Buffon, the philosophers of his youth, Haüy strove to combine "details under the most general conceptions" in a constant search for "the idea of greatest simplicity . . . the basis of the most successful explanations." According to Buffon, all animals were constructed according to a single plan, sustained almost without change from man on down to fish. Similarly, Haüy saw a unity of plan in the realm of crystals, where, he said, "all secondary forms are no more than varieties of the primitive form . . . which have been produced by excess or deficiency."[3] Haüy had begun as a botanist; he had become interested in fish and even in comparative anatomy and had done a summary of Vicq d'Azyr's later works on the role of the yolk in the growth of the newborn chick.[4] Geoffroy could have found in Haüy's thought the guiding principles of his own later research on embryology and comparative anatomy and on the unity of design in living creatures; Haüy could also have been responsible for his rather mystical sense of the underlying unity of a universe in which like attracts like. Cuvier intimated that Geoffroy and Lamarck had borrowed their biological theories from Kielmeyer. But all of them had access to the same sources: Leibniz, Maillet, Maupertuis, Buffon, Vicq d'Azyr, and even Robinet; and it is possible that Kielmeyer's ideas, if Cuvier introduced them to Lamarck and Geoffroy in 1795, served simply to reinforce theories with which they were already acquainted and which were more commonly held at the time than one would think.

Georges Cuvier was born in 1769 to a Protestant family in the little principality of Montbéliard, which had remained French in language although it was a dependency of the Duchy of Württemberg.[5] His family planned for him to become a minister, but he was fascinated by natural history from his early childhood. In 1784 he was entered as a scholarship student at the Caroline Academy in Stuttgart, where he became one of a group of most brilliant students known as "knights." The following year, when he was sixteen, he became a friend of Kielmeyer, a student four years his senior, who taught him to dissect and perhaps introduced him to comparative anatomy. In April of 1788 Cuvier had to leave the Caroline Academy, and he took a position as

3 Abbé R. J. Haüy, *Essai d'une Théorie sur la Structure des Crystaux* (Paris: Gogue 1784), pp. 229–230, 215, 34.
4 Vicq d'Azyr, *Oeuvres*, 4 (1805), 398.
5 There is no recent study available on Cuvier's work except W. Coleman, *Georges Cuvier, Zoologist* (Cambridge, Mass.: Harvard Univ. Press, 1964), 212 p. There are some biographical details and material on his private life in John Viénot, *Le Napoléon de l'intelligence: Georges Cuvier* (Paris: Fischbacher, 1932), 249 p. There is also a good evaluation of Cuvier in Ducrotay de Blainville, *Cuvier et Geoffroy Saint-Hilaire* (Paris: 1890), 446 p.

Plate I

LOUIS JACQUES THENARD (1777–1857) and RÉNÉ-JUST HAÜY (1743–1822)

"*Haüy de bonne humeur sort de l'Institut et son confrère, le Baron Jacques Thénard, sort de moindre humeur.*" Caricature, probably by Pigal, J. Boilly, or Carl Vernet, between 1809 and about 1820. [*Réné-Just Haüy* (Paris: Soc. Française de Minéral., 1954), Pl. 7.]

tutor for a noble Protestant family in Normandy, where he remained for seven years.

During these seven years, Cuvier extended his research in botany and zoology over the Norman coast and countryside and performed many dissections. Kielmeyer, having completed his training at Göttingen under the naturalist Blumenbach, had returned to the Caroline Academy as a professor, and during this period Cuvier had access through his young friend Pfaff[6] to summaries and even to the texts of Kielmeyer's still unpublished lectures.[7]

In his correspondence with Pfaff, Cuvier engaged in lengthy discussions of Kielmeyer's ideas, which were related, as we shall see, to ideas that Geoffroy was to develop later. Cuvier disputed these on the basis of the two great principles that would direct his thinking until his death.

The first of these principles was that scientific theories are useless and even prejudicial and that it is only the facts that count. As early as 1788, when he was nineteen years old, Cuvier wrote to Pfaff: "I wish that all that experience teaches us were carefully disentangled from all hypothesis." Two years later he repeated to Pfaff: "Science should be based on facts, all systems notwithstanding."[8] And two months before his death, on March 15, 1832, he asserted with authority: "Mere imaginative concepts, theoretical dissertations, hypotheses . . . soon follow into the same obscurity the hypotheses or theories which came before them. . . . this collection [the *Nouvelles Annales du Muséum*] . . . will contain only positive facts."[9]

The second principle was that God had created each species, and even each being, to play a definite role in nature; because this role had been fixed from the very beginning, species must vary but little in the course of time, and the same holds true for their organs. Cuvier wrote to his friend Hartmann: "The entire structure of an animal is in necessary harmony with its way of life."[10] Forty years later he declared to the Collège de France: "Each being is perfect and capable of existence in accordance with its order, species, and individuality."[11] As early as 1788 Cuvier had recognized that his ideas were similar to those of

6 *Lettres de Georges Cuvier à C. M. Pfaff*, trans. from the German by L. Marchant (Paris: Masson, 1858), 314 p.
7 *Lettres à Pfaff*, index under the entry "Kielmeyer."
8 *Lettres à Pfaff*, pp. 73, 160.
9 *Nouvelles Annales du Muséum, 1* (1832), ii–iii.
10 G. L. Duvernoy, *Notice historique* . . . [*sur*] *Cuvier* (Paris: Levrault, 1833), 125 p.
11 Last lecture of 8 May 1832 as reported by Baron Pasquier, *Eloge de Cuvier à la Chambre des Pairs* (J. Viénot, *Le Napoleon de l'intelligence*, p. 210).

Bernardin de Saint-Pierre, who thought that the melon had ribs so that it could be divided into equal parts and eaten by a family.[12]

Cuvier thought that beings, with rare exceptions, had been created at the same time and were dependent upon one another; to believe in the existence of an initial being from which all others had been derived was, he said, "to force metaphysical reasoning to the point of ludicrousness."[13] As early as 1791 he had declared himself in opposition to the idea of a "great chain which links beings to one another in nature," an idea that Pfaff had probably borrowed from Bonnet by way of Kielmeyer. The following year Cuvier wrote to Pfaff:

> I will have no part of the obscure metaphysics with which you conclude this article [on insects] . . . I believe, I see that aquatic animals were created for water and others for air. But that they are the branches, or the roots, or even parts of the same trunk is again something which I cannot understand . . . I do not soar high enough to see terrestrial life thus; my way, although longer, will perhaps lead me more safely to the end while the sun melts your wings.

Thus did the cautious Cuvier threaten his dear friend Pfaff, and Kielmeyer behind him, with a quasi-divine punishment.[14]

In Paris, too, naturalists were concerned with the hypothesis of the chain of beings; Lamarck claimed that his interest had been aroused by a passage in *Le Voyage du jeune Anacharsis en Grèce* (1787), in which Barthélemy showed that certain Greek philosophers had believed in such a chain. In 1793 Lamarck and Geoffroy both became professors at the Musèum national d'histoire naturelle, formerly the Jardin du Roi. Lamarck, a botanist, was appointed to the vast and chaotic field of the invertebrates; Geoffroy, who was especially interested in mineralogy, was given the field of quadrupeds and birds. Both had everything to learn, and they seem to have become friends quickly. At the end of a year Geoffroy published his first scientific work, a memoir on the aye-aye. The manuscript contains a passage in which he referred to the chain of being, calling it a premature and visionary hypothesis; but he cut out this passage before the work went to press.[15]

At the beginning of 1795 Cuvier went to Paris in search of a position. He displayed the notebooks of drawings and descriptions that he

12 *Lettres à Pfaff*, p. 73.

13 Manuscript of lectures at Collège de France, Feb.–Mar. 1832, Bibliothèque du Muséum, MS 1601, fol. 93a.

14 *Lettres à Pfaff*, pp. 214–215, 270.

15 Isidore Geoffroy Saint-Hilaire, *Vie, travaux et doctrine scientifique d'Etienne Geoffroy Saint-Hilaire par son fils* (Paris: Bertrand, 1847), 479 p., with detailed bibliography arranged by subject. See pp. 42, 131–133.

had accumulated over seven years, and Parisian naturalists were amazed at such an accomplishment, all the more so because they knew nothing of what he owed to Blumenbach for the classifications[16] and to Kielmeyer for the anatomy. Geoffroy was the most enthusiastic of his admirers. During 1795 they lived together and published under their combined signatures. In a memoir on tarsiers, they stated that "this genus could be considered the link between quadrumana and chiroptera or bats."[17] and in a memoir on orangutans they spoke boldly of the origin of species from a single initial type.[18] Had Geoffroy led his friend to accept theories that the latter had rejected only a few years before? It is more likely that Cuvier still held to his own ideas and that fixity remained his guiding principle in more senses than one, but that he did not want to advertise his fundamentalist convictions during the anti-religious regime of the Convention. By July of 1795, less than six months after his arrival in Paris, Cuvier had achieved financial and moral independence, for he had been appointed professor at the Écoles Centrales, which had replaced the Universities. Then, in December of that year, the Directory, having reorganized the old Academy of Sciences, enrolled Cuvier in its membership. Geoffroy was not to enter the Academy for another twelve years. Thus, in less than a year, Cuvier had outstripped him in the *cursus honorum;* their close collaboration ceased at the beginning of 1796.

The Degeneration of Species and the Mummified Fauna of Egypt

In 1796, while Cuvier regained his freedom, Geoffroy and Lamarck continued to be preoccupied with the origin of species. Over their joint signatures they wrote to an American correspondent, asking him to obtain for them some of the animals that had been introduced into the New World from Europe. Their purpose was to study the "degenerations" that the change of environment had caused these animals to undergo. Such a study, they added, might give them more accurate ideas about species in general.[19]

In 1798 Napoleon embarked on his famous Egyptian campaign, accompanied by a cohort of scholars. The cautious Cuvier remained in Paris; the impetuous Geoffroy left. During the years 1798 to 1801, Geoffroy encountered a thousand dangers, but he brought back to the

16 Ducrotay de Blainville, *Cuvier et Geoffroy Saint-Hilaire*, pp. 68–69.
17 *Magasin Encyclopédique de Millin, 3* (1795), p. 147.
18 *Ibid.*, pp. 451–463. *Journal de physique*, t. 46, p. 185, *Rapport Général des Travaux de la Soc. Philomatique de Paris, 2* (1798), 107.
19 B. Dean, *Science, 19* (1904), 798–800.

Muséum a most splendid collection of specimens. Lacépède, Cuvier, and Lamarck wrote in 1802 that this collection

> contains animals from every century. For a long time we have wanted to know if species change in form with the passage of time, a question which is seemingly trivial but one which is nevertheless essential to the history of the earth and therefore to the solution of a thousand other questions which touch even upon the highest objects of man's veneration [i.e., upon religious beliefs]. . . . [The] ancient Egyptians . . . left us some almost complete zoological collections in their sacred grottoes;

yet that these animals, mummified two or three thousand years ago, were "identical to those of today."[20] When Cuvier signed this report he no doubt thought that the lack of change in the mummified fauna of Egypt was proof of the fixity of species; estimating as he did the age of the present continents at about ten thousand years, he considered three thousand years an extremely long time. By contrast, Lamarck, one of the other signers of the report, considered the few millennia that the fauna represented to be negligible in comparison to the millions of years that he estimated as the age of the continents in his *Hydrogéologie*, published in this same year; three thousand years, he said, was not enough time for species to change in any perceptible way.

Cuvier Versus Lamarck

In 1804 the Pope was in Paris for the coronation of the Emperor, and the government had become favorable toward the expression of religious beliefs. Cuvier had a reputation for half-hearted piety, and his audience at the Atheneum was therefore amazed when he began his course in 1804 with a profession of faith. He then went on to attack Lamarck, whose materialism was well known; he ridiculed Lamarck's *Recherches sur l'organisation des corps vivants* (1802), joking about the idea that organs were formed by habits. The Italian geologist Pencati, who recorded this incident, goes on to say, and not without irony, that Cuvier would henceforth be a formidable opponent of atheism.[21]

Much has been said of the public insult that Napoleon inflicted upon the unfortunate Lamarck in 1809 when he refused to accept the book — probably the *Philosophie zoologique* — that Lamarck offered him. It seems that Cuvier had carefully arranged this incident.[22] At the

20 E. Geoffroy Saint-Hilaire, *Lettres écrites d'Egypte* (Paris: 1901), 280 p., see pp. 253–258. *Annales du Muséum 1* (1802), pp. 234–241.

21 J. Viénot, *Le Napoléon de l'intelligence*, p. 100.

22 P. Pelsneer, *Annales Soc. Roy. Žool. et Malacol. de Belgique*, t. 50 (1919), 53–81. Bibliothèque de l'Institut, fonds Cuvier, MS 252, pc 47.

end of the *Philosophie zoologique*, Lamarck had represented the genealogy of vertebrates in the form of a tree, which, spreading as it did to encompass antediluvian millennia, stated by its very presence the principle of paleontological evolution.

Buffon had recommended to his followers the study of nature's antiquities, or extinct species, which, he said, were rich in philosophical lessons. In 1795 Cuvier and Geoffroy published a short paper on elephant fossils;[23] but from then on Cuvier was to have no partner in this field, particularly after the discovery of the fauna in the gypsum of Montmartre in 1798. He quickly became famous throughout all Europe for his reconstructions of mammals. His method in this work stemmed from his concept of living organisms; believing that God had created nothing useless in nature and had given each being a role that every one of its parts was suited to fulfill, Cuvier thus thought it possible to assert: "At the sight of a single bone, of a single piece of bone, I recognize and reconstruct the portion of the whole from which it would have been taken. The whole being to which this fragment belonged appears in my mind's eye."[24] In other words, he was reconstructing the intentions of the Creator, a method that led him fairly often to erroneous conclusions. Because of his fame he received paleontological specimens from all parts of Europe; and in 1812 he collected his observations in his *Recherches sur les ossements fossiles de quadrupèdes*. In its pages, he denounced Lamarck's opinions:

> In our time some individuals, more prone to free-thinking than ever . . . say . . . that fluid breeds creatures which are at first very simple, such as monads . . . with the passage of time and the adoption of different habits, the various kinds of creatures become more complicated and diversified . . . M. de Lamarck . . . has expounded this system of thought.

One can sense Cuvier's indignation at the persistence of "individuals, more prone to free-thinking than ever" under a government that was reinstating organized religion.

Geoffroy's First Attacks on Cuvier

Cuvier had become a very powerful personage; in 1803, having been named professor at the Collège de France, professor at the Muséum, and Inspector-General of Public Instruction, and on the verge of becoming Perpetual Secretary of the Academy of Sciences, he wrote to one of his friends: "I stand well enough with those in power to be able to hope for everything."[25] He was to make skillful use of men in high

23 *Bull. des Sci. de la Soc. Philomatique de Paris, 1* (1791–1798), 90.
24 Cited by Geoffroy, *Comptes-Rendus Acad. Sci.*, 7 (1837), 116.
25 Letter to G. L. Duvernoy, in Viénot, *Le Napoléon de l'intelligence*, p. 119.

places to become Councillor of State, baron, Grand Officer of the Legion of Honor, etc., etc., etc. This inordinate accumulation of titles, and of salaries as well, did not win him much liking among his colleagues at the Muséum, who were rather fearful of his power. Nevertheless, his relations with Geoffroy remained courteous, at least in appearance, until 1820. It was then that Geoffroy, carried away by his theory of the unity of structure of living things, suggested that the carapace of insects and other articulata corresponded to the vertebrae of vertebrates—that, in other words, the insect lived within its vertebrae. There was no specific evidence to support this absurd theory, and Cuvier attacked it, but orally and in private. Geoffroy could not reply in writing, and he was exasperated: "Cuvier may attack my theory, he may attack it as vigorously as his convictions dictate, but he should at least do it publicly." Cuvier refused, worse yet, he published some favorable remarks on Geoffroy and his comparative anatomy of vertebrates.[26] Geoffroy was furious. Early in 1823 one of his friends wrote to him: "Since your baron [Cuvier] is taking refuge in amphibology . . . you should take your aim through the underbrush in which he is hiding."[27] The following year, there appeared a good target for Geoffroy to aim at: volume five of the second edition of Cuvier's *Recherches sur les ossements fossiles*. In its pages Geoffroy was to discover the "crocodiles" of Caen, which would provide him with the foundations of a new discipline: evolutionary paleontology.

The "Crocodiles" of Caen: An Intermediate Link

At the end of 1815 a young medical student who had come to Paris from Caen was astonished to see at the Muséum the "skeleton, almost intact, of an animal unknown in the living world." This was Amand Eudes, who was to go by the name of Eudes-Deslongchamps. He read Cuvier's works, and, on returning to Caen in 1817, he discovered in a friend's attic two blocks of limestone containing bone fragments, the gift of a grateful patient.[28] These blocks had come from the quarries of Allemagne, a few kilometers from the city; Eudes and his friends

26 Isidore Geoffroy Saint-Hilaire, *Vie, travaux et doctrine scientifique*, p. 374, note 2.

27 Letter from Félix Frestelle, Bibliothèque du Muséum, nouv. acq., fonds Geoffroy. At the Bibliothèque de l'Institut, there is a letter from Geoffroy to Cuvier, 12 Jan. 1824, that shows the extreme tension existing between them at the time (fonds Cuvier, MS 246, pc 11).

28 A. Bigot, "Jacques Amand Eudes-Deslongchamps, 1794–1867," *Bull. Soc. Linnéenne de Normandie*, ser. 9, vol. 5 (1945–47), 74–93, portrait, 249 ref. J. A. Eudes-Deslongchamps, "Histoire d'une Vocation," *Bull. Soc. Lin. de Normandie*, *10* (1896), 26–49.

went there and found other blocks containing bone fragments. Félix Lamouroux, professor of natural history at the University of Caen and an ambitious man, wanted Cuvier's support;[29] he urged the young people to present the illustrious scholar with their find, and he would act as intermediary. They refused at first, but then one of them agreed to exchange one of the blocks for a rare shell from the Indian Ocean, and Lamouroux hastened to send the block to Cuvier.

The gift seemed small; but when the block had been skillfully hollowed out by Cuvier's assistants, it was found to contain the left half of a "crocodile" skull, well preserved and belonging to a body that Eudes-Deslongchamps and his friends had in large part recovered. After lengthy negotiations, Cuvier obtained casts of the bone fragments that had been kept at Caen; and in December 1823 he informed the Academy of Sciences of this discovery, which Lamouroux had already described in summary fashion. Cuvier, overburdened with business, was working and publishing too fast; in his treatment of the "crocodile" of Caen in volume five of his *Recherches sur les ossements fossiles*, he used mediocre evidence, such as an inexact sketch that Eudes-Deslongchamps had made from memory of a fragment for which Cuvier himself possessed a good cast. Of even greater significance was his inexact classification of the skull as that of an extinct species of gavial on the basis of a hasty examination, which would serve Geoffroy as grounds for criticism.

Since the Egyptian campaign, Geoffroy had become a specialist on crocodiles, and he had just completed an elaborate comparative study of the bones of the nasal canal in the vertebrate series. Now he conceived the idea of examining the nasal canal of the "crocodile" of Caen. He went to the gallery of the Muséum, studied the skull donated by Lamouroux, and ascertained that the canal was quite different from that of the gavial but resembled, he thought, the mammalian structure. This verified one part of Lamarck's genealogical tree, and it could serve as a weighty argument against Cuvier's theory of the fixity of species. Geoffroy named the fossil of Caen *Teleosaurus*, the saurian that announced from afar (teleo), from the Jurassic, the coming of mammals at the beginning of the Tertiary. If modern paleontology no longer places the teleosaurians among the direct ancestors of mammals, it does admit the importance and distinctive character of this group, declaring Geoffroy to be right, as opposed to Cuvier.

29 J. A. Eudes-Deslongchamps, "Notice sur . . . J. F. V. Lamouroux," *Mémoires Acad. Roy. de Caen* (1829), pp. 257–283. Bibliothèque du Muséum, MS 627, pcs 64, 65, 73, 74; MS 629 (26 pcs). Bibliothèque de l'Institut fonds Cuvier, MS 240, pcs 8, 9; MS 244, pcs 41, 47, 48.

Geoffroy was aware of the importance of his discovery; in a publication of 1825, he compared himself to Newton.[30] but he did not forget to mention his venerable colleague and friend at the Muséum, M. de Lamarck. The earth, Geoffroy wrote, has undergone great changes, particularly during the diluvian epochs, and these changes could have acted upon living creatures in accordance with Lamarck's two laws, which stated that the frequent use of an organ strengthens it and that the effects of the external environment on living beings become hereditary when they have operated for a long time. If the mummified animals of Egypt were similar to those of today, said Geoffroy, it was because three thousand years was too short a time for change. He recommended to young people that they read Lamarck's *Philosophie zoologique*, available from the author at the Jardin du Roi. Geoffroy's son tells us that his father was made uncomfortable three times in the course of the Restoration because of his ideas; he was an official under a most Christian monarch who could dismiss him. It was for this reason that, after praising the materialist Lamarck, he hastened to compose a short tribute to the divine intelligence — a standard procedure; and Geoffroy, a deist, was almost sincere.

Cuvier was doubly annoyed, first, by Geoffroy's triumphant intrusion into the field of paleontology, which was his own domain, and second, by Geoffroy's conclusions against the fixity of species. He considered it his duty to reply in the article "Nature," in the *Dictionnaire des sciences naturelles*, published by Levrault in 1825. He first proclaimed the implausibility of a theory that would compel the Creator to supply families of useless creatures simply to fill gaps in a ladder of being "which is pure speculation." He then went on to attack the theory put forth by Kielmeyer, who had published in Tübingen in 1796 a discourse on the gradual development of living creatures; here Kielmeyer claimed that a chick embryo began as a worm and passed through fish and reptile stages. Etienne Serres, among others, had produced specific examples in support of this hypothesis, and, in 1824, had stated his famous law of embryogenic recapitulation: "the embryos of superior classes assume in succession the permanent forms of the inferior classes." Cuvier asserted that this theory possessed some plausibil-

30 E. Geoffroy Saint-Hilaire, "Recherches sur l'organisation des gavials . . . [s'ils] descendent, par voie non interrompue de génération, des gavials antédiluviens . . ." *Mémoires du Muséum d'hist. nat.*, *12* (1825), 97–155, pl. 5–6 (read to the Academy of Sciences on 28 Feb. and 7 Mar. 1825). An undated letter from Geoffroy to Lamouroux, probably written late in 1824 or early in 1825, deals at length with the "crocodiles" of Caen; it was listed in the catalogue of autographs of the firm of G. Morssen, Paris, May 1957, and was sold, but I have been unable to discover its present owner.

ity when set forth in general terms but collapsed as soon as one attempted to apply it specifically.

In his memoir of 1825 on the teleosaurians of Caen, Geoffroy had noted that species change too slowly in nature to allow us to observe such modification, but that observation is possible when modifications, whether natural or artificially induced, result in a freak. In an attempt to produce such freaks, Geoffroy went to the village of Auteuil, near Paris, where chickens were being bred industrially by artificial incubation. He used the method of interrupting the development of the egg, particularly by coating the shell with wax to reduce embryonic respiration, and he claimed he could produce monsters at will. But, he wrote, "the possibility of worrisome interference was ever-present";[31] Cuvier might have attempted to obtain a governmental order against his research. Nevertheless, Geoffroy declared, beginning in 1828, that he was no longer afraid to acknowledge his secret objective: to test the theory of the pre-existence of germs. According to this theory, God had provided in the first pair of each species the germs (seeds) of their descendants, all identical to the original pair; this idea supported the theory of the fixity of species. By demonstrating that he could change, almost at will, the sizes and shapes of organs in the chicken, Geoffroy showed that ancestral "germs" were not as immutable as had been claimed.

Geoffroy was well acquainted with the experiments of his friend William Edwards, who had kept at the bottom of the Seine a cage of tadpoles, which were near maturity and which continued their development while retaining their aquatic respiratory system. He was also familiar with the proteus, a cave amphibian capable of reproduction while in the larval stage. He probably hoped to maintain some chick embryos in a reptile stage while seeing to it that they continued their growth; and perhaps he had visions of a chicken with teeth, as it was he who had discovered in 1821 fleeting evidence of teeth in the embryos of certain birds.[32] One thing is certain: Geoffroy, said Cuvier, was carrying out his experiments "in the interest of questions in antediluvian geology." He also intimated that geological formations could be dated according to some series like "Ichthyosaurus, Plesiosaurus, Pterodactylus, Teleosaurus, Megalonix, Megatherium, Anoplotherium, Paleo-

31 "Mémoire . . . [sur les rapports entre] les animaux des âges historiques et vivant actuellement et les espèces antédiluviennes et perdues," *Mémoires du Muséum d'hist. nat.*, t. 17 (1828), 209–229. Jean Rostand, "E. Geoffroy Saint-Hilaire et la tératogénèse expérimentale," *Rev. d'hist, des sci.*, 17 (1964), 41–50.

32 E. Geoffroy, "Sur le système dentaire des oiseaux," *Annales générales des sci. physiques*, 8 (1821), 291.

therium," and so on. Perhaps Cuvier knew Geoffroy's intentions better than we, for he said, "If he succeeds in modifying even one species, geology itself will be modified greatly in one of its fundamental principles."[33] Cuvier was to continue his attacks in succeeding years and to refer to Geoffroy contemptuously as a poet. "We," he said, "are content to describe." The clan that is "content to describe" flourishes even today.

In February 1830 there began at the Academy of Sciences the famous controversy between Geoffroy and Cuvier, the controversy that was to stir up public opinion and arouse the interest of the venerable Goethe. Geoffroy had put before the Academy an ingenious hypothesis, formulated by Laurancet and Meyranx, that lent plausibility to an analogy of design between fish and cephalopods, and for six months Cuvier had prevented its examination. Geoffroy opened fire on Cuvier, stating the case for the mutability of species; Cuvier, annoyed, countered with arrogance; and the debate continued into succeeding sessions. The public, alerted by the press, came in greater and greater numbers. On April 5, Cuvier denounced the dangerous "pantheistic system" that Geoffroy's theories concealed; the Grand Master of the Protestant Faculties of France was turning into the Grand Inquisitor for the Faith. In the face of this threat to the freedom of thought the public gave its sympathies to Geoffroy, whereupon he, on the advice of his colleagues, brought the controversy to an end and published the debates as reported in the newspapers, adding comments favorable to himself.[34]

In July the Revolution broke out, bringing the more liberal reign of Louis-Philippe. Cuvier hoped to reopen hostilities at the Academy in October; but the meeting hall was filled by a restive audience of common people who had no liking for his conservative opinions, and he deemed it prudent not to take the floor. This incident would explain why, in a moment of anger, he called Geoffroy an imbecile who allowed himself to be led by the rabble.[35] The renowned paleontologist departed to find himself a more peaceful rostrum, his course in the history of science at the Collège de France, where he strove to demonstrate the uselessness of scientific theories. In 1832, when he came to consider

33 G. Cuvier, *Histoire des progrès des sciences naturelles* (Bruxelles: Hauman, 1873), Vol. II, p. 327.

34 E. Geoffroy Saint-Hilaire, *Principes de philosophie zoologique discutés en Mars 1830* (Paris: Pichon, 1830), 227 p. J. Piveteau, "Le Débat Entre Cuvier et Geoffroy Saint-Hilaire," *Rev. d'hist. des sci., 3* (1950), 343–363.

35 E. Geoffroy Saint-Hilaire, "Résumé de quelques conditions générales des rochers," *Gazette Médicale* (1831), No. 43. Bibliothèque de l'Institut, fonds Cuvier, MS 255, pc 7.

the history of zoology in his own time, he found his opportunity to launch a vigorous assault on Lamarck, Geoffroy, and the Philosophers of Nature; on May 8, having followed up the attack, he allowed his thoughts to soar to a consideration of the Supreme Intelligence.[36] That very evening, he experienced the first symptoms of the stroke that would claim him a few days later.

In September 1830, when Geoffroy went to Caen to see his nephew Diey, he struck up a friendship with Eudes-Deslongchamps, "that most courageous, most industrious, most single-minded, and most excellent of friends."[37] Eudes had succeeded in recovering from their gangue large numbers of new skeletal fragments of saurians. Geoffroy, using this find to his advantage, presented two memoirs to the Academy, on October 4 and 11, 1830; and on March 27 of the following year he read a third, which, he said, was received in appreciative silence, a departure from custom among academicians. Its title was "Le degré d'influence du monde ambiant pour modifier les formes animales: question intéressant l'origine des espèces téléosauriennes et successivement celle des animaux de l'époque actuelle." The next day Geoffroy wrote to Eudes-Deslongchamps: "I do not think that I have ever dealt with a topic of greater significance"; in effect, he had outlined the problem of the causes of the evolution of species. Several weeks later, about April 15, he returned to Caen, accompanied by Werner, one of the Muséum's painters on vellum; he visited the quarries, studied the remains of saurians in various collections, and gathered enough material for a fourth memoir, which was presented at the Academy on May 9, 1831, and followed by a fifth on August 29. In these five great memoirs, collected and published in one volume in that same year,[38] Geoffroy allotted more space to theory than to observation. He noted, however, that the new teleosaurus skull recovered by Eudes-Deslongchamps confirmed his conclusions of 1825 on the palatine dome; he demonstrated that the teeth had a characteristic structure and that the skin thickened to form two "shells," dorsal and ventral, composed of curious overlapping scales, an arrangement unknown in the crocodile family; the teleosaurus, he said, was "dressed like a fish" in its scales and must have lived in open water, having extremities

36 See Pasquier, *Eloge de Cuvier* (see note 11).

37 Letters from Diey to Geoffroy, 1830–1833, Bibliothèque du Muséum, nouv. acq., fonds Geoffroy; two letters from Geoffroy to Eudes-Deslongchamps, 1830, 1831, Bibliothèque municipale de Caen, MS F°178, fol. 30 and 92.

38 E. Geoffroy Saint-Hilaire, *Recherches sur les grands sauriens trouvés à l'état fossile* (Paris: Firmin-Didot, 1831), 138 p., 1 pl. Reprinted in *Mémoires de l'Acad. des sci., 12* (1833), 1–138, under the title "Divers mémoires sur les grands sauriens . . ."

shaped like paddles; but vigorous probing into the blocks failed to reveal such extremities.

According to Geoffroy, the suborder of teleosaurians, descended from the ichthyosaurus, could have been the ancestor of the crocodiles of the Tertiary, with the genus Steneosaurus, the ex-gavial of Honfleur that Cuvier had described, serving as an intermediate form.

The Means of Transformation of Species

In his third and fourth memoirs, Geoffroy attempted to define his views on "the mutation of body structure," that is, the transformation of species. Contrary to Lamarck's opinion, he said, the external environment could produce lasting and significant changes only in embryos, but not in adults. The environment, or physical surroundings (*le monde ambiant*), had changed with the passage of geological time, quickly at some points and slowly at others; the continents had risen above the primordial seas, and the temperature and humidity had varied; but most important of all, the amount of oxygen in the atmosphere had been reduced when some became combined in calcareous rocks, necessitating a proportional increase in surface area in the respiratory systems of animals. In the period of the teleosaurs, he thought, birds, mammals, and even crocodiles could not have lived, because they lacked a suitable respiratory system; "the appearance of a warm-blooded animal would therefore have been a kind of mistake, and, in our common phrase, a monstrosity; it would have appeared briefly only to die immediately," just as the freaks of living nature die on contact with air when they emerge from the egg or the uterus.

Geoffroy next attacked a subject that was dear to him, that of the transformation of reptiles into birds. He envisioned such a transformation not as a slow change, as did Lamarck, but rather as the result of an "accident" in the course of embryonic development. As he had said in his *Cours d'histoire naturelle des mammifères* (1829):

> [if] the pulmonary sac of a reptile in the early stages of development should undergo a constriction at its midpoint, in such a way as to separate all the blood vessels in the thorax from the bottom of the pulmonary sac in the abdomen, there would result a circumstance favoring the development of the entire structure of a bird . . . the air . . . forced back by the muscles of the abdomen . . . [and] compressed . . . [would contain] more oxygen in a smaller volume,

resulting in the production of warmer, redder, more fluid blood, more energetic muscular action, and the transformation of tegumentary tufts into feathers.

With a certain amount of audacity, our naturalist-philosopher envisaged for the future artificial environments that would make possible the production of sons different from their fathers. He made reference to his law of the balance of organs — for example, the development of the mole's snout, which it uses to burrow, was accompanied by deterioration of the visual apparatus — an interpretation in which he differed from Lamarck. Finally, he noted that the diversification of living beings would accelerate with time, because of the multiplication of genealogical branches.

In the eighteenth century, Haller had said that the embryo became an adult only after profound transformations, which he called "evolutions"; Kielmeyer had said that the mammalian embryo "recapitulated" its ancestry as fish and reptile in the course of these evolutions; and Girou de Buzareingues had stated the same idea in 1828 when he said that the (embryological) evolution of an animal followed the course of "the very evolution of the animal kingdom."[39] There thus came into being the more modern meaning of the word *evolution,* which is no longer used to designate transformation in the embryo, but rather transformation in the living world as a whole. Geoffroy repeated Girou's formula practically verbatim in his fourth memoir: "In the evolution of a being which has passed through all the phases of its life, you have in miniature, in some respects, the panorama of the evolution of the earth." In a marginal note, he again used the word in this new meaning, but several lines further on he called the theory of evolution the theory of fixity, which he was attacking. Geoffroy's observations are precise, but his prose is ambiguous.

The Addition of the Fossil Mammals of Auvergne to the Chain of Being

On July 18, 1833, Geoffroy wrote to a friend: "I am accompanying my wife to the waters at Vichy; the Faculty wants me to drink some, too. I will obey."[40] He was counting on examining some of the fossil reptiles of Auvergne, but he was to find mammals instead. Feignoux, a geometer and amateur naturalist in Vichy, told him of the quarries of Saint-Gérand-le-Puy, located eighteen kilometers from the town; there lay great karstic fissures of the Tertiary that contained large numbers of bone fragments. In a few hours Geoffroy reaped an abundant harvest: "I gathered with ease and intoxication these curious fruits from antediluvian lands, these products of the life of a world so different

[39] Bourdier, "Trois siècles" (see note 2), pp. 38–39.
[40] Bibliothèque du Muséum, MS 2385, pc 52.

from ours."[41] Among these fauna Geoffroy discovered a ruminant that he named *Dremotherium*, an intermediate form, he said, between the muskdeer (*Moschus*) and the chevrotain (*Tragulus*) — another "connecting link" in the chain of being. The proposed connections with *Moschus* and *Tragulus* were remarkable for the time, and they demonstrated Geoffroy's thorough knowledge of mammals. Another of Geoffroy's discoveries at Saint-Gérand was the rodent *Steneofiber*, which he classified as an intermediate form between the beaver and the muskrat.

When his cure was more or less complete, Geoffroy traveled some sixty kilometers southwest of Vichy to the little village of Neschers, near Issoire. There lived the Abbé Croizet,[42] son of a poor family, who had risen to a professorship but whose "advanced" ideas were considered shocking; his bishop had therefore sent him for his health to the little village of Neschers, where he was to remain for forty years. He spent his leisure time probing Celtic and Roman ruins, excavating caves, and, above all, working the rich deposits of mammalian fossils on Mont Perrier, between Neschers and Issoire, which belonged, as we now know, to the transitional period between the Tertiary and the Quaternary. With his friend Jobert he had published between 1826 and 1828 the first part of a large work entitled *Recherches sur les ossements fossiles du Puy-de-Dôme*, which dealt particularly with the fauna of Perrier. This work, dedicated in pompous language to Baron Cuvier, reached an enigmatic conclusion: The authors claimed that they did not share Lamarck's opinions (for to have said otherwise would have made an opponent of Cuvier), but they intimated that the idea of the transformation of species was not incompatible with religious beliefs. Several weeks before Cuvier's death, Croizet had written to him to ask that the Muséum purchase his collection, not neglecting to mention the fact that the British Museum had made him several favorable offers;[43] and it is possible that Geoffroy had been charged with the task of appraisal.

In any event, "the president of the Academy of Sciences," as

[41] Unpublished letter from Geoffroy Saint-Hilaire to the Academy, 3 August 1833 (Archives de l'Académie, dossier de la séance du 12 août). E. Geoffroy Saint-Hilaire, "Palaeontographie. Considération sur des ossements fossiles la plupart inconnus, trouvés et observés dans les bassins de l'Auvergne," *Revue encyclopédique*, t. 59 (1833), 315–346 (read to the Acad. sci., 7 Oct. 1833); reprinted with slight modifications in E. Geoffroy Saint-Hilaire, *Etudes progressives d'un naturaliste* (Paris: Rozet, 1835), pp. 87–123.

[42] F. Grellet, "Eloge biographique de feu M. l'Abbé Croizet," *Mémoires Acad. des sci. de Clermont-Ferrand*, new series, *5* (*Annales, 36*) (1863), 381–406; also, *Bulletin Monumental, 25* (1859), 750.

[43] Bibliothèque du Muséum, MS 1986, pc 400.

Geoffroy called himself, "went to see the rector of Neschers, proceeding by short cuts in order to visit him in his presbytery, a true village hovel, and with the intention of paying homage to a rational spirit." In the priest's bedroom, grouped around a portrait of Cuvier, were more than one hundred extinct species of mammalian fossils, skillfully labeled and arranged; they constituted "testimony," said Geoffroy in characteristic language, "to that many links in a progressive series and to a chain of relationships."

During the summer of 1834, Geoffroy paid a second visit to Croizet, and on his return wrote to the Abbé's sister that her brother could be named canon of Lyons and professor if he would agree to donate his collection to the Muséum. The good Abbé replied in October that he would prefer to be bishop, particularly since there was a vacant bishopric available.[44] As it turned out, he remained rector of Neschers, sold his collection to the British Museum, and then put together a second collection from new excavations, which he sold to the Muséum in Paris for an exorbitant sum several years after Geoffroy's death.

Geoffroy's Last Battles for Evolutionary Paleontology

The life of Geoffroy, the eternal adolescent, consisted of a long series of nervous excitations and depressions, and he often needed the rest of the countryside. At the end of the Egyptian campaign, the mathematician Fourier thought that he had gone mad;[45] and during the last years of his life, nervous exhaustion was an almost permanent condition. One of Cuvier's admirers, Dr. Quoy, had no hesitation in declaring: "M. Geoffroy, that poor man! Loving to make noise so that people will talk about him . . . with his extravagant idealism, his unbearable elocution, and the style of a common cook . . . now, on certain subjects, he is looked upon as half-mad."[46] Geoffroy was not mistaken when he wrote that innovators, like Christ, wear a crown of thorns. Even his best friends dreaded his scientific communications, as lengthy as they were verbose.[47] But in his lectures, and especially in his conversations, the "Grand Marshal of the Grand Army of Philosophers," as Balzac had dubbed him,[48] was less affected and less vague than he was in his writ-

44 Bibliothèque du Muséum, nouv. acq., fonds Geoffroy.

45 *Lettres d'Egypte* (see note 20).

46 J. Quoy, "Notes Intimes sur Georges Cuvier Rédigées en 1836," *Archives de l'Academie Navale* (Dec. 1906), and *Bull. Scientifique de France, 41* (1907), 459–484.

47 A. Moquin-Tendon, *Un naturaliste à Paris sous Louis-Philippe. Journal de Voyage* (Paris: Mercure de France, 1943), 251 p.; see p. 116.

48 Bibliothèque du Muséum, MS 2385.

ings; Moquin-Tendon speaks of the "bizarre, witty, and often profound sentences" of "that devil Geoffroy."[49] He did not hesitate to tackle orally such dangerously materialistic subjects as the origin of life in an "animalizable" sea;[50] and when he was sixty he still had little respect for men who had "arrived," an attitude that appealed to the young and made him receptive to new ideas. "Come, next Saturday [to the Académie des sciences morales et politiques]; I'll show you around; it's fun to see mummies when they're angry."[51] To judge Geoffroy simply on the basis of his publications would be to underestimate his contributions, particularly during the last fifteen years of his life, when he was extremely active as a speaker and a correspondent among philosophers, social reformers, and even theologians.

From 1833 to 1840 evolutionary paleontology remained his major interest, but his tired and seething brain produced no extensive work. He had planned a publication on the fossil fauna of Saint-Gérand, and Werner had made several drawings of *Dremotherium* for it which were to remain unused.[52] In hopes of finding the paddle-like extremities of teleosaurus, he asked the Academy to finance excavations in the geological "charnel-houses" near Oxford. Naturally, the Academy refused, and Geoffroy set out for England, using his own funds, in September of 1836; but he fell ill in London and returned to France, his purpose not achieved.[53] The year before (1835), Eudes-Deslongchamps had obtained from the vicinity of Caen an enormous block containing the remains of a giant saurian, and for two years, Geoffroy tells us, worked away on this block with a hammer; he recorded his results in a "Mémoire sur le *Poekilopleuron Bucklandii*, grand saurien fossile intermédiaire entre les crocodiles et les lézards." Buckland was quick to point out that the skeleton, that of a megalosaurian, was the remains of the first dinosaur to be found in France. But Eudes-Deslongchamps, although far from sharing Geoffroy's philosophical ideas, was obliged to confess that his *Poekilopleuron* "had five digits on its hind foot, like a lizard, but had the physiognomy of a crocodile" and that "this curi-

49 Moquin-Tendon, *Un naturaliste à Paris*, p. 151.

50 R. Van der Elst, *Michelet naturaliste* (Paris: Delagrave, 1914), 345 p.; see pp. 123 and 131, note 2.

51 Moquin-Tendon, *Un naturaliste à Paris*, p. 67.

52 Bibliothèque du Muséum, collection des vélins.

53 Geoffroy Saint-Hilaire, *Mémoires de paléontologie* (Paris: J. Didot, n.d.), 12 p. This curious brochure, which has hitherto gone unnoticed, exists in two copies at the Bibliothèque de l'Institut (Catalogue sur registres, t. 25); cast in the form of a letter to the members of the Academy and distributed to them on 2 Oct. 1837, it contains attacks on the Perpetual Secretaries, which is probably why Isidore Geoffroy Saint-Hilaire does not mention it in the bibliography of his father's works.

ous fossil constituted a link which was well suited to connect the two families."[54] This news filled Geoffroy with enthusiasm, and he left for Caen in September of 1837 to examine the members of the newly found *Poekilopleuron.* He exclaimed in his grandiloquent manner: "God has been good enough to grant me the opportunity of observing crocodilian feet, an opportunity which . . . my partners in scientific endeavor denied me."[55]

"Geoffroy," wrote Eudes-Deslongchamps, "was planning a great work devoted to the description of my teleosaurs, for which he requested my collaboration. But . . . our specimens were no more than ornaments or examples for the pursuit of his theories on the composition of the heads of vertebrates and other lucubrations which were more to his taste." Geoffroy had seen to the execution of beautiful engravings of the fossil reptiles of Caen based on Werner's drawings, but both drawings and engravings have disappeared.[56]

Instead of pursuing extensive research, Geoffroy made several speeches in defense of evolutionary paleontology. Speaking before the Société Géologique de France in November 1833 on the subject of the mammalian fossils of Auvergne, he aired his theories on the advent of living beings:

> This coming must have been slow, gradual, and smooth, and, above all, it must have been produced by changes in the environments in which organized beings lived, changes which seem slight if one examines the life of neighboring periods, but greater if one compares the beings of more distant ages.

At the Society's next meeting, the malacologist Deshayes reproached him for his failure to cite Lamarck.[57]

Moquin-Tendon describes a session of the Academy of Sciences in 1834, in the course of which most of the members, including Geoffroy's own son, left the room as a gesture of protest when Geoffroy began a ponderous and grandiloquent reading of a memoir entitled "Puissance du monde ambiant gouvernant l'engendrement des causes auxquelles se rapportent les formes différentes et la multiplication des espèces végétales."[58] Only the title was published, but the Archives of the Academy recently acquired the twenty-four-page autograph manuscript, throughout which Geoffroy scattered examples of the "unceasing mutability of plant forms." He called particular attention to the "successive muta-

[54] Eudes-Deslongchamps, "Mémoire sur le *Poekilopleuron,*" *Mémoires Soc. Lin. de Normandie, 6* (1838), 36–146; see p. 102.
[55] Geoffroy Saint-Hilaire, *Mémoires de paléontologiste.*
[56] Eudes-Deslongchamps, "Histoire d'une vocation" (see note 28); Isidore Geoffroy Saint-Hilaire, *Vie, travaux et doctrines scientifiques* (see note 15), see p. 445.
[57] *Bull. Soc. Géologique de France,* ser. 1 t. 4 (1833), 90, 99–100.
[58] Moquin-Tendon, *Un naturaliste à Paris,* pp. 89, 105, 116.

tions" of a new species of strawberry plant, which had appeared at Versailles in 1761 and which Duchesne had succeeded in growing from seed.[59] About 1790 this new species was observed, at least in the cultures of Versailles, to reassume certain characteristics of the parent species; "such," said Geoffroy, "was the course of events in antediluvian ages, when species were produced which did not last. Before disappearing, they resulted in a line which retained many, but not all, of the former traits." It was true that this new species had lasted only thirty years; but, he said, whether thirty years or centuries upon centuries, these are but infinitesimal fragments when compared with eternity; and, he added, it is not time that counts, but the essence of the action. His metaphysics takes us rather far from Lamarck, who, as Sainte-Beuve said, "constructed a world with as few elements, as few crises, and as much time as possible."[60]

Falconner and Cauteley had discovered in the foothills of the Himalayas a strange mammalian fossil, which they named *Sivatherium*, in honor of the god Siva, and which they believed to be a link between pachyderms and ruminants. Blainville, disagreeing with them, said that *Sivatherium* was related to the camel. Geoffroy, perceptive as always in a question involving the classification of mammals, pointed out that Falconner and Cauteley were misusing Bonnet's chain of being and that Blainville was equally wrong; *Sivatherium*, he said, belonged to the giraffe family. The future would prove him right, but at the time the question resulted in a controversy between Geoffroy and Blainville that was unfolded at a meeting of the Academy on January 16, 1837. On that day the green cloth of the great table in the meeting hall was literally covered with bone specimens, both living and fossilized, that Geoffroy had brought. He launched a heated attack, a giraffe's head fell noisily to the floor, and Geoffroy's excitement increased as he proceeded. He proclaimed that the principle of the immutability of species was tottering, and that the era of Cuvier, the era of mere descriptions and classifications, was coming to a close. But the collapse of the theories of Cuvier and his students would, alas, be delayed until Darwin published the *Origin of Species* in 1859. In connection with *Sivatherium*, Geoffroy pointed out in the course of his argument that the animals could have been modified through a change of diet resulting from the appearance of dicotyledons.[61]

59 Bourdier, "Trois siècles" (see note 2), pp. 23, 24.
60 Sainte-Beuve, *Volupté, 1* (1835), 227.
61 *Comptes-rendus Acad. sci., 4* (1837), 53–60, 77–78, 80, 113, 168; article in *La Presse,* 20 Jan. 1837 (MS copy in the Bibliothèque de l'Institut, recueil Huzard, no. HR 49*).

From the time of his first works on the classification of apes, Geoffrey had been concerned with the all-important problem of the origin of man. In January 1836, he did a comparative study of the skulls of young and adult orangutans. While the adult had "features that were truly frightful and of a revolting bestiality," as he put it,

> in the young orangutan, childlike and gracious features predominated, except for the fact that the muzzle projected too far; there is the same forehead, high, broad, and well developed; the . . . same gentleness and affection; and also the same ways of indicating sulkiness and unruliness in the face of contradiction [as can be seen in man].

Geoffroy also hinted that the orangutan was one step in the development of the caveman; the encephalo-rachidian system, he said, was similar in the young orang and the infant human; although the development of this system stopped early in the orangutan, it continued in man, where it might be desirable to prolong it artificially.[62] Thus Geoffroy was the first to state the theory of foetalization, or neoteny.

Cuvier had stated, with typical dogmatism, that there could be neither men nor apes in a fossilized state (in the case of man, biblical chronology made it not possible). But on January 19, 1837, Edouard Lartet reported the discovery of the jawbone of an ape in the Tertiary formations of Sansan (Gers). The Academy was skeptical and demanded drawings; the drawings in hand, it demanded the jawbone itself. There was no doubt about it: it was that of an ape. Geoffroy called this discovery the beginning of a new era of "humanitarian" knowledge, and Lartet then had the audacity to declare that the paleontological existence of man was not at all improbable.[63]

As early as 1834, Geoffroy had espoused, "among other oddities . . . the idea that Adam was not the first man; and that there had existed before him an entire race of men with elongated heads, while Adam belonged to a type with spheroidal skulls."[64] Four years later, and one year after the discovery of the ape fossil, Geoffroy visited the University of Liège during a trip to Belgium; the professors there showed him a human skull found by the late Schmerling in the cave of Engis, among the remains of animals of extinct species. It was indisputably the discovery of the caveman, *Homo troglodites;* and as Geoffroy had predicted in 1834, the skull was quite elongated, more than it was shown to be in the illustrations that Schmerling had published (for it was a Neanderthal skull). Despite the fact that Geoffroy had abandoned paleontology the year before, he sent a letter to the Academy telling of these human

62 *Comptes-rendus Acad. sci., 2* (1836), 94–95, 584–585, 601; t. 3 (1836), p. 7.
63 *Ibid., 4* (1837), 85, 981; t. 5 (1837), 41, 159, 185–194.
64 Moquin-Tendon, *Un naturaliste à Paris,* p. 66.

remains "which they refer to at the University of Liège as the bones of antediluvian man. This term implies a theory which is accepted, here at Liège, by the forty-five members of the university." He promised the Academy a more accurate drawing of the skull at a later date.[65]

In another communication to the Academy in 1838, some months after his visit to Liège, Geoffroy proposed the conversion of the family of marsupials into a class that would be called "marsupiaires" and that would include the so-called mammals found in beds antedating the Tertiary.[66] This was probably his last publication on paleontology, but he was still to share with the Academy a few more of his ultimate visions, writing: "There is but one physics in the universe, where worlds weigh upon one another, communicating by means of an immense molecular diffusion. . . .[67] His body survived his mind by three or four years, and he died peacefully on June 14, 1844. The great embryologist Etienne Serres said over his grave that paleontology was confirming every day the theory that beings were linked to one another, from the infusoria to man, in a progressive advance that spans the geological ages. Then, inspired by Kielmeyer and Schelling, he added: "The entire animal kingdom is now understood to be like a single creature which, as it proceeds, stops in its development at various stages [corresponding to various species]."[68]

For some fifteen years Geoffroy had striven to prove from the fossil evidence the reality that lay behind Lamarck's genealogical tree of vertebrates. His extensive knowledge of classification and comparative anatomy had made him a more accurate paleontologist than Cuvier. What would remain of his work after his death?

His son Isidore, a cautious character, admitted to no more than a slight variation in species, a position to which Cuvier himself could almost have subscribed. Geoffroy had another, bolder disciple in the botanist Frédéric Gérard, a man of originality and erudition who was forced, probably by poverty, to devote his time to the writing of popular science and to commercial publishing. He had played a large part in the preparation of the first volumes of the *Dictionnaire universal d'histoire naturelle*, published under the direction of Charles d'Orbigny; this work, one of the best of the period, was sold all over the world, and

65 *Comptes-rendus Acad. sci.*, 7 (1838), 13 (meeting of 3 July); the original letter (dossier of the meeting of 25 June, Archives of the Academy), like the published text, reads *Schermidt* instead of *Schmerling;* my correction is obviously the intended meaning.
66 *Ibid.*, p. 629.
67 *Ibid.*, 9 (1839), 159.
68 Serres, "Discours aux Obsèques de M. Geoffroy Saint-Hilaire," *Gazette Médicale*, 29 June 1844.

Gérard's articles did much to spread the theories of Lamarck and Geoffroy; they also did much to acquaint the world with the word "evolution," which Gérard was perhaps one of the first to use as a matter of course in its newer meaning. His article "Espèce," published in 1844, has an almost Darwinian ring to it. As he saw it, life was nothing but destruction: All the possible combinations among species came into being only to disappear, and the individual did nothing but eat and reproduce. In an article entitled "Géographie zoologique (1845), which stretched over eighty pages of double columns, he described seven successive periods in the evolution of flora and fauna over the geological ages, and explained the geographical distribution of living creatures by stating that each continent functioned as a "center of evolution."

After 1840, bold ideas like Gérard's would give way to the theory of multiple creations, which was incompatible with that of evolution. Multiple creations left something to be desired in the way of biblical orthodoxy; but the idea permitted extraordinary growth in the realm of geological science, for geologists abandoned the idea of lithological formations and adopted that of *stages* represented by their characteristic organic fossils. Each of these stages had begun with a creation of living beings and had ended with a catastrophe that had wiped them out, and so on in a cycle.

But the very progress of geology necessitated an increase in the number of stages and, consequently, of creations. In 1849, Alcide d'Orbigny set the number of successive creations at twenty-nine; and the more geologists added to the figure, the more they came to realize that two successive creations had produced a large number of species that were closely related or common to both. They ascertained, also, that the more recent the stage, the greater the number of complex, improved—or, as they put it, perfect—species. With the blurring of paleontological breaks between the stages and the idea of increasing complexity among species, the geologists who had believed in the fixity of species slowly and inadvertently moved closer to evolutionary ideas. One would not suspect after reading Marcel de Serres' paleontological work, *Du perfectionnement graduel des êtres organisés* (1851–1852), that the author believed in the fixity of species.[69]

Among the paleontologists of the era, Louis Agassiz perhaps came closest to carrying on Geoffroy's work. An indirect disciple of Kielmeyer, he placed great importance on the study of embryos and young forms. In his *Recherches sur les poissons fossiles* (Introduction, 1844), he spoke of the "progressive sequence of the four classes of vertebrates," and the following year he presented at the Academy of Sciences a brief

69 *Actes de la Société Linnéenne de Bordeaux,* années 1852, 1853.

paper, which seems to have undergone some cutting, on "the progressive development of echinoderms."[70] Later, in a communication sent from Cambridge, Massachusetts, in February 1853, he declared it possible to assert that "the embryos and the young of all living animals, of no matter which class, are a living image in miniature of the fossil representatives of the family to which they belong"; thus, for example, the young elephant of today resembles the extinct mastodon.[71] It was a demonstration of the theories of Kielmeyer, Serres, and Geoffroy.

Like Marcel de Serres, Louis Agassiz moved farther and farther toward an evolutionist stance, all the while retaining his belief in a pre-established harmony between the creative thought of God, in which all beings originated, and human thought made in the image of divine thought. When Darwin proclaimed his theory of natural selection, Agassiz was profoundly shocked by its materialistic aspects and stated flatly that there was no proof for the mutability of species.[72]

Etienne Serres, Geoffroy's faithful friend, would also turn to an anti-Darwinian spiritualism, seeing man as a creature apart and the supreme goal of all creation.

Geoffroy remarked: "It is only the doctrines which have been debated and have emerged from the struggle completely victorious that can definitely be considered to belong to the realm of science."[73] Darwin had the honor of winning the last battle for evolutionary theory; but I believe that if Darwin had neved lived, the idea of paleontological evolution would gradually have been imposed by the facts themselves, as is shown by trends in paleontology before 1859.

This new paleontology, based on the idea of evolution, was responsible for the fame of many scholars. Their work has overshadowed that of Geoffroy, their forerunner; but he introduced an idea that seems bold to us even today, that of creating an experimental paleontology that might re-create extinct species, or create new ones, by acting upon the embryos of modern species. Tomorrow our biologists may perhaps make semi-apes out of human embryos and semi-humans out of ape embryos, and they may bring about a continuation of the evolutionary process by creating supermen. As Geoffroy said in this connection, "one need only point out the possibility of such experiments to give an inkling of the new destinies which await natural history."[74]

70 *Comptes-rendus Acad. sci.*, *23* (1846), 276–295.
71 *Bull. Soc. Géologique de France*, séries 2, *12* (1854–55), 353–354.
72 L. Agassiz, "Rapports fondamentaux des animaux entre eux et avec le monde ambiant," *Revue des cours scientifiques*, *5* (1868), 345–354.
73 Geoffroy Saint-Hilaire, *Mémoires du Muséum*, *11* (1823), 143.
74 Geoffroy Saint-Hilaire, *Recherches sur les grands sauriens* (Firmin-Didot 1831), p. 84 (cf. *Comptes-rendus Acad. sci.*, *3* (1836), p. 7).

Mineral Classification in the Early Nineteenth Century

John G. Burke

While on a lengthy visit to Paris in 1824, Eilhard Mitscherlich wrote a letter to his friend and mentor, Berzelius, giving him tidbits of gossip about French scientists as well as the latest scientific developments at the French capital. In one passage he wrote: "As for the chemical system of mineralogy, no one thinks any longer in any other terms; everyone, however, is developing a system of mineralogy of his own, according to his own method; I do not expect much good will result from this."[1]

Mitscherlich, in the first part of his message, was, in effect, announcing victory to the author of a mineralogical system based exclusively on chemistry. His own recent research into isomorphism and polymorphism had vitiated Haüy's insistence that the form of the integrant molecule, derived usually by continued cleavage of a mineral specimen, be given equal if not higher priority than chemical analysis in the definition of a mineral species. Haüy had dominated the sciences of mineralogy and crystallography for over a quarter of a century; at the time of Mitscherlich's letter he had been in his grave less than two years. Yet not one of his many students appeared willing to assume his mantle and assert that crystallographic criteria were of equal importance to chemical characteristics in the classification of minerals. However, Mitscherlich's declaration was not entirely accurate. Although chemistry did become the basis of much systematic mineralogy in the next twenty-five years, Frederick Mohs, the successor of Werner at Freiberg, adhered to a system based on a variety of mineral characters. Another prominent German mineralogist, Gustav Rose, insisted upon

[1] H. G. Söderbaum (Ed.), *Jac. Berzelius Lettres: Correspondance entre Berzelius et Eilhard Mitscherlich (1818–1847)* (Uppsala: Almgvist, 1932), p. 40.

the importance of the findings of crystallography by deriving genera from crystallographic criteria, though basing species on chemical composition, and further, by classifying separately all mineral species according to the crystal system to which they belonged.

However, as Mitscherlich predicted in the latter part of his communication, the majority of mineralogists fashioned their own individual systems according to their preferences and with their own nomenclature, and, in the process, were critical of the shortcomings of the systems of the predecessors. Although the primary purpose of classification was supposedly educational, the plethora of systems must have served rather to confuse than to instruct the novice, especially one who might have possessed the intellectual curiosity to study texts other than that written by his professor. Why did this situation occur? What were the reasons for the development of so many diverse classifications? What was the thinking of the mineralogists involved? If a given classification were widely accepted, what was it hoped that such an accomplishment would perform? It is my intent to answer or to suggest answers to these questions.

It may perhaps be informative to investigate the last queries first: Why should and why do scientists classify things — stars, snowflakes, birds, planets, or minerals — and what do they hope to accomplish in so doing? Aristotle was the first we know of who attempted to give a clear answer to this question. The first five chapters of Book I, *On the Parts of Animals*, presents a reasoned exposition of Aristotle's biological system. Therein, Aristotle states that the method adopted should "attempt to recognize the natural groups following the indications afforded by the instincts of mankind."[2] Thus, the names originally employed by men to designate groups — for example, fishes or birds — could be used as a beginning, but Aristotle insisted that the multiplicity of further differentiae eventually applied should be rigorously defined at the same time. A bipartite division, the method of dichotomy suggested by Plato — for example, flying and nonflying — was worthless, he said; in the example cited, some ants fly and others do not, and thus this method involves the breaking up of natural groups. Similarly, a privative term such as "unfeathered" would serve no useful purpose as a generic differentia. The course of exposition, Aristotle continued, must first be to state the attributes common to whole groups of animals, and then to attempt to give their explanation. In accordance with the philosophy of Aristotle, of course, the explanation was to be framed in terms of final

2 Richard McKeon (Ed.), *The Basic Works of Aristotle* (New York: Random House, 1941), p. 653.

causes. As Aristotle put it, "every bodily member subserves some partial end," and "the resultant end of her [nature's] generations and combinations is a form of the beautiful."[3]

Allowing for the fact that the number and kinds of differentiae employed in modern systems, such as botanical or zoological classifications, have multiplied far beyond Aristotle's most remote imaginings, it is true that he could rightly claim to be the originator of systematics. The chasm between Aristotle and modern taxonomists results from their diverse purposes or reasons for the systematic ordering of the products of nature. It was Aristotle's expectation to be able to derive a satisfactory explanation of the divergence of his groups from their mere ordering. Differentiated as they were by the attributes of their bodily members or structure, the reasons for the variety, he believed, might be elicited by a study of the completed classification. At least since the era of Sir Francis Bacon and René Descartes, scientists have recognized that Aristotle's hope was in vain. But according to the Baconian philosophy of science, a hypothesis might be inferred from classified facts, which further investigation might validate or falsify. For example, the inference might be made from a classification of the occurrence or absence of heat in terrestrial and celestial phenomena that heat was the violent motion of the internal particles of bodies. However, the hypothesis must remain tentative until such time as additional empirical facts substantiated it. Hence, with Bacon, the reason or purpose of classification altered.

Few scientists or philosophers of science at present discuss at any length the reason for the continued, detailed classifications of kinds in any science. There appears to be common agreement that an extremely simple type of explanation can be offered after a classification has been prepared in any science, and that classification is necessary in order to be able to handle properly a vast amount of data. Thus, C. L. Lastrucci states that classification "systems are logically necessary if the large masses of facts are to be handled in any kind of manageable order."[4] He continues: "The development of a comprehensive, precise, and functional set of classifications is a significant index to the degree of maturity of any particular science."[5] Similarly, Ernest Nagel writes:

> When a system of inclusion between kinds is achieved, it is possible to explain (even if only in a crude fashion) why some individual thing is a member of a specified kind by showing that the individual is a member of a subordinate

3 *Ibid.*, pp. 657–658.
4 Carlo L. Lastrucci, *The Scientific Approach* (Cambridge, Mass.: Schenkman, 1963), p. 41.
5 *Ibid.*, p. 78.

kind (for example, the family pet is a mammal because it is a cat and cats are mammals). Such explanations are obviously far removed from the sort of explanations to which the modern theoretical sciences have accustomed us; nevertheless, they are early steps on the road which leads to the latter.[6]

Nagel recognizes that classification is a necessary preliminary step toward the development of comprehensive theoretical science, but he maintains that there is a distinct difference in kind between explanations derived from systematic classifications and the type of explanation considered scientific at the present time.

Still, there appears to be a divergence of views on the subject. Jean Orcel writes: "A good system of [mineralogical] classification is no mere catalogue of facts; it is an instrument of research, the very basis of all theoretical generalizations."[7] In contrast, Lastrucci states: "The function of a classificatory system can be rationalized only with the framework of a given theory . . . it is the theory which both justifies and gives meaning to the particular classification of facts."[8]

The history of mineralogy since the eighteenth century supports, I believe, Lastrucci's contention. Classification has been based upon theory, and new systems of classification have developed from theoretical advances emanating from refined observation and experiment. The editors' prefaces to the seventh edition of *The System of Mineralogy of James Dwight Dana and Edward Salisbury Dana* offer the clearest proof of this fact.[9] That the massive fund of information contained in these volumes is an indispensable aid to the mineralogist is indisputable; that the system of classification adopted will lead to theoretical advance is highly unlikely.

Mineralogical classification, consciously based on theory, commenced in the early nineteenth century. Certainly, there were attempts in the eighteenth century to classify minerals on rough, qualitative chemical tests. Cronstedt, for example, based his classification on the action of fire, water, and oil upon specimens, dividing minerals into four classes: (1) *Erdarten* — those that do not dissolve in water or oil and that retain their qualities when placed in glowing charcoal; (2) *Erdharze* — those that dissolve in oil but not in water and that burn in fire; (3) Salts — those that dissolve in water; and (4) Metals — those that become fluid in fire.[10] The same rationale was used later by Kirwan,

6 Ernest Nagel, *The Structure of Science* (New York: Harcourt, Brace and World, 1961), p. 31.

7 René Taton (Ed.), *Science in the Nineteenth Century* (London: Thames and Hudson, 1965), pp. 318–319.

8 Lastrucci, *The Scientific Approach*, p. 116.

9 Charles Palache, Harry Berman, and Clifford Frondel (Eds.), *The System of Mineralogy of James Dwight Dana and Edward Salisbury Dana* (2 vols.; New York: John Wiley & Sons, Inc., 1944, 1951), Vol. I, pp. vii–xi; Vol. II, pp. v–vii.

10 A. Cronstedt, *Versuch einer Mineralogie* (Gratz: Brünnich, 1775), p. 9.

whose four classes were earths or stones, salts, inflammables, and metallic substances. Kirwan defined an earth, for example, as a tasteless, odorless, dry, brittle, uninflammable substance, with a specific gravity not over 4.9, which gives no tinge to borax on fusion, and which is not perceptibly soluble.[11] Similarly, Romé de l'Isle classified mineral crystals into four divisions according to the results of exposure to fire or water. In the class of stones, however, one genus consisted of seven species of gems; namely, diamond, ruby, sapphire, topaz, olivine, hyacinth, and emerald. From the modern point of view, or even that of the early nineteenth century, this genus is a hodgepodge, consisting of a native element, two identical oxides, and four silicates, three of which are members of the garnet group and the other of the beryl group. It seems obviously based upon rarity, commercial value, color, and brilliance. But Romé de l'Isle explained that there was another reason for grouping these gems as a separate genus. He thought that the gems were composed of small, hard platelets applied directly on top of one another, which resulted in an intimate union of the composing particles and which was responsible for the interior refraction of light and the brilliance.[12] There was, then, in the eighteenth century the recognition that qualitative chemical criteria and structure might be useful differentiae, but there simply was no body of sophisticated theory available either in crystallography or chemistry upon which a systematic classification might be based until the last decade of the century. One major reason for the diverse mineral classifications in the early nineteenth century, when chemistry and crystallography had advanced to a more successful theoretical stage, was the division of opinion as to how theory was to be applied in the construction of mineralogical systems.[13]

An avowed goal of mineralogists proposing systems built on theory was that the classifications be "natural." In the eighteenth century, the number or degree of sophistication of the proposed mineral systems had not provoked much attention to the distinction between "natural" and "artificial" classifications. By the late eighteenth century, however, the early discussions by Carolus Linnaeus and Bernard and Antoine de Jussieu in taxonomic botany were quite familiar to mineralogists. The more natural was, to a certain extent, identified with the comparison of the more remote or hidden forms of organization: in botany with the

[11] R. Kirwan, *Elements of Mineralogy* (2 vols.; London: Nichols, 1794), Vol. I, pp. 1–3.

[12] J. B. L. Romé de l'Isle, *Essai de cristallographie* (Paris: Chez Didot jeune, 1772), pp. 6–7, 148–195.

[13] A. Dufrénoy, *Traité de minéralogie* (4 vols.; Paris: Carilian-Goeury, 1844), Vol. II, pp. 6–7.

organs of plants, and in mineralogy with the chemistry and forms of the atomic or molecular units. Haüy, it should be remembered, was an amateur botanist before he was a crystallographer or a mineralogist; in fact, he was elected to the Académie des Sciences in 1783 as an associate in the botanical class, because no place in mineralogy was vacant. Throughout his works, one finds constant references to botany, and although Linnaeus is responsible for the introduction of the system of ordering minerals into species, genera, orders, and classes, Haüy should certainly be credited with the system's perpetuation. Strangely enough, while botanical taxonomy was a major influence in the development of Haüy's thought concerning mineralogical classification, Haüy proved to be influential in the future development of botanical systems, since his student in crystallography, Auguste de Candolle, applied Haüy's ideas of the regularity and symmetry of crystal forms to the field of botany.

In the first edition of his *Traité de minéralogie*, published in 1801, Haüy made a clear distinction between the systems of Linnaeus, Wallerius, and Daubenton, who employed in the determination of species, genera, and orders certain characters that were visually apparent, and those of Cronstedt, Bergman, and Kirwan, who attempted to classify minerals according to qualitative chemical criteria. Haüy did not call the employment of external characters an artificial method nor the use of chemistry a natural method.[14] These appellations were introduced into mineralogy during the 1820's, when the most devastating criticism of another mineralogical system was to label it artificial. Thus, in 1826, Berzelius termed the classification of F. S. Beudant, published in 1824, artificial, because it was based on two chemical characters: (1) the degree of volatility; and (2) the color of oxidized combinations, such that by a successive transition from one property to another, minerals composed a series.[15] Understandably offended, Beudant immediately rose to his own defense, pointing out that Berzelius had based *his* newly proposed classification on a single, unique principle (the most electronegative element), a practice, he declared, that all naturalists recognized as an indelible stamp of an artificial system.[16] The meanings of the words "artificial" and "natural" were thus protean; despite a chemical basis, a system might still be termed artificial.

[14] R. J. Haüy, *Traité de minéralogie* (4 vols.; Paris: Chez Louis, 1801), Vol. I, p. *vii*.

[15] J. Berzelius, "Des changemens dans le système de minéralogie chimique, qui doivent nécessairement résulter de la propriété que possèdent les corps isomorphes, de se remplacer mutuellement en proportions indéfinis," *Annales de chimie et de physique*, *31* (1826), 34–35.

[16] F. S. Beudant, "Sur la classification des substances minérales," *Annales de chimie et de physique*, *31* (1826), 183.

A. Dufrénoy, who published his four-volume *Traité de minéralogie* in 1844, explained that artificial classifications were employed in the first mineralogical systems in order to establish well-entrenched characters without heed to their relative importance. This practice, he said, resulted in the union of very dissimilar substances according to certain characters but simultaneously separated them from other substances that resembled them closely according to other criteria. It was only later, Dufrénoy continued, that the totality of properties was compared and substances were grouped according to their more or less greater analogy. Natural methods, he concluded, are the résumé of all observations made in a science, and if they could be extended to the last degree of perfection, they would become the exact and complete exposition of all of nature.[17] In this expression, we have an epitome of the Baconian conception of natural science.

While complimenting Haüy and Beudant for their earlier work, Dufrénoy believed that Alexandre Brongniart had proceeded further toward developing a natural classification because he had based it on several principles. Brongniart united stony substances by the electronegative element, under the names of silicates, and the metals by the electropositive element. "This method of proceeding," Dufrénoy said, "also conforms to that which occurs in zoology, where different characters are invoked for the classification of each order. The teeth and organs of nutrition present a natural principle of classification in the mammals, which is abandoned for the reptiles and the fishes, where it does not have the same value."[18] According to Brongniart, it was not necessary that a classification be based on a unique character, and in Dufrénoy's mind, this principle appeared to resolve the main difficulties of mineralogical classification. The choice of bases as the principle on which to classify the metals, he thought, gave birth to true natural groups in which there was one common element; for example, all those of lead, all those of copper, and all those of iron. On the other hand, the silicates formed a distinct class because their hardness, luster, and other characters were similar. Dufrénoy believed, then, that classification by acids ought to be limited to the silicates and viewed his resulting system as more natural and philosophical than those of his predecessors.[19]

As evidenced by Dufrénoy's reference to zoology, there was among mineralogists a feeling akin to envy that biological classifications apparently could be designed relatively rapidly and with such compelling

17 Dufrénoy, *Traité de minéralogie,* Vol. II, p. 1.
18 *Ibid.*, Vol. II, p. 13.
19 *Ibid.*, Vol. II, p. 18.

logic. One can sense frustration among the classifiers because the widely diverse characters of minerals appeared to defy their orderly and rational systematization. Haüy felt this baffllement keenly, but he attempted to overcome it in an optimistic statement that mineralogy had advantages over zoology and botany in that there were fewer mineralogical species and the study of these could be undertaken in all seasons of the year and at any spare moment.[20] William Whewell, however, announced the inferiority of mineralogists to the scientific and educated world in this statement:

> Indeed it will be found, that it is with regard to these latter objects, to organized beings, that the process of scientific classification has been most successfully exercised; while with regard to inorganic substances, the formation of a satisfactory system of arrangement has been found extremely difficult . . . The best exemplification of these branches of knowledge . . . will, therefore, be found in the organic world, in Botany and Zoology; but we will . . . take a brief view of the science which classifies inorganic bodies, and of which Mineralogy is hitherto the very imperfect representative.[21]

A fundamental and continual nagging problem, central to the establishment of a mineralogical system, was the definition of species.[22] In the 1801 edition of his *Traité de minéralogie*, Haüy defined a mineral species as a collection of substances whose integrant molecules were identical in their forms and composed of the same chemical principles united in the same proportion.[23] In essence, he identified the chemical molecule of a substance with the integrant molecule resulting from cleavage. But the fact that the integrant molecules of many substances were found to be cubes clouded this definition; in these cases, chemical analysis alone afforded evidence of specificity. Nevertheless, in his 1822 edition, despite the increasing number of substances found to belong to the cubic system and despite the intervening elucidation of the phenomena of isomorphism and polymorphism by Mitscherlich, Haüy reiterated this definition of species. For one thing, he did not accept isomorphism, and he attacked Berzelius and Mitscherlich for their ideas that "accidental" elements in isomorphous substances were in chemical combination. Also, he emphasized correctly that there were wide divergences

20 Haüy, *Traité de minéralogie* (1801), Vol. I, p. *xxi.*

21 William Whewell, *History of the Inductive Sciences* (2 vols.; New York: D. Appleton & Co., 1884), Vol. II, p. 313.

22 The committee on Nomenclature and Classification of Minerals of the American Mineralogical Society in 1923 declared: The definition reads: "A mineral species is a naturally occurring homogeneous substance of inorganic origin, in chemical composition either definite or ranging between certain limits, possessing characteristic physical properties and usually a crystalline structure." *American Mineralogist, 8* (1923), 50.

23 Haüy, *Traité de minéralogie* (1801), Vol. I, p. 162.

in the values of the angles of the cleavage molecules of substances that Mitscherlich had termed isomorphous.[24] "If all minerals were composed only of molecules proper to them, without any addition," Haüy declared, "the results of analysis would be as invariable as these molecules. But it is otherwise."[25] Geometry, he continued, makes an abstraction of these accidental principles, and although the results of chemical analysis are indispensable for denoting genera, orders, and classes, the limits of species ought to be traced by crystallography. Considered only with reference to chemical analysis, he pointed out, it would be very difficult to get a clear idea of what constitutes the species pyroxene.[26] Thus, to Haüy, the form of the integrant molecule was the principal determinant of species; consequently, in his classification, stilbite and analcime, for example, were listed as individual species, separate and apart from chabasie (chabazite) and the other zeolites.[27] Similarly, actinote (actinolite) was distinguished from the other amphiboles.[28]

In the interval between the publication of Haüy's two editions, Berzelius attempted to introduce a mineralogical classification based exclusively on chemistry. Mineralogy, Berzelius asserted, is merely a branch of chemistry and must take chemistry as its scientific foundation. Chemistry in a state of perfection, he insisted, would transfer its perfect qualities to mineralogy.[29] A mineralogical species to Berzelius, then, was nothing more than a collection of substances having the same composition and the same proportion of the composing elements.[30] The discovery of isomorphism, of course, caused a revision of Berzelius' ideas and led him to propose a new scheme of classification based on chemistry in 1826. If his former definition of species was adhered to, Berzelius stated, the changeable analyses of pyroxene, amphibole, and garnet, resulting from isomorphous substitution, would result in an inordinate multiplication of the number of species. Hence, Berzelius suggested that a species be defined by the more electronegative element present in its composition.[31] But his resulting system still had its disadvantages, since isomorphous substances might and did appear in several places in the

[24] R. J. Haüy, *Traité de minéralogie* (4 vols.; Paris: Bachelier, 1822), Vol. I, pp. 37–43.
[25] *Ibid.*, Vol. I, p. 33.
[26] *Ibid.*, Vol. I, pp. 35–43.
[27] Haüy, *Traité de minéralogie* (1801), Vol. III, pp. 151–185.
[28] *Ibid.*, Vol. III, pp. 58, 73.
[29] J. J. Berzelius, *An Attempt to Establish a pure and scientific system of mineralogy by the application of the electro-chemical theory and the chemical proportions*, trans. John Black (London: Baldwin, 1814), pp. 7–8.
[30] *Ibid.*, p. 51.
[31] Berzelius, *Annales de chimie et de physique*, *31* (1826), 6–7, 14.

classification. For example, garnet was listed eight times, augite six, and hornblende five times in the order of silicates.[32]

Beudant disagreed with Berzelius. A mineral species, he declared, was "the collection of bodies which, by the form or regular structure, the colors, the type and kind of refraction, the specific weight, the chemical composition, have between them analogies which are not found in any other."[33] He granted that chemical composition merited the most attention, and that form as determined by cleavage did not serve to distinguish species, as Haüy had claimed. Nevertheless, chemical mixtures, as Beudant persisted in calling isomorphous substitutions, presented difficulties in the way of a classification based exclusively on chemistry. This condition, he confessed, was one of the most embarrassing for the health of the science of mineralogy, since it raised the question whether there actually are species in the mineral kingdom or only individuals.[34] At first, Beudant believed that every substance having a different chemical formula ought to be regarded as a separate species, isomorphous or not, with groups of species so indicated.[35] Somewhat later, he took the view that chemical mixtures were analogous to mules in the animal kingdom, and that they should be placed in an appendix following the description of the dominant species.[36]

In his introductory chapter, Dufrénoy gave lip service to Haüy's definition of species. Mitscherlich's discovery of dimorphism, he stated, showed that Haüy's definition was too absolute, but it was not erroneous and required only a slight modification, inasmuch as only ten minerals of three hundred and fifty known species displayed two forms. "As to the theory of isomorphism," Dufrénoy declared, "it is entirely favorable to the [Haüy's] laws, because it unites under the same species, minerals that Haüy united by crystallization, for example, pyroxene and diopside." The minerals composing a species, he continued, do not have to have the same content of the elements, but must only present an identical relation between the bases or the acids they contain, or between their isomorphs.[37] Here, Dufrénoy stretched Haüy's definition of species to its maximum limits, and it is doubtful that Haüy would have accepted this distortion of his meaning despite the good intentions

[32] Gustav Rose, *Das krystallo-chemische Mineralsystem* (Leipzig: Engelmann, 1852), p. 3.

[33] F. S. Beudant, *Traité élémentaire de minéralogie* (2 vols.; Paris: Verdière, 1830), Vol. I, p. 482.

[34] *Ibid.*, Vol. II, pp. 483–491.

[35] Beudant, *Annales de chimie et de physique, 31* (1826), 194.

[36] Beudant, *Traité élémentaire de minéralogie*, Vol. I, p. 494.

[37] Dufrénoy, *Traité de minéralogie,* Vol. I, p. 18.

of the author. Later in the work, however, despite the fact that Dufrénoy still believed Haüy's assertion that the chemical and integrant molecules were identical, he declared that chemical composition was the only true basis of species.[38]

By the mid-nineteenth century, however, the accumulating mass of mineralogical literature had not served to clarify the criteria of a species nor promoted unanimity in the expression of a suitable definition. Gustav Rose described the impasse clearly:

> Whereas no difficulty arises with the combination of isomorphous bodies with respect to the determination of the degrees of combination, still the feature of their being able to combine with one another in all proportions presents a great difficulty for the determination of species. Because of this, between two species there exist a more or less great number of members, which form a more or less complete transition of the end members into one another. Where should one draw the boundary line between one or another species, and how many species ought one to accept between the end members? This question . . . has in more recent times very much concerned mineralogy, and the desire to achieve a somewhat precise determination has called forth several proposals . . . all of which have clearly proved not to be practicable . . . It appears to me that precise determinations for the definition of species in these cases are still not available, and that it must be left to the opinion of the individual how far to proceed in this direction. I am of the opinion that the definition of species cannot be dealt with uniformly. The combinations of isomorphous substances are frequently distinguished, even if only slightly, by distinctions in the angles of the crystals, in refraction, color, hardness, and specific weight; distinctions that are produced by the difference in chemical combination.[39]

Inverting Haüy's method, Rose designated the genus within certain classes according to the crystal system. He said:

> The crystal form tells us far more than the mere combining formula of the mineral; it tells us also the way in which the atoms are united; and this variety in the manner of combination of the atoms establishes more often the exterior differences of substances than the variety of the atoms themselves. Thus diamond and graphite are composed of the same atoms, and it is probably not possible to think of a greater difference in the physical condition between different elements.[40]

Thus, in Rose's view, the crystal form reflected the interior structure of the composing atoms of a substance and merited close attention and a significant role in the scheme of classification. In his "crystallo-chemical" system, however, the species depended upon the composing elements and their combining formula.

The attempts to employ the new chemical theories in classification led to difficulties in the establishment of genera, orders, and classes, as

[38] *Ibid.*, Vol. II, p. 17.
[39] Rose, *Das krystallo-chemische Mineralsystem*, pp. 11–12.
[40] *Ibid.*, p. 7.

well as species. The discovery of isomorphism, as noted previously, demonstrated the impossibility of constructing a classification based on the electropositive elements, and quite soon classification based on electronegative elements came under suspicion when it was determined that the arsenic and phosphorus radicals, for example, were also isomorphous.[41] In addition, there was no certain definition of a metal. For example, for a period after the isolation of silicon by Berzelius, there was some question as to whether it should be classed as a metal.[42] Also, in Dufrénoy's system, titanium, tantalum, selenium, arsenic, vanadium, antimony, mercury, molybdenum, tungsten, chromium, osmium, and rhodium were placed in his first class, apart from the fourth class, which comprised native metals and their combinations.[43] Further, before the advent of the concept of valency and structural theories, the imperfect knowledge of chemical combination created obstacles to systematic classification, particularly of the complex silicates. Nevertheless, there was no paucity of efforts to classify minerals on a chemical basis. Probably the most ambitious scheme was that proposed and elaborated by Nils Nordenskiöld in 1827, 1833, and 1849, which Gustav Rose used as a basis for his system in 1852. This system comprised seven classes, as follows:

1. Haploites — comprising the chemical elements
2. Diploites — comprising the union of elements with one another
3. Bidiploites — comprising the union of diploites with one another
4. Tridiploites — comprising the union of bidiploites with diploites
5. Tetradiploites — comprising the union of bidiploites with one another
6. Pentadiploites — comprising the union of tetradiploites with diploites
7. Hexadiploites — comprising the union of tetradiploites with bidiploites

The classes were divided into orders, genera, and species. The orders and genera depended entirely upon chemical formulas, the orders arising from the variety of combinations and the genera from the variety of numerical relations. Species within genera depended upon the numerical relations between the elements or radicals comprising the genus.[44] Berzelius was aghast at Nordenskiöld's temerity, declaring that the system

41 Whewell, *History of the Inductive Sciences,* Vol. II, p. 350.
42 *Ibid.*, Vol. II, p. 308.
43 Dufrénoy, *Traité de minéralogie*, Vol. II, pp. 25–26; 29–36.
44 Rose, *Das krystallo-chemische Mineralsystem*, pp. 2–6.

assumed a perfect knowledge of the composition in every case.[45] It was obvious that contemporary chemical analysis and the understanding of chemical combination did not allow such mathematical precision.

The names that Nordenskiöld adopted for his seven classes illustrate another contemporary problem involved in mineral classification, that of nomenclature. Haüy had made nomenclature revision fashionable, in the first place by renaming many minerals in an attempt to make their designations more rational. Also, each variety in the form of a crystal had a separate name in Haüy's system; for example, calcite *mixte, cuböide, unitaire, binaire, imitable, contractée.* Though Haüy handled the introduction and revision of nomenclature with consummate ease, he was severely criticized for his pains. Thomas Thomson objected strenuously to Haüy's having discarded all the well-known mineralogical terms and to his having introduced over one hundred names of his own invention.[46] Beudant stated:

> Haüy knew hardly one sixth of the crystal varieties of carbonate of lime, and yet he presented one hundred and fifty different names for this substance; from which it resulted that the crystal nomenclature of the varieties of this substance was much more extensive than the general nomenclature of species.[47]

But this criticism did not deter Beudant himself from introducing novel nomenclature. Apart from his new class names — *gazolytes, leucolytes,* and *chroïcolytes* — Beudant's family names were unusual; for

[45] Whewell, *History of the Inductive Sciences,* Vol. II, p. 350.
[46] Thomas Thomson, "Some Observations in Answer to Mr. Chenevix's Attack upon Werner's Mineralogical Method," *Annals of Philosophy, I* (1813), 241–258.
[47] Beudant, *Traité élémentaire de minéralogie,* Vol. I, p. 496.

TABLE I. Names Applied to the Six Crystal Systems

	Weiss and Rose	*Mohs*	*Naumann*
I.	Regular	Tessular	Tessular
II.	Two and one-axis	Pyramidal	Tetragonal
III.	Three and one-axis	Rhombohedral	Hexagonal
IV.	One and one-axis	Orthotype	Rhombic
V.	Two and one-membered	Hemi-orthotype	Monoclinic
VI.	One and one-membered	Anorthotype	Triclinic

example, silicides, carbonides, hydrogenides, and phtorides. Also, Beudant attempted to change the recently established chemical symbols. This step was too much for Berzelius, who complained about it to Brongniart in 1825:

J'ai une grande envie de critiquer Beudant un peu pour la gallomanie dans sa manière de changer les symboles de diverses substances, p. ex. celui de Sn en E, pour l'étain (parce qu'il a voulu prendre l'initiale du mot francais), celui de K en Po pour la potasse etc. Si nous commencons, par un sentiment puéril d'esprit national, à changer la langue universelle des symboles en autant de langues spéciales qu'il y a de différentes nations qui s'en occupant, qu'est-ce que la science y gagnera?[48]

In the face of such attacks, the more outrageous novelties in nomenclature gradually fell by the wayside. Still, it was impossible to gain a consensus for the names of the generally recognized six crystal systems. The diversity of this nomenclature is shown in Table 1.

Although the names designating the six crystal systems varied among mineralogists, it was common to note in the classifications the system to which a mineral belonged. However, this practice was not universal; for example, Francis Alger, who prepared the fifth edition of Phillips's *Treatise on Mineralogy* in the United States in 1844, preferred to refer minerals to fifteen primary forms rather than to the six crystal systems, because he thought the former "afforded a true and natural explanation of the character and relations of form among crystallized minerals."[49]

[48] H. G. Söderbaum (Ed.), *Jac. Berzelius Lettres: Lettres Détachées (1809–1847)* (Uppsala: Almqvist, 1919), p. 27.
[49] Francis Alger (ed.), *William Phillips, An Elementary Treatise on Mineralogy* (5th ed.; Boston: W. D. Ticknor & Co., 1844), p. *xxxii.*

Miller	*Beudant*
Octahedral	Cubic
Pyramidal	Right prismatic with square bases
Rhombohedral	Rhombohedral
Prismatic	Right rectangular prismatic
Oblique prismatic	Oblique rectangular prismatic
Doubly oblique prismatic	Oblique prismatic with bases shaped as oblique-angled parallelograms

By mid-century, structural theory had not progressed much beyond what it had been about 1820, when Weiss and Mohs, revising Haüy's theory, had arrived at the concept of the six crystal systems. The analysis of external symmetry by J. F. C. Hessel in 1830, which demonstrated that only thirty-two point groups could exist, was overlooked by contemporary mineralogists, and Bravais was just beginning to publish his mathematical analysis of point groups in 1850. The notion of spheroidal atoms aggregating to form polyhedral molecules had been advanced repeatedly, by Dalton, Prechtl, and Wollaston, among others, but these speculations did little to influence structural theory. Thus, all that Dufrénoy could say in 1844 was:

> The size of atoms escapes our senses, and matter continues to be divisible until each particle ceases to be appreciable; but at that point also our power to determine the form ceases; however, all probabilities being considered, it is natural for us to represent the elementary bodies by a spherical form, because it is this [form] that matter affects when it is not submitted to the influence of outside forces.[50]

Two lines of investigation were being pursued, however, that pointed in the direction of a sweeping advance in structural theory. The first was the study of optical properties of crystals, initiated by Biot and Brewster and taken up by J. F. W. Herschel and Neumann. These experiments gradually demonstrated that certain optical properties corresponded to the crystal symmetry systems, and further, they stimulated the important mid-century investigations by Pasteur of the crystals of the salts of tartaric and racemic acids. The second line of investigation, closely connected to the first, was concerned with the anomalies in the laws of symmetry presented by such substances as boracite, quartz, and tourmaline. These studies first resulted in the paper of Delafosse, published in 1840, that rejected Haüy's notion of the identity of the chemical and integrant molecule and clearly presented the idea of an internal crystal lattice. Haüy, Delafosse declared, had yielded involuntarily to the tendency of the human mind that leads us to believe that things are really as they appear to our senses, and thus had supposed that the last term of the mechanical division of a crystal, the cleavage molecule, was also a definite physico-chemical entity. This idea, he stated, had caused an arrest in the development of structural theory.[51] Brongniart, Cordier, and Beudant presented a critique of Delafosse's revision without detail-

[50] Dufrénoy, *Traité de minéralogie*, Vol. I, pp. 324–325.
[51] G. Delafosse, "Recherches relatives à la cristallisation considerée sous les rapports physiques et mathématiques," *Comptes Rendus Acad. Sci., XI* (1840), 394–400. See, also, G. Delafosse, "Recherches sur la cristallisation," *Mémoires presentés par divers savants à l'Académie Royale des Sciences, VIII* (1843), 648–651.

ing their objections to it and without accepting it.[52] Dufrénoy, however, while terming Delafosse's hypothesis "ingenious," believed that it was subject to serious objections because his resolution of the anomalies of boracite could not be extended to other substances. Delafosse's suggestions were, in fact, unprofitable at this time; he was still committed to the concept of polyhedral physical-chemical molecules, and he gave no thought, at least in print, to the arrangements of the ultimate particles or atoms. Although these studies foreshadowed the direction that future structural theory would follow, they did not in themselves present sufficient factual data from which a structural theory could be elicited — one that might be useful in the classification of minerals.

Ultimately, then, the confusion inherent in systematic mineralogy in the first half of the nineteenth century resulted from the unsettled state of theoretical chemistry and from the lack of a viable theory of the intimate structure of matter. At the time, Whewell presented this indictment of mineralogy:

> It may appear presumptuous to speak of the failure of those whom, like Berzelius and Mohs, we acknowledge as our masters, at a period when, probably, they and some of their admirers, still hold them to have succeeded in their attempt to construct a consistent system. But I conceive that my office as an historian requires me to exhibit the fortunes of this science in the most distinct form of which they admit, and that I cannot evade the duty of attempting to seize the true aspect of recent occurrences in the world of science. Hence, I venture to speak of the failure of both the attempts at framing a pure scientific system of mineralogy — that founded on the chemical, and that founded on the natural-history principle.[53]

In Whewell's view, "the object . . . of the classificatory sciences is to obtain FIXED CHARACTERS of the kinds of things; and the criterion of the fitness of names is, that THEY MAKE GENERAL PROPOSITIONS POSSIBLE."[54] He continued: "The general propositions to which we hope to be led are such as assert relations between their intimate constitution and their external attributes."[55] It was obvious that none of the proposed systems had accomplished this feat.

But did not this goal assume the perfection of both chemical and structural theory? It did; and hence, we must conclude that a system of classification in any science, as an elaboration of current theory, must remain just as primitive or as imperfect as the theories upon which the classification is founded.

52 *Ibid.*, *XIII* (1840), 205–210.
53 Whewell, *History of the Inductive Sciences*, Vol. II, p. 349.
54 *Ibid.*, Vol. II, p. 314.
55 *Ibid.*, Vol. II, p. 349.

Charles Lyell Is Permitted to Speak for Himself: An Abstract

Walter F. Cannon

Charles Lyell's *Principles of Geology* (1830–1833) was one of the most influential of early Victorian "treatises," works designed to codify a major revision in the way of looking at a body of knowledge. The *Principles* began with a lengthy sketch of the history of geology, showing that only men with views similar to those of Lyell had advanced the science. The chief obstacles to the development of truly scientific geology had been ideas originating in the Christian view of the world: Noah's flood, a limited time span for the earth, a man-centered viewpoint, and a willingness to believe in unique catastrophes and supernatural causes.

In contrast, Lyell featured James Hutton (1788) as the first person to free himself completely from these delusions, and he stressed that the opposition to Hutton had largely been motivated by a fear of radical ideas in the era of the French Revolution. Hutton was the first to work for correct principles in geology, although even he allowed too much importance to catastrophes; and he erred in thinking in terms of cycles of decay and renovation instead of continuously counteracting forces.

Lyell was thus the principal founder of the erroneous historiographical tradition, which presents uniformitarianism as the only correct or scientific approach to geology, and catastrophism as an unscientific retarding force based on religious prepossessions.

Lyell's own system, that of "absolute uniformity," postulated that the same physical laws have always worked in the same system of interactions through the same physical agencies having the same quantities

The complete text of this article will be included in my book, *The Cambridge Network in Victorian History*, to be published by Routledge & Kegan Paul in the near future.

and intensities as at present. He sketched a balanced system, in which the agencies producing unevenness in the earth's crust are continuously counteracted by those producing evenness. In this stable system of endless variation, cumulative development is impossible. Hence, in his second volume Lyell produced a classic refutation of the hypothesis of progressive organic change, as represented most notably by Lamarck's doctrine of the transmutation of species. The subjects Lyell treated in this refutation are quite similar to those in Darwin's *Origin of Species*, which among other things is itself a refutation of Lyell's demonstration that "species have a real and permanent existence in nature."

Lyell contended that individual species have come into existence and passed away without being essentially modified. The same families and orders have persisted throughout geological history; paleontology shows no evidence of fundamental change or progression in fossil forms. A disturbance of the ecological balance — which is a balance of warring species — may kill off one or more species, but the balance is always restored. This view of a self-restoring balance of nature that counteracts the effects of the war among species had already been popularized in English theological thought around 1800 by William Paley.

Lyell's third volume discussed the actual geological monuments from his uniformitarian point of view. Its material on tertiary formations was quite important, but its briefer sections on the earlier formations were soon outmoded by the progress of field work. Whether for this or for other reasons, Lyell dropped it after the edition of 1837, so that the *Principles* as it reached the audience of the 1840's, 1850's, and 1860's consisted of the first two volumes of the original, but with the details continually modified by Lyell to keep abreast of the advancing science.

It was the antiprogressionist "orthodoxy" of his friend and mentor Lyell, not that of some imagined theological orthodoxy, that Darwin had to overcome. Darwin showed that Lyell's geological dynamics, taken by themselves, were compatible with an evolutionary view in geology as well as in biology. The chief retarding effect of natural theology on geology, therefore, was not in influencing catastrophists towards progressionism, but in frightening Charles Lyell away from it.

de Maillet's Telliamed (1748): An Ultra-Neptunian Theory of the Earth

Albert V. Carozzi

Benoît de Maillet's *Telliamed, ou Entretiens d'un philosophe indien avec un missionnaire françois sur la diminution de la mer, la formation de la terre, l'origine de l'homme, . . .* etc., is known by five manuscripts, three printed editions (Amsterdam, 1748; Basel, 1749; The Hague, 1755), and two English translations (London, 1750; Baltimore, 1797). The printed editions were the product of a moderately successful editing done by the Abbé Jean Baptiste le Mascrier in a futile and pathetic attempt to reconcile the proposed materialistic system with the dogma. The details of this editorial venture have been presented elsewhere.[1]

The reaction upon publication of de Maillet's system was extremely violent and lasted for almost a century. As it was so well publicized by its opponents, *Telliamed* became a popular book and also affected all the leading naturalists of its time, particularly Buffon, Cuvier, Linnaeus, Celsius, Pontoppidan, and Bergman. In modern times, *Telliamed* took a well-deserved place as a forerunner of Lamarckism and Darwinism in all books dealing with the history of transformism and evolution. However, the fundamental significance of de Maillet's work in the development of geological concepts has not been fully appreciated, because he himself did not present a synthetical picture of all the facets of his contribution in the form of a theory of the earth. Consequently, historians of science have stressed his spectacular ideas about transformism and largely neglected his amazingly precise geological observations and their fundamental implications.[2]

1 A. V. Carozzi, "De Maillet's *Telliamed* (1748): An Editorial Venture of the Eighteenth Century," *Stechert-Hafner Book News*, XXII (1968), 97–100.
2 Sir Archibald Geikie, *The Founders of Geology* (2nd ed.; New York: Dover, 1962), pp. 84–88. K. A. von Zittel, *History of Geology and Paleontology*, trans. M. M. Ogilvie-Gordon (New York: Hafner, 1962), pp. 32–33.

De Maillet's cosmogonic system is based on the Cartesian theory of vortices combined with Fontenelle's concept of the plurality of inhabited worlds. It assumes that heavenly bodies go through an endless process of renovation consisting of alternating luminous phases, during which they become suns, and dark phases, when they behave as planets. At present, the earth is undergoing a dark phase during which the sea, after having covered the entire globe, gradually diminishes by evaporation until there is total depletion. De Maillet considered the sea entirely responsible for all the physiographical, lithological, and structural characteristics of the earth's crust. These were generated, according to his own estimate, during billions of years, a concept completely unheard of at the time, which demonstrates his acute perception of the duration of geological time. Plants, animals, and mankind originated from the sea and, after emergence of the continents, became terrestrial through a generalized and continuous process of transformism.

In spite of its extreme originality, de Maillet's system is well related to the works of his predecessors and contemporaries. The association of the concept of eternal matter with a transformism of the universe can be traced back to Lucretius and Epicurus. The influence of many Arabic authors is present everywhere in the system. Although we do not know the source of all the precise data given by de Maillet on marine processes at a time when oceanography was nonexistent, there is no doubt that the fundamental concept of the diminution of the sea was assumed by Herodotus and many other ancient Greek and Roman historians. This idea was also the subject of a work by Omar-al-Khayyam,[3] taught at great length by him around A.D. 1100. However, de Maillet was the first to describe in great detail the present-day processes of marine erosion and deposition and to project them into the past, according to uniformitarian principles, in order to explain the features of ancient rocks as a function of the diminution of the sea, which allegedly lasted for two billion years.

The idea of transformism may also be traced in a vague form to classical antiquity, but de Maillet was certainly the first to advocate a general application of transformism and to present a comprehensive and rational discussion of its consequences. Fontenelle suggested to de Maillet a review of the Cartesian theory of vortices and of the concept of the plurality of inhabited worlds; and the importance of volcanoes in the luminous phases of the globes has unquestionably been borrowed from the famous work of Thomas Burnet. De Maillet's own observations on

[3] Charles Lyell, *Principles of Geology* (London: John Murray, 1830), Vol. 1, pp. 21–23, and also in many later editions.

the occurrence and interpretation of fossils were combined with those of many naturalists, such as Antoine de Jussieu, John Woodward, Martin Lister, Nicholas Langius, J. J. Scheuchzer, Bernard Palissy, and even J. B. Adam Beringer, but Agostino Scilla had by far the greatest influence. It is certainly to be regretted that the only major author apparently unknown to de Maillet was Nicholas Steno, whose *Prodromus* contained even more modern ideas than those discussed in *Telliamed*.

Since the deeply modified printed text of de Maillet's system alone had a remarkable influence on scientific thinking, one may wonder what would have been the impact of the original text, which has never been made available. In order to fill this unfortunate gap, F. Neubert[4] undertook in 1920 an elaborate comparison between the five available manuscripts and the printed editions, presenting all the documents necessary for the reconstitution of the original text without attempting it himself. With the help of two new additional manuscripts at the University of Illinois, and on the basis of Neubert's fundamental contribution, I have established, for the first time,[5] the original text of de Maillet's new system of the world, the fundamental features of which are analyzed in this paper.

The Geological Significance of de Maillet's System

1. EVOLUTION OF CELESTIAL BODIES A thorough study of de Maillet's cosmogonic system is necessary for a correct understanding of his ideas concerning the geological evolution of the earth's crust. His system is Cartesian in essence and postulates that the celestial bodies undergo an eternal cycle of dark and luminous phases, which corresponds to a constant transmigration of matter within the various vortices and between them.

This cyclical evolution of the celestial bodies is better explained by using arbitrary stages:

Stage 1. The existence of a vortex is assumed, consisting of a burning star (sun) that rotates on itself and is surrounded by a certain number of opaque bodies that gravitate around it, arranged in order of radially decreasing density. The currents of subtle matter corresponding to the sun's rays are responsible for the rotation of the opaque bodies.

Stage 2. Through its burning process, the sun gradually loses its matter

[4] Fritz Neubert, "Einleitung in eine kritische Ausgabe von B. de Maillets *Telliamed*. Ein Beitrag zur Geschichte der französischen Aufklärungsliteratur," *Romanische Studien, 19* (1920).

[5] B. de Maillet, *Telliamed, or conversations between an Indian philosopher and a French missionary on the diminution of the sea*, trans. and ed. Albert V. Carozzi (Urbana, Ill.: University of Illinois Press, 1968), 496 pp.

in the form of ashes. These are carried away radially by the sun's rays, which on their path collect additional dust and, particularly, water from other globes. This association of ashes, dust, and water is transported toward the periphery of the vortex and deposited over light and porous opaque bodies (extinguished suns), which, because of their low density, are also located in the marginal parts of the vortex. By this process these extinguished suns regain the weight and humidity lost during their previous phase of burning, and become covered by a thick layer of silty materials that will settle at the bottom of the general water envelope deposited at the same time.

Stage 3. When, after a gradual decrease of its intensity, the sun is finally extinguished and becomes itself a light dark body, two situations may occur:

1. The sun is carried to the periphery of its own vortex and its place is taken by the nearest planet that becomes inflamed.
2. The vortex is dispersed and the opaque bodies including the former sun wander away at random, as comets, to be incorporated in other vortices.

Stage 4. When an extinguished sun with its envelope of water is captured by another vortex and stopped inside it at a particular place determined by the activity of its sun, the water envelope may increase or decrease according to its distance from that sun. If the waters decrease by evaporation (as on the earth at present, according to de Maillet), the primitive mountains formed by the currents on the floor of the general ocean will eventually emerge.

Life develops in the shallow waters surrounding these mountains and, through generalized transformism, will eventually populate the continents. Secondary mountains are formed by the marine erosion and reworking of the materials forming the primitive mountains to which numerous fossils are added.

In conclusion, it should be stressed that all the materials forming — either by themselves (metamorphic rocks), as the Primitive mountains, and later by reworking and the addition of fossils (sedimentary rocks), as the Secondary mountains — actually derive from the various kinds of extraterrestrial ashes originally deposited at the bottom of the universal ocean.

If an inhabited globe changes vortex and reaches a new position in which its waters will increase again, the living organisms on it will be destroyed and buried under new layers of silt. This is the other process by which life is destroyed on a globe without the globe passing through the state of fire of the next stage.

Stage 5. While evaporation proceeds toward completion, combustible organic matters corresponding to concentrations of fossils buried in the layers of the secondary mountains begin to burn, generating volcanoes. These gradually spread over the entire globe, destroying all life and changing the opaque body into a new sun.

Stage 6. If the opaque body becomes a new sun before the sun regulating the vortex to which it belongs is extinguished, it will create a vortex of its own, taking away some of the planets of the earlier vortex. This process brings the system back to the conditions of Stage 1, and the cycle begins again.

The duration of a complete cycle undergone by a celestial body, as assumed by de Maillet, may be considered to be of the order of five billion years. This figure is based on his assumptions that the longest time the largest star could remain inflamed would be two billion years, and that the diminution of the sea has been in progress on the earth for at least two billion years.

These alternate dark and luminous phases recall modern ideas of the evolution of stars; when they near depletion of their hydrogen supply, they swell into red giants, to cool and shrink afterward into white dwarfs after complete conversion of their hydrogen into helium.

2. THE DIMINUTION OF THE SEA The idea of the diminution of the sea on the earth is introduced through the observation made by Telliamed's grandfather, of a rock that gradually emerged from the ocean on the neighboring coast during his lifetime. Since the observed diminution of the sea could only be the continuation of a more general process, he concluded that the mountains had also been formed in the sea when its level was much higher in the past. In order to understand the formation of the mountains, he wisely decided to undertake an extensive oceanographical survey of the present-day shorelines and shallow waters in order to use the present as the key to the past. In this introductory section, erosional and depositional processes along rocky coasts, beaches, and deltas are discussed in great detail as a function of wave action. The investigation is even carried out, by means of an intricate diving apparatus, to the bottom of the continental shelf, and the distribution of muds there is discussed on the basis of the pattern of bottom currents. The entire spectrum of coastal sediments and their organic communities is reviewed with a precision that implies the personal knowledge of some unidentified earlier (Arabian?) or hired observer,

if not de Maillet himself. The picture of present-day marine processes is cleverly projected into the past according to uniformitarian principles, and all its facets will be used to explain the petrographical, paleontological, and structural features of ancient rocks.

De Maillet's system, which may be qualified as "ultraneptunian" since it is based essentially on processes included today under the designation of sedimentology, excludes all other geological and geomorphological agents, except perhaps some minor aspects of weathering. This exclusive approach gives to the proposed system the extremely unusual character of being very modern when applied to appropriate subjects, such as sedimentary rocks and fossils, and utterly fantastic when used in other, inappropriate fields.

3. CLASSIFICATION AND TERMINOLOGY OF ROCKS De Maillet's terminology of sedimentary rocks relies on the observation of recent marine coastal sediments and consists essentially of a classification based on the variation of grain size, with no reference to mineralogical composition. All rocks consist of angular blocks, cobbles, pebbles, sands, silts, and muds, and their induration is attributed to the action of marine salt, an idea probably derived from the observation of beachrock. As a consequence of this mechanism of cementation, freshwater deposits are not supposed to indurate and therefore cannot participate in the building of mountains. Because of this peculiar approach, de Maillet did not recognize freshwater deposits and consistently misinterpreted them as marine formations.

It is important now to enumerate the rock types described by de Maillet and his interpretation of their genesis in comparison with recent coastal deposits. The term *marble* always refers to any rock that can take a polish and be used for ornamental purposes; it has no implication of composition, as it does in the modern technical terminology. The following recent marine sediments and indurated ancient equivalents are distinguished:

1. Nonstratified and massive deposits formed at the foot of steep cliffs, consisting of angular blocks in a muddy matrix with very few organisms. The indurated rocks interpreted in this way (*pierres et marbres variés*) are the different kinds of sedimentary breccias and many tectonic breccias misinterpreted as sedimentary deposits. Indurated talus, scree deposits, and alluvial fans are also included in this group because de Maillet failed to recognize their continental subaerial origin.

2. Well-stratified deposits, formed along beaches, which receive an abundant supply of detrital particles and display rich organic communities. These sediments correspond to gravels, cobbles scattered in sands, sands, and silts with abundant shells. The indurated rocks belonging to this group are the different kinds of conglomerates (*pétrifications de cailloutages*), sandstones (*grais, sables durs, sables pétrifiés*), siltstones (*pierres unies, pierres fines*), and claystones (*terre argile*). Shales or slates as such (*ardoises*), characterized by their fissility, are mentioned only once, and their importance apparently is not appreciated. Also included in this group are chert nodules of a color different from the enclosing limestones, erroneously identified as cobbles scattered in sand or muds, and granitic rocks interpreted as massive coarse sands of variable grain size.

3. Massive to stratified deposits formed along protected beaches and consisting of concentrations of corals, shells (with emphasis on oysters), and algae (*mousses*), mixed with a variable amount of mud or sand and cemented into beachrock by salt spray. The indurated rocks interpreted in this manner are the different types of coral- and oyster-bearing limestones, coquinoidal limestones, fine to coarse fossiliferous calcarenites, or organodetrital limestones, called *tufi* or *tufi calcari* (*Miocene-Pliocene*) by the Italian geologists (*pierres de roche, pierres de tuf*).

4. Poorly bedded sands, silts, and muds, with rare scattered pebbles, deposited as bars and mud banks at the mouths of streams. These sediments often display a greenish color, supposedly due to enclosed remains of transported grass. Since such sediments are partially deposited in freshwater conditions, they will remain soft for a certain length of time and consequently will show wavy structures due to their reworking by storms and floods. Most of the indurated rocks included in this group (*marbres ondés et bigarrés*) are fine-grained variegated reddish and greenish limestones and alternating limestones and shales, with structurally controlled flow structures, and they are used extensively in France for ornamental purposes under the names of *marbres griottes* and *marbres Campan*. In a few instances, greenish subgraywackes, green shales, and basic greenstones, particularly serpentine breccias, have also been included in this group.

5. Sands, muds, and silts originally deposited in the sea at the mouths of streams, as in the preceding group, and exposed after induration to weathering. This process is supposed to have split the rocks, and the cracks have become filled with the overlying soil. The indurated rocks interpreted in this manner are several types of sandstones, lime-

stones, and marbles, with tectonically induced veins of white or yellow calcite (*marbres bigarrés à rayures*).

6. Debris of the preceding rocks redeposited at the foot of steep cliffs and cemented by a muddy matrix. The indurated rocks included in this group are tectonic and sedimentary breccias containing fragments that themselves display an internal brecciated structure due to networks of white or yellow calcite veins (*marbres à marqueterie ou mosaïque*).

7. Horizontally bedded quartzose sands, deposited in quiet lagoons by a combination of fresh and marine waters and therefore not indurated. The deposits interpreted in this manner are the blankets of fluviomarine to continental quartzose sandstones and sands (*sables durs non pétrifiés*).

8. Widespread layers of ferruginous muds and muddy sands rich in ammonites, very uniform in composition and thickness. The rocks interpreted in this manner are the Mesozoic oolitic iron ores, in particular the *minette* of Lorraine.

9. Accumulations made by the sea, in the places where they occur, of rotten weeds and fish fat. Such an origin of coal (*charbon*) is demonstrated by its combustible property and bad smell. This applies particularly to cannel coals and bogheads, which are soft and bituminous types characterized, respectively, by a predominance of spores and microscopic freshwater algae. Both kinds of coal yield a high percentage of volatile matters, ignite easily, and burn with a luminous, smoky flame.

10. Horizontal beds of simple and uniform substance, corresponding to sands and muds devoid of fossils or containing very rare ones, which were deposited on the floor of the ocean when it covered the earth completely. The indurated equivalents included in this group are the metamorphic rocks, the gneisses being interpreted as sands and the mica-schists as muds.

11. All minerals and metals, such as gold, silver, copper, lead, tin, iron, sulphur, alum, vitriol, and quicksilver, have been deposited at first by the flames of volcanoes on the sides of their vents like the soot of wood and coal burned in furnaces. If volcanoes are taken in the sense of general magmatic activity, the above statement is certainly a correct understanding of the genesis of mineral veins of hydrothermal processes.

This comprehensive list clearly shows that de Maillet observed, but did not always correctly understand, all the major types of rocks forming the earth's crust (acidic to basic igneous, metamorphic, and sedimentary). However, the above list of rock types given by de Maillet shows the demanding task that he set for himself in trying to establish a general lithological classification biased by an excessive emphasis on unusual building and ornamental stones. He did not realize the irrational character of classifying common rock types on the basis of lithological rarities, because early geologists were not yet doing systematic field work. According to his rigid sedimentological approach, he interpreted all of them as marine deposits formed, according to their grain size, at the foot of steep cliffs, along beaches, in lagoons, in deltas, or in deeper waters.

4. EVOLUTION OF THE EARTH'S CRUST — PRIMITIVE AND SECONDARY MOUNTAINS While the sea was surrounding the earth completely, its bottom currents circulated for countless ages, reworking and transporting ashlike materials consisting of sands, silts, and calcinated stones of extraterrestrial origin. On the floor of this deep ocean, the currents, by their combined depositional and erosional action, generated the so-called *Primitive or primordial mountains.* These are, today, the highest on earth and consist of a simple and uniform substance, either massive or in beds of variable thickness and almost always horizontal. These rocks are devoid of fossils or contain a very small number of them, since at the time of their deposition the ocean was too deep to allow any development of life (Fig. 1).

This description applies to metamorphic rocks, which commonly form the central and highest portions of many mountain ranges and

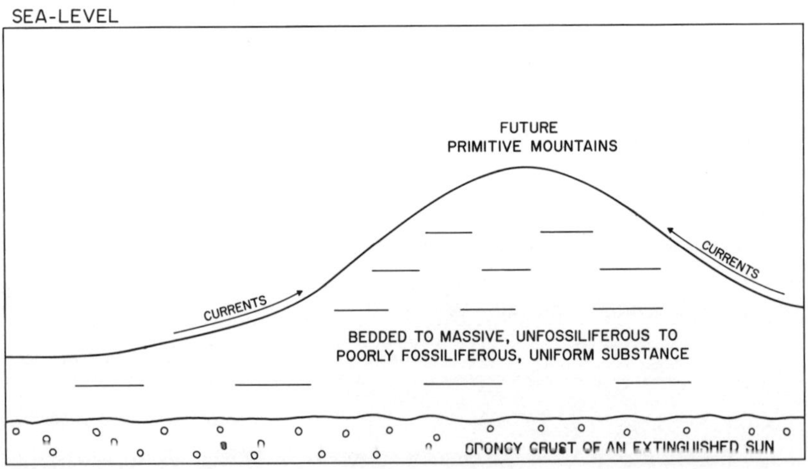

FIG. 1. *Theoretical diagram illustrating the evolution of the earth's crust according to de Maillet's system. Stage 1: Formation of the Primitive mountains by the action of currents on the bottom of the ocean.*

are surrounded by marginal belts of lower mountains consisting of sedimentary rocks (Pallas, Law of 1777).

Because of his purely sedimentological approach, de Maillet interpreted all mountains and valleys — in other words, the entire landscape — as the result of the depositional and erosional action of marine currents. In his mind, there is continuity in all respects between the mountains exposed by the diminution of the sea and those still under water and in the process of being formed.

The formation of mountains is explained by considering them as equivalents to submarine bars generated by currents on the sea floor or in front of deltas. Several types of mountains are described, distinguished on the basis of their morphology. A simple, elongate mountain results from the deposition of a bar at the meeting point of two opposite currents or between a stream and a marine current in front of a delta. A long chain of mountains results from two currents intersecting each other obliquely, causing the weakest current to deposit materials alongside the strongest. The same situation may also occur when two currents flow parallel to each other but in opposite direction and form a bar between them. Two long chains of mountains separated by a longitudinal valley result from one powerful current cutting across two opposite ones, forming a double bar by eroding a channel through the original bar.

The size of mountains is considered a function of the size of the generating current, the amount of materials transported, and the depth of the sea. All the valleys, with their steep slopes or cliffs, the passes, or any other interruptions between mountains were also generated at the same time as the mountains through the truncation of the deposits by contemporaneous transverse currents. Gentle slopes are interpreted, however, as depositional in origin and due to the normal termination of currents. In such a context, the correlation of beds from one mountain to another across a valley or a basin is considered by de Maillet as a very logical procedure.

In summary, all the positive and negative features of a given landscape result from erosional or depositional mechanisms and are consequently generated during the sedimentation of the soft materials. These are gradually indurated by the action of the marine salt and eventually exposed through the diminution of the sea as a completed and "congealed" scenery that will undergo only minor changes through weathering.

Although de Maillet understood quite well the origin and properties of soils and the small-scale effects of weathering, and even described the erosion, transportation, and deposition due to streams, in his theory rivers play no role in the shaping of the scenery. He clearly states that stream valleys are pre-existing channels due to marine currents and subsequently used after emergence by streams to concentrate and evacuate rain waters to the sea. Such channels are deepened, kept open, and extended downstream by the tidal currents during the gradual retreat of the sea, as shown today by estuaries. In a general way, de Maillet did not understand the geomorphology of folded mountain ranges consisting of alternating resistant and soft formations. Consequently, prominent and repeated rocky cliffs, either straight or with the shape of amphitheaters, corresponding to resistant limestone formations and separated by gentle slopes or plateaus due to softer shaly beds, are always interpreted as ancient marine coves representing the steps of the diminution of the sea and compared to wave-cut terraces and notches visible along present-day coasts.

De Maillet's negation of the role of running waters in forming landscapes is another striking illustration of his rigid attitude of attributing all the features of the earth to the action of the sea. Such a position is even more incredible in the light of the fact that Avicenna[6] as early as A.D. 1021–1023, Agricola[7] in 1546, and of course Steno[8] in 1669 had understood perfectly the role of rivers as fundamental agents in geomorphology.

After a certain amount of evaporation takes place, the upper parts of the primitive mountains are finally exposed to the air. This emergence initiates the processes of weathering, and in the shallow waters surrounding the newly emerged lands life appears (Fig. 2). This appearance of life on earth must be considered in the light of de Maillet's general concept that the entire space of the universe is full of seeds of everything, sensitive or vegetative, that can live. These seeds are so minute as to be invisible and consequently cannot perish in essence. They are more abundant around the opaque globes, in thick airs, and in waters than in interplanetary spaces.

These seeds are available at all times for participation in the operations of nature that make the waters around the globes at certain times and under certain circumstances proper for fecundity and for the crea-

6 Ibn Sīnā (Avicenna), *Avicennae de Congelatione et Conglutinatione Lapidum*, being a section of the *Kitâb al-Shifâ'*, trans. E. J. Holmyard and D. C. Mandeville (Paris: Librairie Orientaliste, P. Geuthner, 1927), Section L, "Upon Mountains," pp. 18–32.

7 G. Agricola, *De Ortu et Causis Subterraneorum*, Book II, in *De Re Metallica*, trans. from the first Latin edition of 1556 by H. C. Hoover and L. H. Hoover (New York: Dover, 1950), pp. 595–596.

8 N. Steno, *The Prodromus of Nicolaus Steno's Dissertation*, trans. J. G. Winter (London and New York: Macmillan Co., 1916), pp. 231–234.

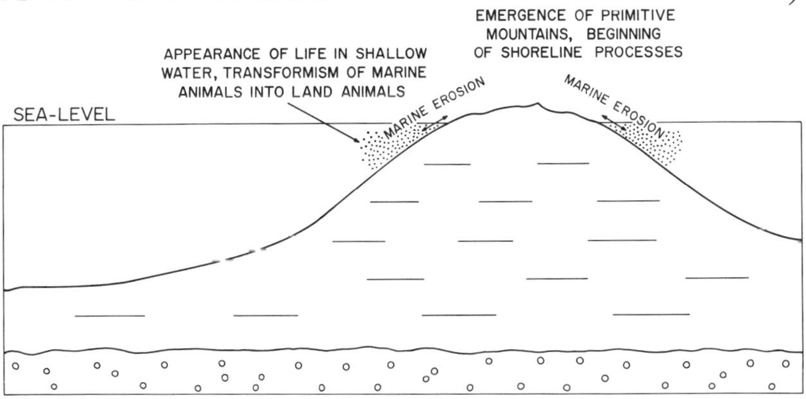

FIG. 2 *Theoretical diagram illustrating the evolution of the earth's crust according to de Maillet's system. Stage 2: Emergence of the Primitive mountains, appearance of life in shallow waters, transformism of marine organisms into terrestrial ones, and population of the continents.*

tion of new species. On the Earth, such favorable conditions did not occur until the existence of shallow waters around the newly emerged primitive mountains. Indeed, this situation provided a supply of food not available before and a good mixing between the agitated and shallow waters warmed by the sun and the air containing the seeds. These seeds find on the bottom deposits of the retreating seas, particularly in the rich silts, the conditions appropriate for their initial phase of development, after which they go out in the waters in search for additional food to bring them to the final size characteristic of a given species.

Under the effect of the diminution of the sea and the continuously increasing continents, all forms of life undergo a generalized transformism to adapt themselves to the new environments. Marine plants slowly change to land types; sea animals master flight and walking, giving rise to birds and land animals; mermen and mermaids abandon their aquatic life and become terrestrial human beings.

While this spectacular transformism is taking place, the generation of the *Secondary mountains*, daughters of the Primitive ones, begins. They consist of horizontally bedded sands and muds derived from the erosion of the primitive mountains by wave action along the shorelines, to which are associated abundant remains of marine organisms (Fig. 3). Continental areas now being exposed, these sediments also receive a certain amount of mineral and organic components derived from mechanical weathering, wind action, and stream erosion on land areas, but since in de Maillet's system the streams play no appreciable role in the shaping of the scenery, such materials are considered to be negligible.

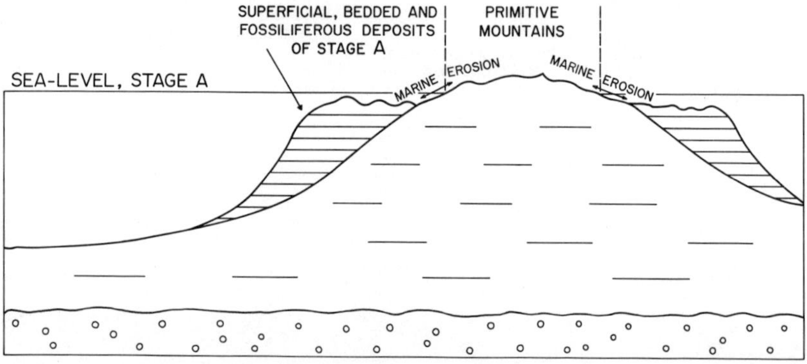

FIG. 3. *Theoretical diagram illustrating the evolution of the earth's crust according to de Maillet's system. Stage 3: Beginning of the formation of the Secondary mountains by shoreline erosion of the Primitive mountains.*

Consequently, the secondary mountains result essentially from the erosional action of waves along shorelines and will display all the sedimentary rock types equivalent to the coastal and shallow-water sediments described by de Maillet in his oceanographic survey.

The Secondary mountains differ from the Primitive ones not only by the presence of abundant fossils but also by their formation through the combined depositional and erosional action of bottom currents and waves in shallow conditions along retreating shorelines, and no longer on the floor of the ocean. However, following the same concept applied previously to the Primitive mountains, all the details of the aspect of the Secondary mountains and even the shape of single outcrops are also considered as original features dating back to the time of deposition. It should be stressed again here that all the lithic materials forming the Primitive mountains (metamorphic rocks) and the Secondary mountains (sedimentary rocks) by reworking of the Primitive ones are of extraterrestrial origin. Actually, only the fossils are indigenous to the earth, having developed from the seeds ever present in space.

The formation of the Secondary mountains is actually a generalized process of reworking of earlier deposits comparable to the generation of terraces by an entrenching stream (Fig. 4). As the marine erosion of the Primitive mountains provides the materials for the first set of Secondary mountains, these, in turn, are the source for the sediments of the second set located at a lower elevation. This process, expressed by de Maillet as "higher mountains will give birth to lower ones," naturally excludes the action of streams, which would rework all previous higher deposits, not only those located immediately above a given one. The process will continue as long as the sea exists to receive the materials

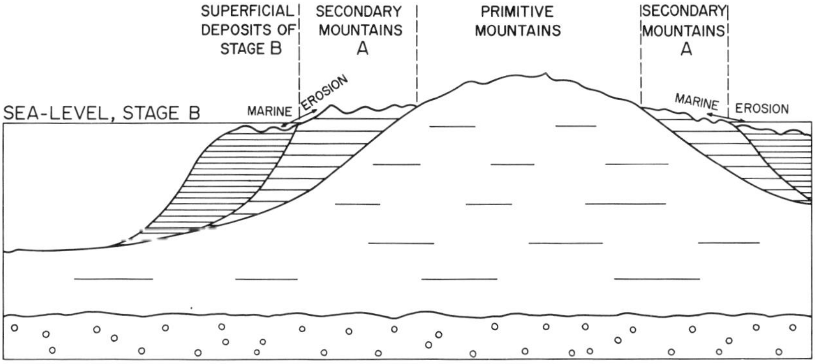

FIG. 4. *Theoretical diagram illustrating the evolution of the earth's crust according to de Maillet's system. Stage 4: Formation of the lower Secondary mountains by erosion of the upper Secondary mountains.*

from its shoreline erosion. Such a mechanism explains all the rock types previously described, particularly those that contain debris of other rocks that themselves may have gone through previous phases of deposition and reworking (*marbres à marqueterie ou mosaïque*). The horizontal bedding of the Secondary mountains corresponds to the tabular appearance of many of the chains along the Mediterranean coasts, and de Maillet interpreted their steplike morphology as ancient sea cliffs, with their notches alternating with wave-cut surfaces extending down to the present shorelines (Fig. 5).

Since sea level is decreasing, the Secondary mountains will be younger with decreasing elevation, but this age relationship does not contradict the fact that in a given geological section, the beds are super-

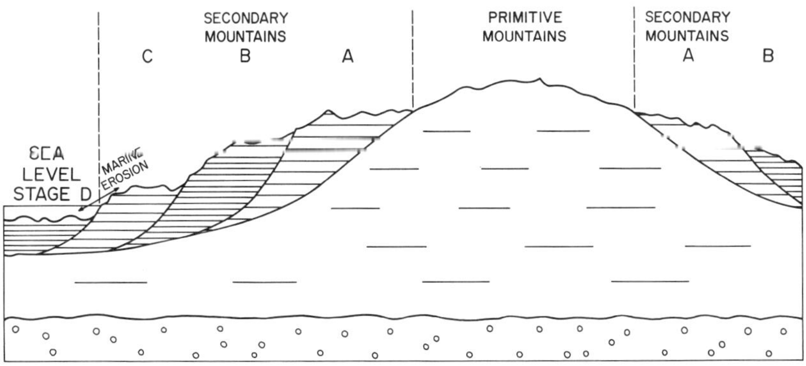

FIG. 5. *Theoretical diagrams illustrating the evolution of the earth's crust according to de Maillet's system. Stage 5: Formation of present-day marine sediments by shoreline erosion of the lowest Secondary mountains.*

posed from bottom to top, and the law of superposition is still valid (Fig. 5). De Maillet was well aware of this fundamental principle when he stressed the idea that each horizontal layer is considered to have kept its original elevation when deposited on the bottom of the diminishing sea. This attitude shows that the proposed system is devoid of any orogenic mechanism that would uplift, tilt, or fold strata. Since de Maillet's geological experience was limited to sedimentary rock areas with essentially horizontal or mildly deformed strata, he considered tilted and folded beds as exceptional features of mountains that should normally consist of horizontal or gently inclined beds deposited by marine currents.

De Maillet's explanation of tilted and folded beds follows again his logical approach and is based on the action of marine bottom currents as allegedly observed along present coastal areas. Tilted or even vertical beds result from the erosional action of marine currents that undermine previously deposited horizontal beds, causing blocks of them to fall back on the sea floor. This process compared to the action of a stream undermining its banks, is considered to be relatively rare.

Sets of low- or high-angle tilted beds due to the effects of faults were not understood by de Maillet who did not recognize faulting in general because he considered such geometrical relationships between beds as primary depositional features resulting from currents coming from different directions. It may be recalled that he likewise confused tectonic breccias with sedimentary ones.

Small-scale regular folds were interpreted as ripple marks whose subaqueous origin was perfectly understood by de Maillet, whereas large-scale tilted and regularly folded beds were explained by assuming that currents deposit layer after layer of sediments with even thickness over irregularities of an indurated substratum, like coats of paint or plaster applied with a brush on an irregular wall. As a consequence of this process, tilted beds were always considered to be inclined in the direction from which the depositing current originated.

Complicated forms of folding were interpreted as the results of violent storms during which currents pushed against rocky obstacles flow in complex patterns, eroding and depositing sediments at the same time until highly contorted beds are eventually generated.

This "plastering" process, also advocated by J. G. Lehmann[9] in 1756, is not possible, of course. The idea must have originated from the observation that rocks along muddy coasts may indeed show, at low

[9] J. G. Lehmann, *Versuch einer Geschichte von Flötz-Gebürgen* (Berlin: Klüter, 1756).

tide, a slick coating of mud mostly trapped by algal mats or settled directly from the water but not deposited by the action of bottom currents. The bottom currents only deposit horizontal beds (with possible internal cross-bedding), even on an irregular substratum, as pointed out as early as 1669 by N. Steno in his *Prodromus* (pp. 228–230).

5. DE MAILLET'S FIELD OBSERVATIONS AND THE SIGNIFICANCE OF FOSSILS

A few typical examples will be sufficient to demonstrate the amazing accuracy of de Maillet's field observations, which represent an almost incredible feat for a man who was a career diplomat, not a trained naturalist.

The following account is part of the numerous data he collected in Egypt between 1692 and 1708:

> Three miles away from Cairo, there is a long valley which with a gentle slope leads the traveller in three days to the Red Sea. This valley, which is a mile and in some parts two miles wide, is filled with a layer several cubits thick consisting of all kinds of shells. This deposit extends from the upper end of the valley to the shoreline where new shells are daily added in great number. (Vol. 1, pp. 118–119)[10]

This east-west trending valley, which lies to the east of Cairo and has been used by pilgrims to go to Suez and the Red Sea, is called Wadi Digla. It displays wide outcrops of very fossiliferous sand and calcareous beds of marine Miocene, which increase in thickness from west to east (30 m. to 180 m.). The road passes by the Bir Gendali well, which is still a classic collecting locality for oysters, pecten, gastropods, corals, echinoids, and bryozoans.

From the observations collected by de Maillet while he was Consul in Livorno between 1708 and 1714, come the following three:

> I have not seen anything more remarkable or worthy of attention, or more instructive, than the banks of oyster shells with which several hills of Tuscany are capped, especially those in the vicinity of Pisa. Some of these banks are two to three miles in extent, and overlain by three to four feet of earth or sand, accumulated by the winds since these hills came out from the sea. The shells, broken loose from these banks by the rain and transported by any other means to the neighboring fields, cover all the adjacent areas, just as ours are strewn with stones and pebbles . . . (Vol. 1, pp. 114–115)

There are indeed large oyster banks in the hills south of Pisa which consist essentially of *Ostrea edulis* belonging to the Calabrian (Lower Pleistocene). The Calabrian begins with transgressive marine clays overlain by brackish yellow sands with frequent oyster banks. It terminates with a regression surface and yellow-reddish sands largely of eolian ori-

[10] This quotation and the ones following are from the final edition of *Telliamed*, in two volumes, The Hague, 1755.

gin. It seems as if the wind-blown character of these terminal deposits has even been recognized by de Maillet.

I have seen a whale rib in the steep rock, on which the fortress of Porto-Ercole is built. This rib was shown to the present King of Spain when his galleys called at that port. . . . (Vol. 1, pp. 108–109)

Porto Ercole, on the south side of Monte Argentario has two fortresses built on the so-called "vuggy limestone" of Norian-Rhetian age. By analogy with other localities in Tuscany and in the Monte Argentario itself, the vertebrate bones and ribe (not of whale but probably of *Elephas*) belong to bone breccias filling fissures and cavities of karstic origin and of Pleistocene age.

In the island off Portovenere where black marble with yellow streaks approaching a gold color is quarried. . . . (Vol. 1, pp. 63–64)

The black marble with gold-yellow spots and veins mentioned here occurs in beds six to seven meters thick intercalated among the dolomitic limestones (Lower Liassic) that form the western promontory (Portovenere) and the islands of the Gulf of La Spezia. This marble is called locally *portoro* (from *oro*, gold).

This last example dates probably from de Maillet's stay in Paris (1720–1721):

In the white stone used in building the cathedral of Rouen and in a hundred other places of Normandy, we find large pieces of black stone. . . . How can we account for these puzzling occurrences, if we do not admit that while the sea was forming that white stone, currents or a storm carried into the mud of that color which was being deposited along a coast, some fragments broken loose from a black rock. . . . (Vol. 1, pp. 96–97)

The chalky limestones used in the construction of the cathedral of Rouen contain, indeed, irregularly shaped nodules of black chert replacing sponges or located along burrows of scavenging organisms. They were quarried at Caumont (Eure) and belong to the *Micraster cor testudinarium* zone of the Senonian Chalk. Many churches and other buildings in Normandy are similarly built with chert-bearing chalky limestones (Turonian-Senonian) of local origin.

In his field descriptions, de Maillet recognized and demonstrated at great length the organic origin of fossils, although, as discussed later, his definition of extraneous bodies also included the remains of human beings and the products of their activity, such as ships, weapons, tools, and earthenware. In general, de Maillet, following his strictly oceanographical approach, conceived only marine faunas and almost never recognized continental or freshwater fossils. This attitude is particularly well shown, for instance, by his persistent interpretation of elephant bones as remains of whales. In a general fashion he noticed that fossils occur frequently oriented parallel to bedding and are concentrated preferentially in certain layers. Applying the same principle as for the enclosing beds, the fossils follow the law of superposition and are supposed to occur at the same elevation at which they originally lived and died on the sea floor. Consequently, said de Maillet, "They are superposed to each other like the materials of a wall being erected in front of us."

Fossils are supposed to become more abundant upwards because of the increasingly favorable conditions afforded by the shallowing seas. Although this statement is incorrect, de Maillet noticed, probably for the first time[11] that the lower layers contained animal and plant fossils unknown along present coasts or on present continents, whereas the upper layers contained fossils comparable to modern coastal or terrestrial forms. In other words, he made the fundamental observation that ancient faunas and floras were different from modern ones and noticed the existence of faunal successions through geological time. In the absence of any concept of organic evolution and geological changes of climate, de Maillet suggested three possible explanations for this important observation, all essentially of oceanographical nature. According to a first explanation, the unknown character of the faunas of the lower beds is due to their deposition in deep water. Indeed, the fauna living at such depths is poorly known, a situation emphasized by the observation that even in relatively shallow conditions, many modern shells live buried in the mud or attached to rocks and rarely occur along coasts and beaches. In a second explanation, the species found only in the lower layers are supposed to have become extinct when the particular basins they characterized dried up as a consequence of the diminution of the sea. The third explanation assumes that when the sea was deeper and more extensive, long-distance flotation by marine currents could freely operate. Since the unknown species of plants found in the lower layers of Europe are supposed to be typical of Asia or North America, such a process could explain their occurrence. Emerging lands gradually reduced and finally stopped the flotation, thus explaining the differences observed between ancient and modern floras.

6. THE ORIGIN OF MANKIND AND THE "PETRIFIED" SHIPS In the framework of his idea that the diminution of the sea had been in progress for at least two billion years, de Maillet considered that mankind came out of the sea and populated the continents many thousands of centuries after the emergence of the Primitive mountains, probably when the earth had reached half of its present age. According to him it took man-

11 K. T. Mather and S. L. Mason, *A Source Book in Geology* (New York: McGraw-Hill, 1939), pp. 47–48.

kind the other half to develop navigation from a simple plank to the "fossilized" ships allegedly found in rocks. He furthermore postulated that cities and harbors already existed when sea level was 6,000 feet above its present elevation. This statement, combined with the present rate of diminution of 3 inches per century based on the observation made by Telliamed's grandfather in wells connected with the sea, would give to these harbors an age of 2.4 million years before present. Such a figure is actually in default, because de Maillet believed that the rate of diminution of the sea increases with time accordingly as it becomes shallower, smaller, filled with additional sediments, and interspersed with more numerous and larger islands and continents.

It is fundamental to understand how de Maillet reached this incredible idea of the very early appearance of mankind. In the absence of any concept of evolution, this major event did not present any particular problem for him, being the normal result of the emergence of mermaids and mermen from the ocean when conditions on the continents became favorable for such a transformism. De Maillet's alleged proofs result essentially from the fact that, although he realized the immensity of geological time, he had no concept of relative age. Therefore, he unfortunately combined a correct understanding of prehistoric artifacts of the Iron and Bronze ages with the erroneous interpretation of Cenozoic vertebrates as human skeletons, fragments of red shales reworked in conglomerates as debris of earthenware, and silicified logs in fluvio-continental deposits as petrified ships.

This last erroneous interpretation, shared, in fact, by N. Steno is of great interest. It is based on several occurrences, but particularly on the following observation of de Maillet:

> It is very common for those who travel across the sandy deserts of Libya and Africa to find, in digging wells for water, petrified hulls of small ships which were undoubtedly wrecked when the sea still covered these areas. . . . About a day and a half, or at most two days journey west of Cairo, there is in the middle of a desert of sand quite a long valley bounded and strewn with rocks at present partly covered with sands. This place is called by the Arabs *Bahar-Balaama*, that is, sea without water, because that plain is indeed dried up. In it a great number of small boats and ships are found which formerly had been wrecked and are now petrified. Masts and yards are particularly abundant, many of which are still in one piece. . . . (Vol. 1, pp. 91–92)

The Bahar-Balaama is marked on modern maps as Wadi-el-Faregh, located south of Wadi Natrun, 60 to 70 kilometers northwest of Cairo. It is a large outcrop of Oligocene sands and gravels, vividly colored and displaying abundant silicified tree trunks, some of which reach 30 meters in length, and remarkably aligned by stream currents during their final deposition.

There is no doubt that the preferred orientation of the huge silicified logs associated with smaller debris suggested the appearance of ships with their broken masts and yards, confirming de Maillet in his belief in the early appearance of mankind.

This final discussion of the origin of mankind integrates all living organisms into their geological framework, both resulting from the gradual diminution of the sea, a process that is itself a fleeting phase of the endless rejuvenation of celestial bodies according to the theory of the vortices. The system is therefore complete, satisfying to the mind, and a perfect illustration of rigid Cartesian logic.

Le Mascrier was right, for once, in his praise of the geological portion of the system that resulted essentially from de Maillet's lifelong personal observations. Even today, his theory of the earth remains a remarkable achievement for a career diplomat and a traveler who could hardly be called a scientist. Nevertheless, in addition to being a cosmogonist and a forerunner of transformism, de Maillet's "ultra-neptunian" or sedimentological approach makes him a marine geologist of the eighteenth century.[12] His everlasting reputation extends also to geophysics, since S. Günther considered him, together with Emanuel Swedenborg, as the promoter of the Desiccation Theory,[13] the process by which the earth slowly dries up as expressed by the diminution of the sea. However, Günther erroneously stated that de Maillet assumed a disappearance of the waters in cavities inside the earth, a concept that de Maillet refuted at length (Vol. 1, pp. 188–191).

This well deserved fame, spanning the biological and earth sciences, was certainly not anticipated by de Maillet's most sarcastic detractor, Voltaire, who, under the cover of M. de Morza, said in a footnote to the *Cabales* of 1772:[14]

> Ce consul Maillet fut un de ces charlatans dont on dit qu'ils voulaient imiter Dieu, et créer un monde avec la parole. C'est lui qui, abusant de l'histoire de quelques bouleversements avérés, arrivés dans ce globe, prétend que les mers avaient formé les montagnes, et que les poissons avaient été changés en hommes.

12 Although J. Thoulet recognized de Maillet as a marine geologist rather than an oceanographer, like his contemporary, L. F. Marsigli, he erroneously attributed to de Maillet the idea of oscillations of the sea under astronomical causes, which would make the ocean waters alternatively cover and uncover certain areas of the globe. De Maillet only postulated a worldwide decrease of sea level through generalized evaporation into space. See J. Thoulet, "L'étude de la mer au XVIIIème siècle: De Maillet, Buache et Buffon," *Mémoires de l'Académie de Stanislas, 6* (1908–09), 214–256.

13 S. Günther, *Handbuch der Geophysik* (2nd ed.; Stuttgart: F. Enke, 1897–1899), Vol. 2, pp. 577, 663, and 676.

14 Voltaire (J. F. M. Arouet), *Œuvres complètes* (Paris: Firmin-Didot, 1875), Vol. 2, p. 749.

Edward Jorden and the Fermentation of the Metals: An Iatrochemical Study of Terrestrial Phenomena

Allen G. Debus

Recent research on Paracelsian and iatrochemical texts has emphasized the insistence of these authors on a new science led by men who were willing to question the ancient maxims taught at the schools — a new science to be founded on fresh observations and experiments. Paracelsus had insisted that his works were based on "the light of Nature, and experience,"[1] and his chief sixteenth-century interpreter, the Dane, Peter Severinus, had written in 1571 that honest students of nature should sell their possessions and burn their books. With the proceeds they must buy sturdy clothes and set out to examine and observe everything with their own eyes. Above all, they should purchase coal, "build furnaces, watch and operate with the fire without wearying. In this way and no other you will arrive at a knowledge of things and their properties."[2] For such men there seemed no question of the need for a new experimental science — a science that the universities were not providing.

The Paracelsians had a definite goal in mind.[3] Above all, living in an

This paper was prepared while the author was an Overseas Fellow at Churchill College, Cambridge, during the tenure of a Guggenheim Fellowship and with the aid of a research grant from the National Institutes of Health (LM 00046).

[1] Paracelsus, "Of the Nature of Things," in Michael Sendivogius, *A New Light of Alchymy* (London: A. Clark for Tho. Williams, 1674), sig. L8 recto.

[2] Petrus Severinus, *Idea Medicinae Philosophicae* (3rd ed.; Hagae Comitis, 1660), p. 39.

[3] The reader should refer first to the many works of Walter Pagel. Among them we might single out specifically his *Paracelsus: An Introduction to Philosophical Medicine in the Era of the Renaissance* (Basel: Karger, 1958), *Das Medizinische Weltbild des Paracelsus, seine Zusammenhänge mit Neuplatonismus und Gnosis* (Wiesbaden: Franz Steiner Verlag, 1962), and "Religious Motives in the Medical Biology of the XVIIth Century," *Bulletin of the Institute of the History of Medicine, 3* (1935), 97–128, 213–231, 265–312. See, also, A. G. Debus, *The English Para-*

intensely religious age, they sought a true understanding of their Creator. Our observations of nature can only reveal to us the divine truths of God's creation, and we may rightly seek corresponding truths in the Bible, where we should turn first to the creation account in Genesis. The Paracelsians commonly pictured this as a divine chemical separation in which special emphasis was placed on the elements from which all other substances derive. Whether the particular author centered his discussion around the ancient Aristotelian elements or the newer Paracelsian principles, or even if he set up an independent system of his own, he was able to think of this as a basis for all created substances as well as a cosmological system. Indeed, the significance placed on this biblical account by the Paracelsians goes far to explain why there exists such a voluminous literature relating to element theory in the Renaissance, and why attack on the Paracelsian principles should have been considered essential by opponents of the iatrochemists.[4]

Chemistry then had a divine significance. Since the creation might be understood as a chemical process, it was thought that nature must continue to operate in chemical terms. It was the true key to nature — all created nature. Most of these men were convinced also of the truth of the macrocosm-microcosm relationship. The universe was conceived to be relatively small, with all of its parts interconnected. In particular, man was seen as a small copy of the great world, and analogies were frequently drawn between him and the earth or the universe as a whole. The latter was considered to be a vast system of living nature on all levels — one to which we might rightly apply the tools of chemistry as a guide.

Human, earthly, and universal phenomena were thought of as chemical processes and interpreted in this fashion. In the literature we find examples of this being brought forth on all occasions. Robert Fludd gave a chemical description of the circulation of the blood that parallels the circular motion of the sun in the great world,[5] and Joseph Querce-

---

celsians (London: Oldbourne Press, 1965), pp. 9–48; "Renaissance Chemistry and the Work of Robert Fludd," included in *Alchemy and Chemistry in the Seventeenth Century* (Los Angeles: Clark Memorial Library, University of California, 1966), pp. 3–29 (reprinted in revised form in *Ambix, 14* (1967), 42–59); and *The Chemical Dream of the Renaissance* (Overseas Fellowship Lecture Series, Churchill College, Cambridge) (Cambridge, England: Heffer, 1968).

[4] This problem has been discussed in greater detail in Debus, "Fire Analysis and the Elements in the Sixteenth and the Seventeenth Centuries," *Annals of Science, 23* (1967), 127–147.

[5] Debus, "Robert Fludd and the Circulation of the Blood," *Journal of the History of Medicine and Allied Sciences, 16* (1961), 374–393. On the relation of Fludd to William Harvey, see Pagel, *William Harvey's Biological Ideas: Selected Aspects and Historical Background* (Basel/New York: Karger, 1967).

tanus compared the phlegm associated with a running nose with similar condensation effects observed in the macrocosm — winds, sleet, and snow.[6] Burning diseases are explained as reactions of sulphureous and nitrous vapors in the body, which are analogous to gunpowder explosions or even the effects of thunder and lightning.[7]

Applying this chemical philosophy to problems that we would call geological, Gabriel Plattes offered a chemical experiment as visible proof of his concept of the formation of the earth (1639).[8] Rejecting the common belief that rocks and mountains were growths on the land like warts and tumors on man, he argued instead that they were features formed from the vapors of bituminous and sulphureous substances. The veins of metals he explained were engendered subsequently in the cracks and crannies of the mountains. Hills and valleys then resulted from the gradual action of the sea. In support of his views he suggested the following:

> Let there bee had a great retort of Glasse, and let the same be halfe filled with Brimstone, Sea-coale, and as many bituminous and Sulphurious subterraneall substances as can bee gotten: then fill the necke thereof halfe full with the most free earth from stones that can be found, but thrust it not in too hard, then let it be luted, and set in an open Furnace to distill with a temperate Fire, which may onely kindle the said substances, and if you worke exquisitely, you shall finde the said earth petrified, and turned into a Stone: you shall also finde cracks and chinkes in it, filled with the most tenacious, clammy, and viscous parts of the said vapours, which ascended from the subterraneall combustible substances.[9]

In a similar fashion, mountain streams were understood as the end products of internal distillations presided over by the Lord God who was pictured as the eternal Distiller. Thus, Sendivogius explained that springs result not from starry emanations or virtues, but rather from a divine earthly process:

> . . . there are in the bowels of the Earth places in which Nature distills and separates a sulphureous Mine, where, by the Central Fire it is kindled. The Waters running through these burning places, according to the neerness or remoteness are more or less hot, and so breaks forth into the superficies of the Earth, and retains the tast of Sulphur, as all broth doth of the Flesh that is boiled in it. After the same manner it is, when Water passing through places where are minerals, as Copper, Allum, doth acquire the savour of them. Such therefore is the Distiller, the Maker of all things, in whose hands

6 Joseph Duchesne (Quercetanus), *Traicté de la Matière, Preparation et excellente vertu de la Medecine balsamique des Anciens Philosophes* (Paris, 1626; first edition in Latin, 1603), p. 183.

7 Debus, "The Paracelsian Aerial Niter," *Isis, 55* (1964), 43–61.

8 This is discussed in greater detail in A. G. Debus, "Gabriel Plattes and his Chemical Theory of the Formation of the Earth's Crust," *Ambix, 9* (1961), 162–165.

9 Gabriel Plattes, *A Discovery of Subterraneall Treasure* (London, 1639), p. 6.

is this Distillatory, according to the example of which all Distillations have been invented by Philosophers; which thing the most High God himself out of pity, without doubt, hath inspired into the Sons of Men: and he can, when it is his holy Will, either extinguish the Central Fire, or break the Vessel, and then there will be an end of all.[10]

If the chemists taught that chemistry was the key to the universe, fire, in turn, seemed to be the key to chemistry. Heat was the universal analyzer of all substances, and chemical processes were generally applications of fire. For many, "pyrotechny" was another name for chemistry, and chemists proudly signed their works "pyrophilus." It was only natural, then, that these men should turn to the old belief in a great central fire as an explanation for earthly phenomena.[11] The effects of volcanic action seemed to be visible proof of such a fire, while the high temperatures of many mineral water spas and deep mines gave further evidence in support of this thesis. Yet, how might such a fire be kindled and maintained?[12] Some said that this fire originated through the celestial bodies. Rays of heat from the stars and the planets focused on the earth and there generated fire by friction on one another. These celestial virtues then meeting with otherwise inert matter gave rise to the metals. Thus, Sendivogius spoke of the influence of celestial rays and of the internal distillations and sublimations that produced the metals within the earth,[13] while Glauber wrote of the sidereal rays that strike the earth and "are contracted into a streight room, and (driven back) from thence are sublimed and distilled throughout the whole Orb, from which all kinds of Metals and Minerals (by the help of the Earth and Water corporifying them) are produced."[14] For Glauber the internal fire differed greatly from the superficial fires that brought forth volcanoes, but all would not agree. The question of fuel was an important one. Agricola could not accept a sidereal explanation, and he insisted on more natural causes. Bitumen and sulphur deposits beneath the surface were ignited and gave the only possible natural explanation of the ob-

10 Sendivogius, *New Light of Alchymy*, pp. 94–95. On seventeenth-century geology and the central fire, see Robert Lenoble, *La Géologie au Milieu du XVIIe Siècle*. Les Conférences du Palais de la Découvrte, Serie D, No. 27 (Paris: University of Paris, 1954).

11 See Francis M. Cornford's discussion of the concept of the central fire among the Pythagoreans in his *Plato's Cosmology* (New York: Liberal Arts Press, 1957), pp. 126–130.

12 See Frank Dawson Adams' discussion of the fire at the center of the earth in *The Birth and Development of the Geological Sciences* (New York: Dover, 1954), pp. 279–286.

13 Sendivogius, *New Light of Alchymy*, pp. 40–47.

14 John Rudolf Glauber, *The Works . . . containing, Great Variety of Choice Secrets in Medicine and Alchymy. . . .* trans. Christopher Packe (London: Thomas Milbourn, 1689), p. 116.

served phenomena. As Adams has summarized this argument, these deposits existing beneath volcanic centers "were ignited by intensely heated vapors which, in their turn, derived their heat from friction which was set up within the gaseous mass itself or by its contact with the walls of narrow spaces through which it was forced when in rapid movement within the earth."[15]

For all chemists the generation of the metals was a matter of considerable concern. Yet, at the same time, many of the most learned chemists of the sixteenth and seventeenth centuries were primarily neither metallurgists nor alchemists — rather, they were physicians who studied the earth to learn by analogy of man and supernature. Of special importance to these men was the study of mineral water spas, which became the subject of an extensive literature in this period. To these springs physicians commonly sent their most hopeless cases for relief, and elaborate procedures were developed to identify their contents in the hope that the active substances might be utilized in the preparation of artificial waters.[16] Thus, John French speaks of the making of artificial Tunbridge and Epsom waters,[17] and Quercetanus writes that we know that spas contain niter, alum, sulphur, pitch, antimony, lead and other minerals. Because of this there could be no doubt that minerals were helpful for man[18] — a point of no small importance to Paracelsians, who were continually attacked for their use of minerals by the Galenic herbalists. And as these scholars were concerned with the identification of the substances of these wells, they were no less interested in the origin of the minerals themselves and in the abnormal heat associated with the waters.

It is against this chemical, medical, vitalistic and religious backdrop that we may best view the work of the subject of this paper, Edward Jorden (1569–1632).[19] As a physician-chemist, he fits in well with the influential group of chemical philosophers who loom large in early seventeenth-century intellectual thought. As a man who questioned celestial influences on earthly matter, he fits in well with reformers who

15 Adams, *Birth and Development*, p. 281.

16 Debus, "Solution Analyses Prior to Robert Boyle," *Chymia*, *8* (1962), 41–61.

17 John French, *The Art of Distillation* (4th ed.; London: E. Cotes for T. Williams, 1667), p. 182.

18 Duchesne (Quercetanus), *The Practise of Chymicall, and Hermeticall Physicke for the preservation of health*, trans. Th. Tymme, Minister (London, 1605), sig. Q3 verso.

19 On the life of Jorden, see the biographical sketch by Joseph Frank Payne in the *Dictionary of National Biography*, and the note in William Munk, M.D., F.S.A., *The Roll of the Royal College of Physicians of London* . . . (3 vols., 2nd ed.; London: Published for the College, 1878), Vol. *1*, p. 113.

were appalled by the excesses of the Paracelsian mystics of his time. And as a man who sought a new explanation of the generation of metals and the source of the internal heat of the earth, he perhaps forms an interesting subject for those who are interested in the problems of geology in the period of the scientific revolution.

Born at High Halden, Kent, in 1569, Edward Jorden was the younger son of a gentleman of good family. After studying at Hart Hall, Oxford, he proceeded to Cambridge, where he obtained an M.A. at Peterhouse in 1586. Like so many other sixteenth-century English students who were interested in medicine, he then went to the continent, where he obtained an M.D. at Padua (c. 1591). On his return to England he became first a Licentiate (November 7, 1595) and then a Fellow (December 22, 1597) of the London College of Physicians. This organization was surely one of the most impressive scientific societies in existence at this time. Chief among its members was the renowned William Gilbert, who had been admitted as a Fellow in 1573, had already served in almost every post the College had to offer, and was soon to become president of the august body (1600). Yet there were others whose work and views may have been no less influential on the young Jorden. Walter Bayley, an authority on eyesight who had died as recently as 1592, had authored a short but important work on the analysis of mineral waters. Mark Ridley (F.C.P. 1594) published work on the magnet second in importance only to Gilbert. And although these men were not far removed from the time when John Geynes had been suspended from the College for questioning the infallibility of the writings of Galen (1559),[20] there were already among them a large number of proponents of the new chemical remedies that were commonly thought to be Paracelsian in origin. Thus, we find that when an official pharmacopoeia was proposed by the College (1589), a special committee on chemical medicines was set up. On this committee was Thomas Muffett (F.C.P. 1588) who was one of Europe's leading proponents of the new medicine. Nor need we be surprised to learn that the widely respected Sir William Paddy (F.C.P. 1591) was a close friend of both the mystical alchemist Robert Fludd (F.C.P. 1609) and the practical iatrochemist Sir Theodore Turquet de Mayerne (F.C.P. 1616). Similarly, we find the noted surgeon John Banister, an enthusiast of chemical medicines, licensed to practice medicine by the College (1594). And in the early years of the new century, in addition to the work of Robert Fludd and Turquet de Mayerne, we find a deep understanding

20 W. S. C. Copeman, *Doctors and Disease in Tudor Times* (London, 1960), p. 14.

of the work of contemporary chemists in the writings of Matthew Gwinne (F.C.P. 1599), Thomas Rawlins (F.C.P. 1605), and even William Harvey (F.C.P. 1607). Traditional characterizations of the College as a stronghold of Galenic conservatism are surely in error, for tangible proof of the Fellows' willingness to entertain innovations in medicine may be found in the publication of the official *Pharmacopoeia* (1618), which gave full support to chemically prepared medicines.[21]

Jorden must have been at ease among these chemically oriented members of the Royal College of Physicians. He had surely read widely in all fields of natural philosophy, and his writings show that he was well acquainted with all of the more important chemical authors of the period. Among metallurgists he frequently quoted Biringuccio, Agricola, and Ercker, but perhaps even more interesting for our purpose is the fact that Jorden represents a direct English link with the continental iatrochemical tradition. Although his detailed knowledge of the refutation of Paracelsus by Thomas Erastus is evident throughout, and he spoke with some distaste of the "Platonicall grandiloquence" of Peter Severinus,[22] Jorden referred often to Penotus, Thurneisser, Muffett, Dorn, and Crollius, as well as Paracelsus. Among chemical authors, however, it is evident that he leaned most heavily on Andreas Libavius, who had evidently befriended him while he was a student on the continent. In Libavius' most important work, the *Alchemia* (1597), we find Jorden listed first among a group of friends — among them Tycho Brahe — who had offered him information for his book.[23] It was in the *Alchemia* that Libavius included a lengthy and important chemical discussion of mineral waters, and it is for his investigation of the baths at Bath that Jorden was best known in the seventeenth century. Similarly, we note a common rejection of the works of the mystical alchemists by Libavius and Jorden.

For many reasons we may classify Jorden as one of the more enlightened English scholars of the early seventeenth century. Constantly rejecting the supernatural as a valid form of explanation, he repeatedly sought causes in natural events and forces. Thus, he wrote, "we must

21 Georg Urdang, "How Chemicals Entered the Official Pharmacopoeias," *Arch. Int. Hist. Sci.*, *33* (1954), 304–313. See, also, Urdang's historical introduction to his reprint of the first 1618 edition of the London *Pharmacopoeia*. Additional points are given in Debus, *English Paracelsians*, pp. 137–156.

22 Edward Jorden, Dr. in Physick, *A Discourse of Naturall Bathes, and Minerall Waters* (2nd edition "in many points enlarged"; London: Thomas Harper, 1632), p. 69. Unless otherwise noted, all citations will be made from this edition, which is considerably expanded from the first edition of 1631. All later editions (1633, 1669, 1673) are based upon the revision of 1632.

23 Andreas Libavius, *Alchemia* (Frankfort, 1597), sig c.

not imagine that the government and ordering of the world and nature in a constant course, is performed by miracle, but that naturall effects have naturall causes, and must be performed under the same *genus*."[24] This approach, common to most Paracelsians and natural magicians, is clearly evident in Jorden's first publication.[25] Having been called in to investigate a case of demonic possession, Jorden proceeded to reject that explanation and to attribute the attack rather to natural causes. Nor did he share the belief of many of his contemporaries that there had been an appreciable decay of the pristine knowledge of Adam. Instead, he felt that we could not help but know more now than the ancients did, since "*Dies Diem docet: Alpham Beta corrigit.*"[26] Still, even the great minds of his own age were subject to error, and as they felt free to publish their views, so, too, Jorden thought he might be excused for presenting to the public the results of his research. He stated, "I trauell where no path is to be seene of any learned foot that here hath beene,"[27] yet his work constantly reveals the mark of a cautious scholar. Thus, he diligently sought the advice and criticism of others prior to publishing his views. In his discussion of the flow of mountain springs, he tells us that he described his views in a letter that he sent to Henry Briggs (the first professor of geometry at Gresham College), "mine ancient friend, for his advice in it, being a point wherein he was well studied: but before my Letter came to Oxford, he was dead. But not I have aduentured to publish it, to stir up others to search out the causes hereof, better then hath yet beene discouered."[28] Again, in addition to the numerous personal observations recorded in his pages, we find a persistent interest in the observations of others who had traveled abroad, an interest strongly reminiscent of the early Fellows of the Royal Society of London. Thus, it was through a correspondence with sea captains and merchants who knew the East Indies that Jorden corrected his early belief that camphor was a bituminous substance.[29]

Jorden is an attractive figure to the twentieth-century mind for many reasons. We find him highly skeptical of the evidence commonly

24 Jorden, *Discourse of Naturall Bathes*, p. 109.

25 Jorden, *A briefe discourse of a disease called the Suffocation of the Mother, written upon occasion which hath beene of late taken thereby, to suspect possession of an evill spirit or some such like supernaturall power. Wherein is declared that divers strange actions and passions of the body of man. which . . . are imputed to the Divell, have their true natural causes, and do accompanie this disease* (London, 1603). This short work, which was never reprinted, and the *Discourse of Naturall Bathes* are the only two works written and published by Jorden.

26 Jorden, *Discourse of Naturall Bathes*, p. 6.

27 *Ibid.*, p. 103.

28 *Ibid.*, p. 20.

29 *Ibid.*, p. 34.

given for transmutation in nature. In his discussion of vitriol, he referred to springs in Cyprus, in Transylvania, and the famous Hungarian waters at Smolnicium, places in which

> . . . Copper is ordinarily made out of iron by infusing it with these waters. I will not determine whether this be transmutation of one species into another, as some doe hold, or rather a precipitation of the Copper which was formerly dissolved in the water by means of the sharpe Vitriol; which meeting with the Iron, corrodes it, and imbibeth it, rather than the Copper, and so lets the Copper fall, and imbraceth the Iron in place of it. We daily see the like in Aqua fortis, which having imbibed one metall, will readily embrace another that is more familiar to it and let fall the first.[30]

On this point Jorden referred to Libavius and Ercker. The latter, noting that iron precipitates copper from coppery solutions, was first convinced that "it is only the copper in such solutions that is precipitated by the iron and the iron itself does not change in copper." Later, however, he observed that where copper ore is mined, iron objects appear to become copper. Noting further that in solutions the precipitated copper was greater in weight than the metal in solution originally, he reversed his earlier opinion. "I have therefore been forced to the conclusion that iron turns into copper." Libavius, concerned whether or not this was a true transmutation, came to the same conclusion and cited Ercker in support of this view.[31]

> Ferrum in aes mutatur per aquas atramentosas in Hungaria & Saxonia iam vulgari arte, ut videatur utrumque metallum tantum chalcanthi copia multa vel pauce distare, Aes enim est fere totum vitriolum, fitq; ex ferro recepta huius natura. Corroditur ferrum in pultem viridem, quae ex coctione abit in aes Cyprium melius quam conflatile ex sua vena. Erkerus aliquando in ea fuit sententia, que à ferro tantu deflueret, descenderitq; cuprum: Sed postquã vidit clavos, & ferramenta in fodinis chalcanthi in aes constans mutari, fassus est veram transmutationẽ fieri.

No less susceptible to criticism was the contemporary use of sympathy and antipathy as a mode of explanation of natural phenomena. On this Jorden commented, "if we examine aright what this sympathy and antipathy is, we shall finde it to bee nothing but a refuge of ignorance, when not being able to conceiue the true reasons of such actions and passions in naturall things, wee fly sometimes to indefinite generalities, and sometimes to this inexplicable sympathy and antipathy: attributing voluntary, and sensitiue actions and passions to insensible substances."[32]

30 *Ibid.*, pp. 49–50.

31 See the chapter titled "Whether Iron Turns into Copper," in Lazarus Ercker, *Treatise on Ores and Assaying*, trans. from the German ed. of 1580 by A. G. Sisco and Cyril Stanley Smith (Chicago: University of Chicago Press, 1951), p. 223. Referring to Ercker on this point, Andreas Libavius wrote in his *Syntagma Selectorum undiquaque et Perspicue Traditorum Alchymiae Arcanorum* (Frankfort: Excudebat Nicolaus Hoffmannus, Impensis Petri Kopffi, 1611), p. 280:

32 Jorden, *Discourse of Naturall Bathes*, p. 121.

For Jorden heat could generally be thought of as a form of motion of the "seeds or formes" of a substance. Thus, he explained that a solid substance may be heated or set on fire through various forms of motion, such as collision, concussion, dilation, compression, putrefaction, fermentation, or reflection. However, some results led Jorden to question the universality of this rule. Anticipating the procedure of Joule two centuries later, Jorden studied the temperature of water in violent motion. At the cataracts of the Rhine near Splug, "the agitation and fall of water upon rocks is most violent, and make a hideous noise; yet it heats not water."[33] His incorrect conclusion was that here we have an exception to the general rule. In short, air and liquids are not heated but cooled by attrition.

His view of heat as a form of motion clearly affected his concept of the system of the universe. Concerned with the origin of the heat of some mineral water spas, he referred to the thoughts of those authors who held that this, of necessity, must derive from the sun. If this is so, Jorden continued, it must be due to the motion of the sun. And if the sun has motion, we must think in terms of either motion on its own axis, or the circumgyration of the sun on a fixed sphere. The latter view he called the common opinion,

> which if it be so, he must be carred more swiftly than a bullet out of a peece of Ordnance. I read in the Turkish History at the siege of Scodra, of a bullet of twelve hundred weight shot out of a cannon called the Prince, and it seems a great matter. But to haue a bullet as the globe of the Sunne, which is held to be 166 times bigger than the globe of the earth, to bee carried in a swifter course, and that perpetually, is a monstrous, furious, and mad agitation, *insanus motus*, as one termeth it.[34]

Unwilling to discuss this subject in greater detail, Jorden referred his reader to the sixth book of Gilbert's *De magnete* for a more detailed confutation of the motion of the spheres.

In his method for the analysis of mineral waters, Jorden shows himself to have been one of the outstanding analytical chemists of the early seventeenth century. Well aware of the earlier literature on this subject, he praised Agricola, Fallopius, Baccius, Solenander, Thurneisser, and Libavius for their contributions to this subject. These were men who had described detailed analytical procedures for examining the contents of mineral waters, and because of their work Jorden felt confident of the inadequacy of other authors who gave results of the contents of springs simply by distilling a sample to dryness and then tasting the residue.[35] He suggested, rather, that the first check for im-

33 *Ibid.*, p. 90.
34 *Ibid.*, p. 91.
35 *Ibid.*, p. 122.

purities in a water should be comparing the weight of a sample of that water with a similar sample of pure water. Still, he complained that "it it hard to have great balances so exact, as a small difference between them may be discerned by them."[36]

Jorden divided the ingredients of mineral waters into the following categories: simple earth, stone, bitumen, salts, mineral spirits, mean metals, and spiritual substances.[37] Of these, the most interesting for the historian of chemistry are his tests for salts. He stated that there are four types of salt — niter, salt, alum, and vitriol — but that each of these has diverse species.[38] Jorden's procedure for the identification of salts was based primarily on distillation and the recrystallization of the residue. At this point he depended primarily on crystal form. He noted that "the shooting or roching of concrete iuyces, is worthy to bee observed, seeing euery kinde hath his seuerall manner or fashion of shooting, whereby a man may see the perfection of each kinde."[39] The method was simple: after a stick had been placed in the liquor, it was allowed to evaporate, and

> within a few dayes, the concrete iuyces will shoote upon the wood, in Needles, if it bee Niter; in Squares, if it be Salt; and in Clods and Lumps, if it be Allum, or Coperose; and the other minerall substance which the waters have received, will either incorporate a tincture with them, or if it be more terrestriall, will settle and separate from it, and by drying it at a gentle fire will shew from what house it comes, either by colour, taste, smell, or vertue . . .[40]

Crystal form had been used for identification earlier than this, but it had never been considered so essential as here. Jorden suggested that saltpeter should be tested for purity by crystallization techniques, for only when it is pure will it shoot into needles. When impure, it will appear in "squares or angles or lumps."[41] He also saw that when saltpeter is mixed with salts of ashes, and then allowed to crystallize, the salt of ashes precipitates first as squares and the saltpeter appears last in needles.[42] Jorden found it difficult to find a reason for the characteristic

[36] *Ibid.*, p. 13.
[37] *Ibid.*, pp. 20–67, *Ibid.* (1631), pp. 17–76.
[38] *Ibid.* (1632), p. 37.
[39] *Ibid.*, p. 38.
[40] *Ibid.*, p. 124.
[41] *Ibid.*, p. 38.
[42] *Ibid.* A less detailed method of analysis including crystallization was suggested for English miners at approximately the same time. Gabriel Plattes wrote that the prospector should proceed in the following manner:

> The nexte worke is to finde out the Springs of Water issuing out of the said Mountaynes; and those being found, a quantity of the sayd water is to be boiled in a new cleane pipkin, to the consistency of a thinne Oyle, but no so thicke as a Syrrup, and when it is almost cold, then to put it into a Urinall, and to set it in the coldest place that can be found for 3 daies, then to play the Physitian, and to observe it exquisitely what residence it yeeldeth: if nothing

crystal forms. It would seem, he said, that when different salts are dissolved in the same water, the resultant crystals would appear uniform. But such is not the case, since such a "mixt water will yeeld his seuerall salts distinctly, and all at once."[43] The reason can only rest in the fact that nature has distinguished these species no less distinctly than plants. We need not look for varying proportions of elements or qualities for our answer,

> but only from the forme, *anima*, seed, &c. which frames euery species to his owne figure, order, number, quantity, colour, taste, smell, &c. according to the science, as *Severinus* termes it, which euery seed hath of his owne forme. So also it is in minerals, which haue their seuerall and distinct species in nature, and their seeds to maintaine and perpetuate the Species.[44]

In the earth these salts may not always appear in their characteristic crystalline forms, either because of a mixture with other minerals, because of a lack of space, or because of the lack of water to dissolve them. Yet, through chemical crystallization techniques, we are able to discover these forms, which are clearly made evident.[45]

Jorden suggested also that if the analyst did not wish to wait for the salts to crystallize, they might be isolated in a second way, by precipitation,

> whereby those minerall substances are stricken downe from their concrete iuyces which held them, by addition of some opposite substance. And this is of two sorts: either Salts, as Tartar, Soape, Ashes, Kelps, Vrine, &c. Or sowre iuyces, as Vinegar, Lymons, Oyle of Vitrioll, Sulphur, &c. In which I have observed that the Salts are proper to blew colours, and the other to red: for example, take a piece of scarlet cloath, and wet it in Oyle of Tartar (the strongest of that kinde) and it presently becomes blew: dip it againe in Oyle of Vitriol, and it becomes red again.[46]

> settle but a black earth or mudde, it is a signe of Coales: if some part thereof shoot into Ice, or a substance like Ice or Vitrioll, then to observe the colour thereof; if it be greene or blewish, it is an evident signe of copper; if whitish then it may signifie any other Metall with out exception. (Plattes, *Discovery of Subterraneall Treasure*, pp. 9–10.)

It would appear that Jorden was in error in his discussion of the precipitation of saltpeter and salt of ashes. Based on the solubilities of K_2CO_3 and KNO_3, one might expect the saltpeter to precipitate first. Indeed, in a description of a similar process by Nicolas Lemery, he concluded that "the long crystals that we see Saltpeter shoot into, do proceed from its volatile part, for that which is crystallized last, is fixt like sea-salt, and looks just like it." Nicolas Lemery, *A Course of Chymistry*, trans. Walter Harris, M.D. (2nd edition from the 5th French edition; London, 1686), p. 292.

43 Jorden, *Discourse of Naturall Bathes* (1632), p. 39.

44 *Ibid.*

45 *Ibid.*

46 *Ibid.*, p. 124; *Ibid.* (1631), p. 76. One cannot be certain just what dye was in the cloth Jorden used. However, in a letter to the author, Sidney M. Edelstein has pointed out that the common scarlet dye in Jorden's lifetime was cochineal or kermes and that these "would have acted exactly as Jorden describes the reactions with alkali and acid in his text."

This, then, is an extremely early instance of the use of an acid-base indicator, for in essence, his "salts" here are all basic, while his "sowre iuyces" are nothing but acids. However, Jorden certainly was not consistent in his terms, for although all of his examples here are clear enough, in his other division of salts into niter, alum, salt, and vitriol, such a definition would not hold. These observations with the scarlet cloth and crystallization are important, but Jorden generally failed to apply quantitative methods to his analyses, an approach that by the 1630's was not uncommon with other authors who were interested in the ingredients of mineral waters.

The subjects most important to Jorden in his *Discourse of Naturall Bathes*, however, were not these, but rather his search for the true explanation of the growth of minerals and the source of the heat of mineral water springs. These were major problems for Renaissance natural philosophers and they seemed intimately connected. Jorden, drawing on a varied background of current thought, offered to the reader a fermentation hypothesis that seemed to him to account adequately for all observed phenomena.

Jorden's fermentation hypothesis was based upon his conviction that there is a life force within metals. As a thorough-going vitalist he did not hesitate to compare the growth of minerals with that of plants and animals. In the Renaissance this theory was widespread and frequently referred to by alchemists, but the concept is one that may be traced back much farther; it is found in the work of Anaxagoras, Agatharchides, Theophrastus, and Dionysius.[47] Sendivogius, speaking of the generation of the metals, discoursed at length about the role of the seed or sperm, which "is the Elixir of every thing, or Quintessence, or the most perfect Decoction or Digestion of a thing or the Balsom of Sulphur, which is the same as the Radical Moisture in Metals."[48] For him, the center of the earth was to be compared with the womb of a woman; meeting a proper matrix, the seed engenders the various metals. Similarly, William Gilbert found the earthly fluids that result in metals comparable to the blood and the semen in the generation of animals.[49]

Jorden felt that the book of Genesis might be used as a guide on this point. He was convinced that the Lord taught us that the seeds were created prior to the completed plants. By analogy, therefore, we may assume that minerals

> were not at first created perfect, but disposed of in such sort, as they should perpetuate themselues in their seuerall kindes. Wherefore it hath euer

47 See Adams on "*The 'Seed of the Metals'*" in *Birth and Development*, pp. 289ff.
48 Sendivogius, *New Light of Alchymy*, p. 6.
49 William Gilbert, *De magnete*, trans. P. Fleury Mottelay (New York: Dover, 1958), p. 36.

beene a receiued *Axiome*, among the best Philosophers, that minerals are generated, and experience hath confirmed it in all kindes.[50]

Thus, we find saltpeter regenerated on an earthen floor, and tin regenerated in the exhausted mines of Cornwall. Similar reports of the growth of iron ore were confirmed by Agricola and Baccius, and Erastus and Thurneisser discussed the growth of silver on a beam of wood placed in a pit to support the ceiling.[51]
What more need be said?—"The manner of generation of minerals and metals, is the same in all, as is agreed upon both by *Plato* and *Aristotle*, and *Theophrastus*."[52] All evidence, both sacred and profane, seemed to confirm Jorden in his belief that mineral propagation could quite properly be compared with that of the other two kingdoms:

> Animals are propagated by begetting of their species, the power whereof is in euery indiuiduall, which, no doubt, will not giue ouer this trade as long as the world lasteth. Vegetables are also fruitfull in their kinds, euery one producing 100, or perhaps 1000 seeds of indiuiduals yeerly, to perpetuate their species. Minerals haue no such meanes, but onely haue their seeds in their wombes, whereby they are propagated: and if these generations, being longer in perfecting of their species, were not supplied with a larger extent for their productions; nature had been defectiue in not prouiding sufficient meanes for their perpetuity, as well as for others, and might easily suffer a decay, and a vacuity of the minerall species; which agrees not with the prouidence of nature, and the ornament of the world. The necessity hereof depends vpon the first benediction, (*crescite & multiplicamini*), which; no doubt, belongs as well to minerals in their kinds, as it doth to animals and vegetables, and by virtue hereof wee see that they are propagated daily . . .[53]

Yet how might we explain this growth or increase? The effects we perceive can result only from celestial or from terrestrial influences. Jorden proceeded to discuss each in turn. It seemed most unlikely to him that metallic generation should result from starry emanations. In his *Phisica Genesis*, the learned Paracelsist Gerhard Dorn taught us that plants with their seeds were created on the third day and planets on the fourth.[54] Surely, then, terrestrial substances are not dependent upon planets for their generation. Nor do we find the light or heat of the sun to be necessary for this process. Mineral generation takes place in eternal darkness, so we may rule out light as a factor.[55] Nor does the heat we perceive proceed from celestial bodies. It is not likely that they have the proper degree of innate heat to produce such an effect on earth, and their substance is much too thin for them to acquire addi-

50 Jorden, *Discourse of Naturall Bathes* (1632), p. 67.
51 *Ibid.*, pp. 67–68.
52 *Ibid.*, p. 69.
53 *Ibid.*, pp. 118–119.
54 *Ibid.*, pp. 70–71.
55 *Ibid.*

tional heat by motion.[56] Their beams do not have enough heat to explain the phenomena, and even if they did, we could not explain the effect by reflection on the earth because of the constancy of the observed temperatures in mines and mineral waters. The celestial beams are sometimes obscured and are constantly being directed at us from different angles. We should therefore expect varying temperatures to result from the different intensities.[57] We may concede that heat may aid the generation of metals, but we should not forget that it is only a quality itself, which can breed no substance. At most, the sun may help further the natural process, but Jorden could not consider it the principle cause of the growth of metals or their ores.[58]

If we may rule out a celestial cause, we must then search for a terrestrial explanation.[59] Here, we must examine in particular the role of the elements in metallic generation. As substances that represented a cosmological scheme as well as a chemical system, the elements had been a subject of considerable controversy between the Aristotelians and the Paracelsians for a century. Indeed, many of Jorden's authorities had discussed the matter in great detail. Jorden reflected this dispute in his impatience with those who wished to attribute nearly all things to the elements. Some authors insisted that our souls are elementary and others derived from them tastes, colors, and a host of other properties.[60] All of these views, Jorden asserted, could be summarily rejected, for surely the elements cannot give these properties to other things, since they do not possess them themselves.

Among the elements, only fire could be a possible answer to our problem, since the other three are cold and therefore could not cause generation of any sort. The authorities who stated that the element fire is responsible for the generation of metals did so because they believed that heat must assemble all elements together and then proceed to "ranke them." Not so, replied Jorden. "Heat can but heat," and all chemists know that the result is the separation of different substances and the congregation of similar ones.[61] Indeed, while none would deny the existence of common fire, the existence of the element is a different matter. Jorden rejected the latter, as did many of his contemporaries:

As for the element of fire, which should bee pure, not shining, and therefore invisible, and subsisting without a subject or fewell: let them finde it who

56 *Ibid.*
57 *Ibid.*, p. 110.
58 *Ibid.*, p. 71.
59 *Ibid.*, pp. 72–78.
60 *Ibid.*, p. 75.
61 *Ibid.*, p. 72.

know where to seeke for it. For my part I know no element of fire, vnlesse we should make it to be that which is naturall to all creatures and their seeds, causing their fermenting heat . . .[62]

When we turn to the alchemists we meet with their sulphur-mercury theory of the metals. If they meant common sulphur and common mercury, they were obviously mad, but it would seem that most of them understood by these words "some parts of the seminary of metals which have some analogye with these: and so their opinion may be allowed. For the spirit, which is efficient in these generations, doth reside in a materiall substance, which may be resembled to Sulphur or Oyle, as some other part may be resembled to Mercury."[63] The chemists' views are permissible, then, if they are understood in terms of the seed-sperm concept of generation.

We may find real truth in an application of two axioms of Hippocrates: (1) "everything is dissolved into that whereof it was made"; and (2) "wee are nourished by such things as we consist of."[64] We find that neither natural nor artificial dissolutions give to us the four elements. Hippocrates, speaking of natural dissolution or putrefaction, observed three distinct substances: *calidum*, *humidum* or *fluidum*, and *siccum* or *solidum*.[65] This agrees well with the artificial distillation analyses of the chemists who obtain *vaporosum*, *inflammabile*, and *fixum*—that is, mercury, sulphur, and salt.[66] Thus, if all things are dissolved into their original components, these primal components cannot be the four elements. Furthermore—again following the maxims of Hippocrates—if things do consist of the four elements, they are not nourished by them. Jorden therefore felt it safe to assert that metallic growth is not nourished by the four elements. At best we may consider the traditional elements as matrices or wombs within which the growth process occurs from "the seeds of things, which are agreeable to their species."[67]

Since none of the earlier concepts were satisfactory, Jorden felt free to give his own views on the generation of metals. Here he adapted to his own purposes the concept of fermentation. This phenomenon had long held the attention of alchemists, who saw a similarity in the action of yeast with that of the philosopher's stone. Martin Ruland (1612) quoted Hermes, Geber, Richardus Anglicus, and Raymundus Lullus in his description of the relation of fermentation to the great

62 *Ibid.*, p. 96.
63 *Ibid.*, p. 73.
64 *Ibid.*, p. 77.
65 *Ibid.*, pp. 78–79.
66 *Ibid.*, p. 79.
67 *Ibid.*, p. 80.

work.[68] For Bonus of Ferrara (c. 1330), the ferment was the great secret of the art of alchemy, and as such it was one that the sages spoke of only in the most obscure terms. Its action is "not quite analogous to leaven. For leaven changes the whole lump of dough into a kind of leaven, but our Stone, instead of converting metals into Tincture transmutes them into gold."[69] With Paracelsus we find an interest in fermentation directed more at an understanding of generation in the earth. He suggested that the reaction of salt and sulphur in the earth produces an internal fermentation that results in a liquid that subsequently solidifies into sharp rocks on exposure to air.[70] Van Helmont proceeded to emphasize the role of fermentation in earthly generations and to reject specifically fire and heavenly influences as important factors in the process.[71]

Although the concept of fermentation was not a novel one, Jorden did apply it to the problem with greater internal consistency than had earlier authors. The process is surely a simple one:

> There is a Seminarie Spirit of all minerals in the bowels of the earth, which meeting with conuenient matter, and adiuuant causes, is not idle, but doth proceed to produce minerals, according to the nature of it, and the matter which it meets withall: which matter it workes vpon like a ferment, and by his motion procures an actuall heate, as an instrument to further his work; which actuall heate is increased by the fermentation of the matter.[72]

Indeed, this may properly be likened to the making of malt,

> where the graynes of Barley being moystened with water, the generatiue Spirit in them, is dilated, and put in action; and the superfluity of water being remoued, which might choake it, and the Barly laid vp in heapes; the Seeds

[68] Martin Ruland, *Lexicon Alchemiae* (Frankfort, 1612; Hildesheim: Georg Olms, 1964), pp. 211–213.

[69] Bonus of Ferrara, *The New Pearl of Great Price*, trans. and abridged by A. E. Waite (London: Vincent Stuart, 1963), p. 254. For a late medieval alchemical discussion of fermentation, see, also, Sir George Ripley, "The Compound of Alchymie" in the *Theatrum Chemicum Britannicum*, collected and annotated by Elias Ashmole, introduction by Allen G. Debus (New York/London: Johnson Reprint Corporation, 1967), pp. 107–193, especially pp. 173–178. Walter Pagel discusses the role of fermentation in alchemy in *Paracelsus*, pp. 261–263.

[70] Paracelsus, "*Philosophie de Generationibus et Fructibus Elementorum*," in *Opera Bücher und Schrifften*, ed. J. Huser (2 vols.; Strassburg: Zetzner, 1616), Vol. *2*, p. 57. "Dann das Salz/ und der Sulphur, die da gestanden seind/ haben sich nicht mögen vergleichen in der Globul mit der Erden/ sondern durch die Erden in das widerspiel gebracht/ unnd in ein Wüten kommen/unnd (sich) gebläet/ als sein schaum oder gest von einem Fleisch oder Wein." J. R. Partington noted this passage in the Latin translation in *A History of Chemistry* (Vol. 2; London: Macmillan, 1961), p. 147. More traditional alchemical thought on fermentation is reflected in the pseudo-Paracelsian *Aurora Philosophorum*. See Paracelsus, *Chirurgische Bucher und Schrifften*, ed. J. Huser (Strassburg: Zetzner, 1618), p. 772.

[71] John Baptista van Helmont, *Oriatrike, or Physick Refined*, trans. John Chandler (London: Lodowick Loyd, 1662), pp. 31–34, 155–156.

[72] Jorden, *Discourse of Naturall Bathes* (1632), p. 84.

gather heat, which is increased by the contiguity of many graines lying one vpon another. In this worke natures intent is to produce moe indiuiduals, according to the nature of the Seede, and therefore it shoots forth in spyres: but the Artist abuses the intention of nature, and conuerts it to his end, that is, to increase the spirits of his Malt. The like we find in minerall substances, where this spirit or ferment is resident, as in Allum and Copperas mynes, which being broken, exposed, and moystened, will gather an actuall heat, and produce much more of those minerals, then else the myne would yeeld; as *Agricola* and *Thurneisser* doe affirme, and is proued by common experience.[73]

Thus, Jorden continued, the evident heat experienced in deep mines need not be ascribed to the sun, which clouds can remove, or to an internal subterranean fire. The earth abounds with matter "which attends vpon the species of things: and oftentimes for want of fit opportunity and adiuuant causes, lies idle, without producing any species."[74] Again rejecting any active role of the elements, Jorden insisted that "this matter is not the Elements themselves, but subterraneall seedes placed in the Elements, which not being able to liue to themselues, do liue to others."[75] Here, then, we have the proper explanation of the nutrition of natural things. It is only in this way that we may properly account for colors, tastes, numbers, proportions, distempers, and all the other observed differences of the varying substances of the earth.[76]

Jorden's seminary hypothesis may be pushed somewhat further with additional analogies between the organic and the inorganic worlds. Transplantations occur in the animal and vegetable kingdoms. Thus, we note variations from the norm in the issue of dogs and wolves, horses and asses, and partidges and hens. We find similar changes in plants, since when we mix species the first seeds "doe oftentimes bring forth other fruits then their owne."[77]

In Minerals we find the like transplantations: as Salt into Niter; Copperasse into Allum, Lead into Tinne, Iron into Copper; Copper into Iron, &c. And this is the transplantation whereupon the Alchymists ground their Philosophers stone.[78]

Thus, although common explanations of transmutation are to be frowned upon, the fermentation concept makes it possible to think of the phenomenon in terms of the transplantation of mineral seeds. It is a position not far removed from that of van Helmont, who wrote that

73 *Ibid.*, pp. 84–85.
74 *Ibid.*, p. 86.
75 *Ibid.*
76 *Ibid.*
77 *Ibid.*, p. 87.
78 *Ibid.*

"the ferment is the parent of transmutation."[79] For Jorden this law of natural growth seemed basic and indisputable. The seminary spirit acting as a ferment seemed to have the backing both of philosophical authority (Aristotle) and observational evidence. More than this, Holy Scripture seemed to require the growth of all things from the first benediction, "*crescite & multiplicamini*."[80] Jorden thought that the generation of minerals formed only part of a greater divine law, one that was founded no less firmly than the laws of mathematics and logic.

The answer to Jorden's second major question—the source of the heat of the mineral springs—now seems obvious. It is surely not due to frictional heat of the wind, air, or other exhalations rising from the internal regions, since friction seems to cool rather than heat air and liquids.[81] Nor can the sun heat the internal waters of the earth any more than it does the metallic ores in the mines.[82] Those who apply the doctrine of antiperistasis here are equally wrong, since this would require that the waters must be hotter in the winter than the summer which is hardly the case.[83] And those who postulate large deposits of quicklime within the earth, which are heated on contact with water, may be refuted, since quicklime is a substance that does not exist in the natural state and can only be made in the laboratory.[84]

But surely the most common explanation is based upon a belief in a vast subterranean fire that results in heated waters, volcanoes, and all sorts of other surface effects. For Jorden this could not exist. If it did, we must assume that the center of our planet is hollow. This is not likely, since all heavy things descend toward the center of the earth.[85] Because of this, we might guess that the center of the earth is more compact than the surface. Again, if the earth is hollow and filled with a gigantic fire, we may think in terms of an immense distillation system. Vast receptacles below the surface must act in the form of chemical vessels filled with water, which are then heated to boiling so that the springs we see form the end product of a macrocosmic distillation system. However, laboratory experience shows us that in such systems almost all the heat is lost on condensation, whereas in the hot natural springs the temperatures remain very high.[86]

[79] van Helmont, *Oriatrike*, pp. 115, 207–208. Here, again, note the alchemical heritage expressed in van Helmont's thought.
[80] Jorden, *Discourse of Naturall Bathes* (1632), pp. 87–88.
[81] *Ibid.*, p. 89.
[82] *Ibid.*, pp. 90–93.
[83] *Ibid.*, pp. 93–95.
[84] *Ibid.*, p. 95.
[85] *Ibid.*, p. 99.
[86] *Ibid.*, p. 100.

And what of the subterranean fire itself? Jorden had questioned the very existence of the element, and if we speak not of it, but rather of the common kitchen fire, we are faced with a difficult problem of explanation. If this fire exists, how is it kindled and how does it continue to exist? Those who argue for it may have forgotten that any fire requires a tremendous amount of air, a fact obvious to anyone who has confined a flame with a cupping glass. Here the flame is immediately put out since "fuliginous vapours . . . choake it if there were not vent for them into the ayre."[87] Surely, the required abundance of air cannot reach the central regions under normal circumstances. And what of the problem of fuel? The proponents of the central fire insist that a limitless supply of fuel exists in deposits of bitumen and sulphur. But if this is true, the earth should sink in the areas of these deposits as they are consumed in the internal conflagration.[88] And furthermore, experience teaches us that not all hot baths are sulphureous or bituminous—nor are all bituminous or sulphureous springs hot.[89] Hence, we need not assume that if such a fire exists, these are the substances that maintain it. There are those who argue that the existence of volcanoes overrides all objections on this score. This Jorden refused to admit, and he countered that volcanic action could easily be thought of as confined to the regions close to the surface of the earth. Here, subsurface deposits of sulphur or other combustible material may be kindled by lightning without too much difficulty and proceed due to the proximity of the atmosphere.[90] In short, a subterranean, central, fire could not be maintained. There could not be enough fuel or enough air in the internal regions of the earth to maintain it, and there is far too much water within the earth to permit it. Indeed, previous authors have stretched the facts beyond endurance to account for such a fire, while they might have been explained far more simply by dropping this outmoded hypothesis entirely.

Surely, Jorden explained, we must return to our fermentation hypothesis of the generation of the metals. With this we may account for the heat by natural means, as the result of the motion between the agent spirit and the passive matter.[91] Such a heat is natural rather than violent and will result in generation rather than the familiar destructive powers of the ordinary kitchen fire.[92] Furthermore, it is a slow process

87 *Ibid.*, pp. 97, 101.
88 *Ibid.*, p. 112.
89 *Ibid.*, p. 102.
90 *Ibid.*, p. 101.
91 *Ibid.*, p. 104.
92 *Ibid.*

rather than a rapid combustion. Therefore, it will account satisfactorily for the heating of springs in the same area for long periods of time.[93] If we accept this explanation, we need not exhaust our wits to locate vast sources of fuel to maintain the internal conflagration and we need not rack our minds to think of ways for air to be admitted to the bowels of the earth to permit the continuance of the flames. And what more natural explanation could there possibly be for the presence of minerals in a water, since their generation was the very cause of the heating process.[94] For Jorden fermentation had become the basis of a new system of the earth. If we accept it we may do away with the difficult concept of the internal fire, and at the same time we may account far more satisfactorily than before both for the generation of the metals and for the internal heat of the earth.

Edward Jorden's *Discourse of Naturall Bathes, and Minerall Waters* was well known in the seventeenth century. It went through two editions and five printings (1631, 1632, 1633, 1669, 1673), and it was frequently referred to by many authors.[95] Jorden's rejection of supernatural causes in nature, his disapproval of sympathy and antipathy as a mode of explanation of natural phenomena, and his views on the heliocentric system distinguish him as a widely read and an intelligent scholar. Yet it is as a chemist that he stands forth with most distinction. The twentieth-century historian of science might point first to Jorden's clear description of the color test for acids and salts (bases), but Jorden himself surely considered his fermentation hypothesis much more significant. The fact that it was wrong need not lessen our interest in it, since Jorden's work conveniently offers to us an example of a seventeenth-century chemical philosopher attempting to explain geological phenomena in a new fashion. In his work Jorden clearly emerges as a proponent of new observations and experiments as a guide to nature —an approach that reflects the texts of the Paracelsians, which he frequently cites. Yet most other iatrochemists discussed primarily the application of chemistry to physiological processes. Other than their discussions of the divine chemistry of the creation, we find their explanations of the divine chemistry of the macrocosm relatively sketchy. Jorden, however, centering his work on metallic generation and the internal heat of the earth, gives us an example of a chemical philosopher

93 *Ibid.*, p. 105.
94 *Ibid.*, pp. 106–107.
95 For some seventeenth-century references to Jorden, see Debus, *English Paracelsians*, p. 174 (note 101).

who sought a new chemical explanation of the earth. His views on the growth of metals give us new insight into the problems of early seventeenth-century natural philosophers who relied on observation and experiment as proper guides to nature, but whose work was stimulated by an unshakable belief in now-outmoded systems of thought. At the same time, Jorden's work gives to us greater insight into the breadth of the interests of the chemical philosophers. Clearly dissatisfied with earlier views, he adapted fermentation—traditionally one of the great alchemical mysteries—to a new explanatory role for our understanding of metallic generation and the composition of the earth.

The more we study the work of the sixteenth- and early seventeenth-century Paracelsians and iatrochemists, the more we can see the truth of their assertion that their goal was a comprehensive chemical explanation of the universe. We are only now beginning to realize the extent of the popularity of the chemical system of nature in the period prior to 1660. Because of the influence of these men, we should be encouraged to look further into the writings of the chemical philosophers for their macrocosmic as well as their microcosmic concepts.

James Hutton and the Concept of a Dynamic Earth

R. H. Dott, Jr.

The full significance of James Hutton's contribution to geology has been somewhat eclipsed by a kind of "Lyellian myopia." It is my opinion that the great attention given to such figures as Charles Lyell, William Smith, and others of the nineteenth century has obscured the fact that Steno and Hutton, before them, already had enunciated the most fundamental principles for the budding science, and Werner had forged a new sense of chronology. Hutton extended Steno's basic principles for deciphering earth history, and proceeded to formulate the first modern-appearing, unifying theory of the earth. Pursuing the history of geology with the bias of one interested in the tectonic evolution of the earth, I find that Hutton looms larger than any other figure in terms of early dynamic theories.

It is the purpose of this paper to show that Hutton provided a revolutionary unifying synthesis of geology in the late eighteenth century, the full significance of which has not been sufficiently appreciated and which seems to have been somewhat misinterpreted. I feel that the theory provided by him, though less concise and nonmathematical, was of the same order of importance to geology as were the syntheses of Newton and Lavoisier. As in the latter cases, a continuous thread of thought can be traced from Hutton to modern theory. His dynamic concept, coupled with the espousement and practice of actualistic reasoning, qualifies Hutton as a pre-eminent founder of modern geology.

Hutton's Background

BIOGRAPHICAL SUMMARY Several biographical treatments of Hutton have appeared,[1] so that only a brief summary will be given here.

[1] See John Playfair, "Biographical Account of the Late Dr. James Hutton, F.R.S. Edinburgh," *Transactions of the Royal Society of Edinburgh, 4* (1805), 39; E. B.

Hutton was born in Edinburgh in 1726 and died there in 1797, the year of Lyell's birth. His residence in Scotland was to prove fortunate in developing his geological concepts. As a student at Edinburgh, an interest in chemistry was sparked, and he ultimately determined to pursue medicine as the practical pursuit most closely allied to chemistry. In 1749 Hutton received the M.D. at Leyden, but a partnership in an Edinburgh enterprise for producing sal ammoniac from chimney soot obviated the necessity of conventional employment. Hutton turned to farming, and spent approximately two years in England and the Low Countries learning advanced methods. In 1754 he returned to Scotland to introduce the new agriculture at a family farm, Sligh House, Berwickshire, southeast of Edinburgh.

By 1768 Hutton had developed his farm to a high degree, and he wished to be free to follow more intellectual pursuits. He moved to Edinburgh, where he could enter more actively into the extraordinary society of learned men then concentrated there.

EARLY GEOLOGICAL INFLUENCES All of Hutton's geological publications were produced after his removal to Edinburgh about 1768, but interest in geology was spawned much earlier. While in England learning agriculture, he became interested in the origins of landscapes and soils, and so "was looking with anxious curiosity into every pit, or ditch, or bed of a river that fell in his way." He recognized the importance of weathering and erosion of older rocks to form the ingredients of sediments, thus the cyclic decay of old lands to form new strata. Though not entirely original observations, these were important in stimulating early questions in Hutton's mind, the answers to which formed the foundation of his ultimate theory of the earth. Out of Hutton's agricultural background also came a rudimentary appreciation of heredity and adaptation in plants and animals.[2]

Hutton's guiding philosophy through his geological researches was a devout belief that the earth represents a cyclic system of rejuvenation and decay, the sole *raison d'être* of which was to provide a habitat for life—all perfectly in balance. "A living world is evidently an object in the design of things, by whatever Being those things had been designed,

---

Bailey, "James Hutton, Founder of Modern Geology (1726–1797)," 150th Anniversary Commemorative to James Hutton, *Proceedings of the Royal Society of Edinburgh, 63* (1950), 357; and E. B. Bailey, *James Hutton—the Founder of Modern Geology* (Amsterdam: Elsevier, 1967).

2 Bailey (*James Hutton—the Founder of Modern Geology*, pp. 14–19) ascribes to him a possibly exaggerated anticipation of Darwin through passages from *Principles of Agriculture* (unpublished).

Plate II

JAMES HUTTON (1729–1797)

"*. . . the eminent old geologist, Dr. James Hutton, rather astonished at the shapes which his favorite rocks have suddenly taken.*" Caricature by John Kay of Edinburgh made in 1787. [Thomas Wright, *A History of Caricature* (London, 1865), p. 462, Fig. 221; *A Series of Original Portraits and Caricature Etchings*, by the late John Kay, ed. H. Paton (New ed.; Edinburgh, 1877), Pl. 24.]

and however either wisdom or folly may appear in that design."[3] It followed naturally from this mechanistic notion that Hutton would view the physical world analogically like an organism; the transfer of material by erosion and sedimentation was compared to the flow of blood.

Other obvious influences upon Hutton's geology were his own long-held interest in chemistry and the chemical researches of friends such as Joseph Black and Sir James Hall of Dunglass. The constant stimulation and criticism of the latter were of special importance in the maturation of Hutton's ideas, even though some of his chemical geology proved fallacious. The significance he ascribed to heat in earth processes derived considerably from his chemistry.[4]

ORIGINALITY In reading either Hutton's *Theory* or Playfair's *Illustrations*, a geologist cannot help but be impressed by the great number of basic facts and essentially modern interpretations that were detailed by Hutton. Coupled with the paucity of documents bearing on early development of his ideas, this has raised the question of originality. V. A. Eyles, in his contribution to this book, presents convincing evidence of a much greater eighteenth-century scientific communication over Europe than is generally realized. Also, some fragmentary Hutton correspondence written about 1770[5] shows clearly that a surprisingly complete knowledge of British areal geology existed and somehow was communicated. For example, Hutton at that time classified the rocks of Scotland into freestone, schists, and granite, and he detailed the distribution of each, noting that the structure of Scotland was more complex than that of England. He correlated the coal measures and underlying limestone of the two countries. He cited fossil localities and sent specimens to correspondents, though fossils interested him only for their presence in rocks then generally considered Primary and for their modes of mineralization. Playfair[6] tells us that by about 1785 Hutton had read most books of travel from which anything geological could be learned. Writers referred to by Hutton included Buffon, Burnet, Bou-

[3] See James Hutton, *Theory of the Earth with Proofs and Illustrations* (Edinburgh: William Creech, 1795; London: Cadell and Davies, 1795), Vol. II, p. 546.
[4] Since this was written, Gerstner analyzed the apparent origin and development of Hutton's rather unique concepts of energy, force, and matter. Gerstner feels that Hutton's concept of subterranean heat without fire was particularly important, being unique for the time. See P. A. Gerstner, "James Hutton's Theory of the Earth and His Theory of Matter," *Isis, 59* (1968), 26–31.
[5] V. A. and J. M. Eyles, "Some Geological Correspondence of James Hutton," *Annals of Science*, 7 (1951), 316–339.
[6] See Playfair, "Biographical Account," p. 51.

guer, Cronstedt, de Luc, de Saussure, Dolomieu, de Maillet, Monnet, Palassou, Pallas, Patrin, Raspe, Voigt, and others.

On the local Scottish scene, there were two significant possible precursors, John Walker (1731–1805), first professor at the University of Edinburgh (appointed 1779), and George Toulmin (about 1755–1817), who studied at Edinburgh from 1776 to 1779. Walker, who had such notables as John Playfair, Sir James Hall, and Robert Jameson as students in his class seems at first a likely candidate.[7] But Scott's valuable publication of his lecture notes reveals no hint of the Huttonian theory. Moreover, Walker's temperament and chief interests suggest that he could not have formulated the very intuitive and speculative ideas that appear therein. Walker was a very cautious, conservative individual, repelled by speculation and theorizing. Moreover, his chief geological interest and contributions were in taxonomic mineralogy. Yet it is a mystery that neither Walker nor Hutton mentions the other, for they were contemporaries in the same city. They must have known each other, and two of Hutton's close friends attended Walker's lectures.

From circumstantial evidence, Toulmin would seem a more likely precursor, for as Tomkeieff and McIntyre note, there was a striking similarity of his and Hutton's writings.[8] Moreover, Toulmin's first book on geology appeared in 1783, two years before Hutton's abstract was published. Davies contends, however, that Toulmin "was evidently Hutton's plagiarist, not his precursor."[9]

In Russia, M. V. Lomonosov (1711–1765) developed a concept of internal dynamism of the earth and an evolutionary rather than catastrophic viewpoint. Tikhomirov, in his paper in this book, presents insight into Lomonosov's contributions, which were contained chiefly in a synthesis entitled *On the Earth's Strata* (1761). Lomonosov had ideas similar to Hutton's 22 years before publication of the latter's *Abstract*. Lomonosov is reported to have used the actualistic method, assumed a great age for the earth, and attributed a major role to subterranean heat. Moreover, it is said that translations of his works were circulated during Lomonosov's lifetime. In view of the considerable scientific communication between Russia and Scotland during the

[7] See John Walker, *Lectures on Geology*, ed. H. W. Scott (Chicago: University of Chicago Press, 1966).

[8] See S. I. Tomkeieff, "James Hutton and the Philosophy of Geology," 150th Anniversary Commemorative to James Hutton, *Proceedings of the Royal Society of Edinburgh, 63* (1950), 351; and D. B. McIntyre, "James Hutton and the Philosophy of Geology," in Claude C. Albritton, Jr. (Ed.), *The Fabric of Geology* (Stanford, Calif.: Freeman, Cooper, 1963).

[9] See G. L. Davies "George Hoggart Toulmin and the Huttonian Theory of the Earth," *Bulletin of the Geological Society of America, 78* (1967), 121.

eighteenth century,[10] it appears likely that Hutton could have known of Lomonosov's ideas, though I know of no tangible evidence to verify this.

From 1765 to 1785 is a "dark period" in Huttonian history, for which we have little evidence of the development of his ideas. The great scope of the theory when it first appeared in 1785 implies a long gestation period. Playfair reminds us that Black's research on the causticity of lime was published in 1755.[11] Implications of the possible effects of confining pressure on volatile gases even at high temperature were essential to Hutton's theory; therefore, Playfair concludes "It was not, I imagine, till after the year 1760 that they [Hutton's researches] came to take the form of a theory."

HUTTON'S TEMPERAMENT Playfair has given us a clear picture of Hutton as an unusually enthusiastic, quick-witted individual driven by an insatiable curiosity. His enthusiasm often engendered a certain dashing carelessness. Though a deist, Hutton reflected the age of reason in being little restrained by conventional prejudices. Playfair tells us that he possessed perfect candor, was totally unaffected, and ". . . he was much more delighted with the contemplation of truth than with the praise of having discovered it. . . ." This last trait, together with a difficulty in written expression, delayed publication of and obscured his ideas. There is no doubt that his obscurity of expression has been responsible considerably for the relative neglect of Hutton by subsequent generations. Even Lyell admitted to having relied almost solely upon later sources for access to Hutton's ideas, and most modern scholars must confess likewise.

Hutton emerges as a keen observer, but also as an unusually inquisitive and creative man. His intellectual curiosity drove him, and his enthusiasm and clear oral logic quickly persuaded most who had close contact with him. Apparently, he was of an impulsive and speculative nature, yet he was among the first geologists to base interpretations upon carefully documented field evidence (especially in Volume III of the *Theory*). Unlike more cautious scientists, such as Black and Walker, he was compelled to seek sweeping generalizations, even if they were not completely supported by the fullest possible factual details; his writings reveal a profound appreciation of the role of generalization. He was favored with the luck of usually being right, even if occasionally for the wrong reasons. He was one of those rare intuitive geniuses whose originality of thought and timeliness periodically have

[10] See Eyles and Eyles, "Some Geological Correspondence."
[11] See Playfair, "Biographical Account," p. 59.

led to unparalleled breakthroughs in the simplification of knowledge. Hutton was the right man at the right time when geology in Europe needed a sophisticated unifying synthesis to amalgamate it into a viable discipline.

Development of the Huttonian Theory

LATE EIGHTEENTH-CENTURY GEOLOGICAL ISSUES In Hutton's day there were two paramount geological issues, and there is need to draw a distinction between them, because, at least for purposes of analysis, they represent two nearly independent controversies that too commonly are blurred together. These were the neptunian-plutonian and the uniformitarian-catastrophic controversies.

The first issue, *neptunism versus plutonism* (here including vulcanism), predated both Werner and Hutton, but became focused in the conflicts that raged between their camps. It is important to realize that the neptunism attacked by Hutton himself was largely pre-Wernerian. There was not simply an argument over the origin of basalt and granite, or of the deposition of true sediments from the sea as it is so commonly represented; Hutton agreed completely with the marine origin of most sediments. Much more profound was the little-appreciated underlying neptunian implication of a structurally static solid earth versus the plutonian assumption of a continually dynamic one. In neptunism, only the receding ocean changed significantly as it precipitated the strata of the crust and erosion attacked the emerged land. The remainder of the earth was assumed by many neptunists to be relatively static. In the simplest view, mountains were seen as vestigial protuberances of the "original" surface of the earth.[12] Even volcanoes, when their reality begrudgingly was admitted, generally were assumed of mere superficial origin through combustion of coal or oxidation of sulphide minerals in the uppermost strata. But in plutonism, the entire earth was seen as dynamic and in a constant state of changing. Dynamism was evidenced in forcible emplacement of molten granitic rocks deep in the crust and in the extrusion from depth of hot basaltic lavas, while the bowels of the earth were periodically shaken by earthquakes and intricately contorted to form mountains.

The second great issue was the familiar *catastrophic versus uni-*

12 Besides Lomonosov, other pre-Huttonian figures had at least vague ideas of some internal earth dynamism, most notably the early Italian geologists. In this book, Wegmann shows that Runenberg in Finland (1765) postulated slow tilting of the crust by subterranean forces based on the differential tilting of elevated Pleistocene shorelines. Despite the fact that isolated earlier ideas of internal dynamism did exist, the unifying Huttonian dynamic theory represents a "quantum jump" in geological thought.

formitarian or actualistic views of change in nature, which has received more attention than the neptunism-plutonism issue because of a somewhat greater (or more obvious) significance to philosophy and the development of the concept of organic evolution. While these two issues were not wholly unrelated, it is important to realize that most plutonists of the early nineteenth century also were catastrophists. Hutton contributed heavily to both debates, but it is my purpose to stress the first, as it was almost completely dealt with by Hutton, whereas the latter did not culminate until a generation after him through the persuasion of Lyell.

CYCLIC DEVELOPMENT OF LANDSCAPES The conventional eighteenth-century philosophical belief in a grand design or purpose in nature, namely, the provision of a perfectly balanced habitat for life, appears to have formed a basic assumption behind all of Hutton's geological theorizing.[13] Taking purposiveness as a Huttonian assumption, then, the very first observations (1752–1753) of the weathering and denudation of rocks, of soil formation, and of the erosion of valleys by rivers formed the whole crux of his theory. All subsequent observations and inferences, which deserve so much credit from the geological fraternity, merely supported and provided explanations of the continual rejuvenation of landscapes for the purpose of maintaining plants and animals. The chronology of his discoveries, as we know it, bears out this conclusion (Table 1).

TABLE 1 Chronological Summary of Development of Hutton's Theory of the Earth

1726	Hutton born in Edinburgh
c. 1740–1749	Acquired chemical and medical training
1752–1753	Observed weathering and erosion while studying agricultural methods in England
1754	Traveled to Low Countries to study agriculture
1754	Took up farming in Berwickshire; medical and agricultural experiences undoubtedly strengthened his view of the unity of nature
1755	Black's research on causticity of lime published
1764	First observation of granite and metamorphic rocks in northeastern Scotland (including the famous Portsoy "graphic" granite)
c. 1768	Moved from farm to Edinburgh to pursue intellectual matters full time; reportedly much of the theory conceived by this time

[13] See, for example, Hutton, *Theory of the Earth*, Vol. II, p. 186.

c. 1770	Decided on igneous origin of basalt; made first discovery of sodium in rock material (zeolite); correspondence indicates extensive knowledge of British geology and of fossil localities
1774	Traveled in Wales and western England with James Watt; sought sources of Secondary conglomerates near Welsh border and visited salt mines in Cheshire
1777	Published on controversy over quality of coal and culm
1783	Royal Society of Edinburgh formed; theory surely formulated by now
1785	Discovery of Glen Tilt granite dikes; abstract "Concerning the System of the Earth and its Duration, and Stability" published, contained essence of the entire theory
1786	Galloway granites explored and several intrusive contacts found
1787	Intrusive granite contacts, dike swarms (containing natural glass), and an angular unconformity observed on the Isle of Arran; unconformity exposures also discovered around Jedburgh
1788	Unconformity seen at Siccar Point; visited Isle of Man; found fossils in metamorphic rocks then considered Primary in the Lake Windermere area
1788	Extended statement of the theory published in Volume I, *Transactions of the Royal Society of Edinburgh*
1790	Oral paper presented describing the Glen Tilt dikes
1790–1791	Playfair and Sir James Hall found more fossils in rocks considered Primary; de Luc criticized Hutton in *Monthly Review*
1794	Note on Glen Tilt dikes published in *Transactions of Royal Society of Edinburgh*
1793	Kirwan's attack published
1795	Volume I and II of *Theory of the Earth with Proof and Illustrations* published; *Principles of Agriculture* nearly completed
1797	Died in Edinburgh
1798	Sir James Hall's experiments with crystallization of basalt and glass published
1802	Playfair's *Illustrations of the Huttonian Theory of Earth* published
1899	Volume III of *Theory of the Earth* published posthumously (manuscript of a promised fourth volume never found)

Just as is true of organisms, landscapes mature, then decay and disappear. Therefore, to provide a continual habitat for life, new landscapes must be born. And as new plants sprout in soil formed by the decay of rocks and older plants, so new landscapes are formed from the sedimentary products of the decay of older landscapes. "The spoils or wreck of an older world are everywhere visible in the present. . . ."[14]

[14] Playfair, "Biographical Account," p. 52.

These products were deposited in layers on the sea floor, consolidated (by heat, thought Hutton), and then upheaved by mountain-building processes. Hutton's greatest originality was in inferring this origin of landscapes and in generalizing it through indefinite time. In effect, Hutton had anticipated the important concept of *dynamic equilibrium.*

IGNEOUS ROCKS But what evidence was there, if any, of repeated upheavals of strata from beneath the sea, and what might be the cause? Here it is difficult to establish fully the exact chronology of Hutton's observations and inferences. The quest is compounded by the erratic sequence of publication of his discoveries, for it little reflects his development of ideas.[15]

Hutton perceived the igneous origin of whin or basalt probably about 1770 or 1775, apparently largely from observations made at Salisbury Crags, near his home in Edinburgh. There, a basalt sill forcibly intruded and altered Secondary strata (Old Red Sandstone). It possesses a crystalline rather than clastic texture, contains large baked inclusions of the sediments within it, has deformed the strata along its margins, and shows evidence of chilling at its contacts. The obvious baking of coal cut by basalt at other localities and the development of columnar jointing in many basalts also were used as arguments against the neptunists, as was the apparent impossibility of dissolving all constituents of igneous rocks in sea water. Hutton never interpreted any British basalts as truly volcanic, but ascribed all to a shallow intrusive origin, which he called *subterranean lavas,* considering them to be different from famous French and Italian extrusives. Hutton's genius in observation and reasoning is shown clearly by recognition of analogies between the rate of cooling of man-made glass and the relation between natural glasses and coarser basalts. In 1795 and 1798 these analogies were verified and extended experimentally by Sir James Hall, who determined not only that a high temperature (roughly 800–1200°C.) was required to melt the materials, but also that rate of cooling determined their texture.[16] Meanwhile, Desmarest's classic analysis of the Auvergne District appeared in 1774 (reported orally, 1765), but Werner was announcing in favor of an aqueous oceanic origin for Saxony basalts.

15 See Table 1, and Bailey, *James Hutton — the Founder of Modern Geology.*

16 Ironically, Hutton doubted that the question could be resolved by "looking into the bottom of a little crucible." He believed that the heat to which rocks had been subjected lay beyond the reach of laboratory simulation, "and that the operations of nature were performed on so great a scale, compared to that of our experiments, that no inference could properly be drawn from one to the other" (see the paper by C. S. Smith in this book).

In 1764 Hutton had traveled to northeastern Scotland, and there he saw for the first time crystalline granitic and metamorphic rocks. Exactly when is not clear, but sometime between 1764 and 1775 he extended the igneous concept for basalt to granite and concluded that granite must have formed by slow crystallization from hot, molten material. An originally fluid state was inferred in part from the irregular shapes of the massive granite bodies, as opposed to the more regular geometries of stratified rocks. Igneous origin was extended further to mineral veins, which he noted tended to occur in the same terranes as granite. Moreover, veins contained materials wholly foreign to surrounding rocks, so they "must" have been derived from great depth. Hutton's granite interpretation was based initially upon the peculiar texture of the Portsoy graphic granite (beautifully illustrated in Plate I, Volume I), which has feldspar and quartz grains intergrown in such intimate fashion that he concluded the two must have crystallized simultaneously and the rock could have "been consolidated by fusion, and in no other manner." His chemical background and Black's researches strongly influenced Hutton's interpretations of igneous rocks, which certainly were very advanced for the times. Playfair later acknowledged that the argument for crystallization of massive granite from fluids was more compelling than the inference that it required elevated temperature, the hot origin for basalt being better established than for granite.[17] While recognizing a clear compositional and textural difference between the two, Hutton had believed that granite was gradational to porphyry, which in turn was considered only a hybrid of basalt. Therefore, it was natural to infer a like origin for all three.

HEAT AS ULTIMATE CAUSE OF UPHEAVAL Exactly how and when Hutton became enamored of the importance of subterranean heat has not been fully established. Probably his initial consciousness of internal heat was inherited from the ancients, who assumed a hot interior and a molten origin of the earth. The early suggestion by Strabo that volcanoes functioned as safety valves when the interior became overheated appealed to Hutton. By 1740 the downward-increasing geothermal gradient had been observed in mines. According to Hutton himself[18] he reasoned subterranean heat first from his proof that mineral bodies must have been in a state of fusion. But this appears in a defense against Kirwan's attack, and I question the accuracy of his memory. Internal heat seems simply an a priori assumption, and I sense an element of

[17] See Playfair, "Biographical Account," p. 90.
[18] See Hutton, *Theory of the Earth*, Vol. 1, pp. 235–236.

circular reasoning in Hutton's treatment of igneous rocks and subterranean heat.

If upheaval is to provide land for life, rocks first must be consolidated, or erosion would too soon destroy the new land. So an overenthused Hutton called upon heat to consolidate newly-formed sediments on the sea floor. In a major faux pas, he roundly denied the role of cementation of sediments by simple chemical precipitation, basing his conclusion partly on Black's work and on the effects of pressure on coal when heated. Next, he appealed to heat to soften and then fold strata as it produced uplift by expansive forces. Though this panacea holds no credence today in such simplistic form, Hutton was lucky in hitting upon heat as ultimately the most important form of terrestrial energy.

PROOF OF INTRUSIVE RELATIONS OF GRANITE The neptunians contended that granitic and what we today call metamorphic rocks belonged to a Primitive or Primary zone of the crust, and were formed under unique conditions. Moreover, the early neptunians thought granite represented the original surface of the globe. But two classes of granite were recognized—stratified and massive. The stratified variety (gneiss) seemed to Hutton to be of essentially the same age as the enclosing schists. But because he felt that he had proved an igneous origin for massive granites as well as basalt, he concluded that massive varieties postdate the schists. He considered heat to be a major factor in the formation of metamorphic rocks, and if granites and metamorphism both were produced by heat, and heat were also the main source of energy for crustal upheaval to form new land, then massive granites should provide indirect evidence of upheaval. So, to prove his plutonic thesis, it became desirable to find evidence of forcible intrusion of granitic rocks into older ones, and this he proceeded to do, but *after the theory was formulated.*

The original discovery of intrusive granitic dikes commonly has been ascribed to Hutton alone, yet both Buffon and de Saussure, prior to 1775, had made crude descriptions of dikes in Primitive rocks. So it is not clear if Hutton was completely independent in his prediction that intrusive relations of granite with older rocks could be discovered near the margins of large granitic masses. In any case, search did provide a means of testing his hypothesis, and it allowed Hutton to expound the important extension of Steno's *Principle of Superposition* to the determination of relative ages of igneous bodies by cross-cutting intrusive relationships. He also stated the corollary that inclusions (xenoliths) in plutonic igneous masses must represent rocks older than the igneous intrusion. These were extremely important new tools for the geologist, and subsequently were extended to a host of additional situations (e.g., microscopic studies of sequences of crystallization).

A large mass of granite was known in the southern Highlands, and Hutton reasoned that the Blair Atholl region near its southern margin was a likely place to see the contact. Accordingly, he arranged permission through his friend John Clerk of Eldin to explore in Glen Tilt in 1785. Hutton was elated to discover no fewer than six granitic dikes clearly cutting across layers of schist. This crucial verification of the plutonic hypothesis was not described in print until 1794, though it had been reported orally in 1790. Bailey argues that Chapter 4 of Volume III of *Theory of the Earth* (published posthumously in 1899) must have been written in late 1785, soon after the Glen Tilt expedition and before much of Volumes I and II (published in 1795) was written.[19] Such jumbling of Hutton's discovery and publication sequences has retarded a full analysis of his work.

Hutton was eager to compound his Glen Tilt triumph, and he reasoned that the Isle of Arran should be a rewarding place for search. In 1786 he set out with Clerk to explore Arran, but instead probed the southwestern mainland of Scotland, the Galloway district. Most of the rivers of the region contain granitic boulders, and diligent search of valleys and coastal exposures was rewarded by several discoveries of intrusive contacts between granites and "schistus." In the following summer, Hutton was more generously rewarded on Arran with the discovery again of granite intruding schistus, but also very well-exposed swarms of dikes of varying rock types and ages, as well as his first-recognized angular unconformity. Arran became a bonanza of corroborating evidence! In 1788 he duplicated these observations on the Isle of Man and recognized the original lateral continuity of rocks there with those in Galloway.

THE ANGULAR UNCONFORMITY AND BASAL CONGLOMERATE No finer examples of deduction, prediction, and hypothesis testing can be found in the early history of geology than the successful discovery by Hutton of intrusive granites and of the angular unconformity at the base of the Secondary rocks (i.e., the base of the Old Red Sandstone of the Devonian system). Unconformities almost certainly were being observed earlier, for Pallas (1777) inferred hiatuses between the rocks of

[19] See Bailey, *James Hutton — the Founder of Modern Geology*, p. 1330.

his Primitive, Secondary, and Tertiary mountains, and they were implied in Werner's postulated oscillations and scouring of channels by the sea. De Luc had described (1779) an unconformity between Secondary and Primitive rocks in the Harz Mountains, and at about the same time a clear unconformity also was described by Voigt. But none of these authors appreciated the full significance of their observations, whereas Hutton inferred it from their descriptions alone.

For Hutton, stratigraphic discontinuities provided evidence of repeated upheaval and burial of strata. Exactly when Hutton conceived of the probable existence of angular unconformities in Britain is not clear, but he tells us:

> The immediate connection of the alpine schistus with the strata of the low country (Secondary) is an object which I have long looked for, I may almost say in vain. I expected to have seen it in entering the Grampian Mountains . . . but I was disappointed in my expectations . . . I have in like manner looked for it on both sides of the southern mountains of Scotland, and that in several places without receiving any satisfaction on that subject.[20]

It is clear that Hutton reasoned the existence of an angular unconformity from the striking structural contrast between the Primary and the Secondary strata of the Southern Uplands. He also was influenced by the varied pebble types in the conglomerates of the Old Red Sandstone at the base of the Secondary.

In 1787, a banner year for Hutton, the long-sought contact was finally seen on Arran. In the same year it again was sought successfully around Jedburgh, southeast of Edinburgh, and the following year, at the most famous locality of all, on Siccar Point in Berwickshire, not far from his own farm. In both of these last two areas, clear basal conglomerates were present that contained both angular and rounded fragments of varied older rock types. The *angular unconformity* is so taken for granted today that its enormous significance is easily forgotten. Its meaning represents an extension of Steno's *Principle of Original Horizontality* of strata. The neptunists were aware of tilted and even vertical strata in mountainous areas, but they contended either that these were the first clastic strata deposited upon slopes of initial irregularities of the crustal surface, or that they had resulted from collapse of strata into caverns. To Hutton, on the other hand, tilted and contorted strata provided concrete evidence of crustal upheaval, and basal conglomerates indicated origin of new sediments from old landscapes. *The Principle of Superposition* provides the index of relative age of an unconformity, but Hutton extended this most important principle by expounding the significance of the pebbles in con-

[20] Hutton, *Theory of the Earth,* Vol. III, pp. 230–231.

glomerates above unconformities. These, too, provide an indirect criterion of relative age as well as insight into the nature of ancient landscapes immediately prior to burial. Most important of all is the fact that the angular unconformity provides the chief means of dating mountain building.

CONTINUITY OF CRUSTAL UPHEAVAL Hutton's theory postulated a continued, cyclic decay and resurrection of land with "no vestige of a beginning and no prospect of an end"; thus, in effect, he argued that the earth's interior always had been structurally dynamic. Obviously, this contention flew full in the face of the much more popular neptunian view. But to prove fully the continuity that he advocated would have required a greater knowledge of stratigraphy than yet existed. Moreover, Hutton seems not to have had as much stratigraphic sense as did such geologists as Lehman and Werner. Though he used Primary, Secondary, and Tertiary repeatedly, it is not entirely clear if he envisioned these terms as having some universal significance of synchroneity. In some instances he seemed to believe so, but in others not.

Hutton confined his attention largely to discrediting the common neptunian belief that the rocks of Primitive mountains were unique and had formed by wholly different processes than had those of Secondary and Tertiary mountains, that is, "anterior to the operations of this globe as a living world." He was especially eager to disprove the old notion that granite represents the original surface of the earth, because he believed that *nothing truly Primitive was preserved*. Discovery of granite dikes intrusive into the Primary schistus met this requirement, and though he did not observe any post-Secondary *granites* himself, Hutton knew of many basalt intrusions and mineral veins cutting the Secondary in Britain. To him this was adequate proof of the continuity of *all* igneous processes; later, Playfair (1802) cited examples of granites intrusive into Secondary rocks outside of Scotland.[21]

According to the neptunians, clastic sediments formed only after Primitive time, but Hutton found numerous examples in the schistus of the Highlands (e.g., Ballaculish quartzites cited by Playfair). The neptunists also claimed that Primary rocks were barren of fossil orga-

21 Southern Uplands granites (e.g., those of Galloway) are emplaced in but slightly metamorphosed graywackes and shales representing the Transition series of Werner's modified chronology (Lower Paleozoic of today). Though Hutton recognized the lithological distinctness of "northern and southern schisti" (Vol. III, pp. 57–60), he still regarded all as Primary. Hutton's several unconformity and basal conglomerate localities were of essentially one age — basal Secondary. It was left for later generations to establish unconformities of widely differing ages reflecting a true multiplicity of upheavals.

nisms, so Hutton was elated in 1788 to discover fossils (now known to be Silurian) in rocks then considered Primary near Lake Windermere. Additional ones were found in 1790 by Playfair, and in southern Scotland in 1791 by Hall. Now Hutton felt he could say, "There is nothing formed in one epoch of nature, but what has been repeated in another, however dissimilar may be the operations which had intervened between those several epochs."[22] To Hutton, the "Primary" rocks were simply harder and more deformed than others, and because of this, they formed mountains.

UNIFORMITY IN NATURE The discrediting of neptunian dogma of the uniqueness of different spans of "geologic time" was a corollary to Hutton's assumption of uniformity and continuity of earth processes in the past (more correctly, *actualism*). But trying to demonstrate a nonuniqueness of past time intervals brought Hutton face to face with catastrophism. Hutton's postulated cyclic continuity of landscape birth and decay was, of course, the antithesis of catastrophism, and virtually demanded what came to be called uniformitarianism. Only such a view "can give us any satisfactory idea with regard to that apparent disorder and confusion, which would disgrace an agent possessed of wisdom and working with design."[23] The greatest "catastrophes" permissible in Hutton's theory were phenomena such as floods, earthquakes, and volcanic eruptions. If processes observable today were essentially the same processes that formed the earth as far back as we can study it, "what more can we require? Nothing but time."

Hutton went to considerable lengths to discredit the orthodox view of time. The freshness of ancient Roman roads and walls in Britain was cited as evidence of the slowness of weathering and erosion. With lucky intuition, he correctly reasoned that Alpine glaciers once had been far more extensive and had transported the famous erratic boulders of the Swiss plain, rather than some unknown catastrophic mechanism such as a deluge. Though he never visited Switzerland, Hutton became one of the first men to argue for more extensive past glaciation.

Whereas the turning point in favor of plutonism was essentially realized late in Hutton's life, the combat with the catastrophists was barely beginning. As noted earlier, most of Hutton's observations (e.g., the unconformity) became widely accepted, but their interpretation was to be rendered in a very different light. The final catastrophist confrontation was fought in the nineteenth century by Charles Lyell.

22 Hutton, *Theory of the Earth,* Vol. I, p. 364.
23 *Ibid.*, p. 472.

Evaluation of Hutton's Influence on Modern Geology

Specific geological contributions by Hutton include the igneous interpretation of basalt and granite, elaboration of the significance of angular unconformities, the erosion of valleys by rivers, and the recognition of interrelations among uplift, denudation, and sedimentation. Even more fundamental was the exposition of the important principles for determining relative ages of rocks from the cross-cutting and inclusion relations of igneous bodies and from the pebbles in conglomerates associated with unconformities.

Uniformitarianism commonly is considered Hutton's most significant contribution. In subscribing to the popular view of harmony in nature, Hutton required uniformity and continuity of processes, though in a repetitive or cyclic fashion. But Hutton, unlike Lyell, invoked only the assumption of uniformity of processes. Hooykaas does not distinguish Hutton's uniformity from Lyell's,[24] but Hutton's statements that "we are not to limit nature with the uniformity of an equable progression" and that distinctiveness of the Primary, Secondary, and Tertiary are "in reality nothing but the variable measure of similar operations" indicate that he was closer to *actualism* than strict uniformitarianism. The significance of the uniformity doctrine lies in the historical context of Lyell's reaction to catastrophism. It was *after Hutton* that the implication of a nearly absolute, Lyellian uniformity of the earth was acquired, in which constant rate and intensity of processes were implied. After Lyell, geology returned to a more Huttonian, actualistic conception. But there is no question that the narrower doctrine still imposes subtle confining effects upon the geological psyche; uniformitarianism is a historic artifact that should be abandoned,[25] as evidenced by distortions stemming from it yet today.

A dynamic cyclic earth view has been given great emphasis by Tomkeieff as Hutton's most important contribution. Although some neptunians had limited notions of internal dynamism, development of the concept by Hutton was of a totally different order of magnitude.

The stress given by Tomkeieff to cyclicity seems less valid, for I believe that it became a dangerous dogma in the hands of geologists after Hutton's time. A wholly cyclic view of the earth is illogical, yet ever since Hutton, geologists have sought cycles in nature (climatic, erosional, orogenic, and especially sedimentary ones). They seem to have

24 See R. Hooykas, *The Principle of Uniformity in Geology, Biology and Theology* (Leiden: E. J. Brill, 1963).

25 See S. J. Gould, "Is Uniformitarianism Necessary?" *American Journal of Science, 263* (1965), 223.

been eminently successful, but unfortunately, many supposed cycles are fabrications. The cyclist zeal should be dampened by a reading of Zeller's experiment in the fabrication of a supposedly repetitive stratigraphic pattern by using the Lawrence, Kansas, telephone directory as a random numbers table.[26] Many cycles in the geological literature clearly represent an imposing of order on nature by man. The error is a subtle assumption of perfect transcendent cyclicity, that is, the complete return to a previous configuration as in a sine curve—a kind of *cyclic uniformitarianism.* Long ago, Lord Kelvin regarded such a view of geological history as erroneous and so launched his half-century vendetta with geology. He was basically correct in attacking the popular nineteenth-century Lyellian concept of uniformitarianism as amounting to perpetual motion—a physical absurdity. Lyell's view, which allowed only oscillations within strict limits, amounted to a *steady state* earth; though changes always were occurring, the earth over-all would look the same at any moment in time.

Hutton perceived what we today may term *dynamic equilibrium,* and the concept is of enormous importance in natural science. The key point is one of time scale, a steady state view being a useful concept over the short term, but not for the long term. Kelvin argued that, given a finite internal energy reservoir and the obvious dissipation of energy in geological processes, the earth system must run down, and therefore the earth must change appreciably through time. But Kelvin's time scale was in error by a factor of nearly 200, and it is a classic irony in the history of science that the discovery of radioactivity not only undid Kelvin's most basic assumption about energy source, but ultimately exonerated the older age inferred by geologists.

In effect, Kelvin laid the ground for a modern, *evolutionary view* of the earth. Few geologists have acknowledged the inconsistency between the strictly cyclic, steady state and continually changing evolutionary views of earth history. Tomkeieff suggested something of the latter idea even while stressing Hutton's cyclic concept. He speaks of Hutton having originated the "idea of cyclic evolution" in geology,[27] although he does not clearly define this ambiguous term. Elsewhere, he suggests that Hutton, in effect, extended the dynamic concept of nature, developed earlier mathematically by Galileo, Newton, and Leibniz, to the earth, specifically stating that "integration of small events in their cumulative result lead to great changes [over time]."[28] More

26 See E. J. Zeller, "Cycles and Psychology," in D. F. Merriam (Ed.), Symposium on Cyclic Sedimentation, *Geologic Survey of Kansas Bulletin, 169* (1964), 631.
27 See Tomkeieff, "James Hutton," p. 396.
28 *Ibid.*, p. 390.

recently, Simpson has argued that geological time represents a long series of interrelated cumulative events, no two of which ever have been precisely the same.[29] In other words, history never repeats itself exactly! This is evolution, and though originally conceived in a biological context, it has just as great implications for the inorganic earth. It is more correct to think of geological time in analogy with a helical spiral propagating through time, rather than as the sine curve of the cyclist. From this point of view, there is no violation of actualism, but Lyellian uniformitarianism has no place.

A dynamic evolutionary earth view, to bring the issue up to date, is the most satisfactory model. We can think of the earth as a gigantic chemical system that has been changing in a generally unidirectional, evolutionary ("helical") fashion through time. The earth has undergone a great chemical physical evolution from a homogeneous protoplanet stage through an early differentiation of core, mantle, primeval crust, ocean, and atmosphere. Then a much longer stage of further evolution has followed, marked by obvious changes in the crust and atmosphere (even life can be thought of as a part of this great chemical evolution). The necessary energy derives largely from radioactive decay, the now-known major source of Hutton's postulated subterranean heat. Historical events recorded in volcanic eruptions, earthquakes, granitic and other igneous intrusions, angular unconformities, and intense deformation and metamorphism of rocks are all effects of dissipation of terrestrial energy. Zones of the crust marked out as mountain (or mobile) belts, both modern and ancient, have been the special sites of this dissipation, and such a zone was the source of Hutton's crucial Scottish observations. Because all such phenomena can be seen happening today, it follows that the earth still is structurally dynamic and energy still is being dissipated. As chemical evolution occurs, the earth is becoming more and more differentiated from its original, assumed homogeneous state, and total entropy is increasing.

Because the structurally dynamic nature of the earth is a manifestation of the pervading processes of over-all chemical evolution, first full recognition of that dynamism and elaboration of its implications in a bold synthesis, even though speculative and wrong in certain particulars, must stand as one of the fundamental unifying signposts along the path of geological knowledge. Such was the greatest contribution of James Hutton.

29 See G. G. Simpson, "Historical Science," in Claude C. Albritton, Jr. (Ed.), *The Fabric of Geology* (Stanford, Calif.: Freeman, Cooper, 1963).

William Smith: Some Aspects of His Life and Work

Joan M. Eyles

Many short and a few long papers have been written about William Smith and his work, but apart from the long-out-of-print biography by his nephew, John Phillips, a comprehensive account has not yet been published, and it is necessary to consult a number of different sources to get a full picture of this remarkable man. It seems worthwhile, therefore, to indicate here what are, at present, the principal sources of information, both published and unpublished, about Smith's life and work.

I propose to follow this with some account of Smith's early years in Somerset, where he first made out his two fundamental laws of stratigraphy, the regular superposition of strata and the identification of strata by the contained fossils. The regular succession of certain strata had already been recognized, of course, by some geologists, and had even been pointed out in 1669 by Steno, but William Smith made the discovery independently, and it provided a necessary foundation for his recognition that fossils could be used to identify the stratigraphical succession. This latter discovery, for which he is now famous, was made in 1795, when he was employed by the Somersetshire Coal Canal Company. No account has ever been published that relates the progress of the canal construction to the advances in Smith's geological knowledge, so an outline of the principal stages will be given here, based on an examination of contemporary advertisements and notices in local papers and the reports of the Canal Committee of Management.

Finally, I intend to describe briefly the great variety of work carried out by Smith during his busiest years, when he was sought after as a successful mineral surveyor and civil engineer. It was to some extent the burden of this work that prevented him from completing for publication his geological map of England and Wales and his description of the stratigraphical succession; on the other hand, the variety of this work enabled him to acquire a far wider knowledge of British geology than he had in 1801, when he first issued a prospectus for a proposed work on the strata. Yet had this book appeared immediately, accompanied by the map and sections that he promised, Smith might have achieved a fame among contemporary geologists comparable to that of Werner. It was his misfortune that as knowledge of his discovery spread, without any real recognition that he was its discoverer, so its novelty diminished, and anything he chose to publish about it seemed less remarkable. As Conybeare and Phillips wrote in 1822, he "had freely communicated the information he possessed in many quarters, till in fact it became by oral diffusion the common property of a large body of English geologists, and thus contributed to the progress of the science in many quarters where the author was little known.[1]

Sources of Information About Smith

William Smith has certainly received less attention from biographers than his work would justify. After his death in 1839 his nephew, John Phillips, himself a prominent geologist, went through Smith's large collection of papers, which must have included diaries, letters, reports, attempts at autobiography, and preliminary drafts for books on geology. From this assemblage Phillips prepared a 150-page biography of his uncle, with the title *Memoirs of William Smith, LL.D.*[2] This work is now very rare. It was the main source for all information about Smith for nearly a century. It is not indexed, nor even divided into chapters, but it does contain an immense amount of information about Smith.

In 1873, a book about William Smith and the Somerset Coal Canal, by W. S. Mitchell, LL.B., F.L.S., F.G.S., was announced in the January number of the *Geological Magazine* as ready for publication by subscription.[3] Mitchell had been an assistant in the library and museum of the Geological Society from 1867 to 1869. This book never appeared, and all that is known to remain of it is a volume of photographs, taken along the line of the canal, with a geological key showing the outcrop of the strata. This is now in the library of the Paleontology Department of the British Museum (Natural History).

In 1917, Thomas Sheppard of the Hull Municipal Museum in York-

1 W. D. Conybeare and W. Phillips, *Outlines of the Geology of England and Wales* London: (W. Phillips, 1822), p. xlv.
2 John Phillips, *Memoirs of William Smith, LL.D.*, London: (John Murray, 1844).
3 W. S. Mitchell, "A Book about William Smith and the Somersetshire Coal Canal," *Geological Magazine, 10* (1873), 31.

shire published a lengthy paper entitled "William Smith: his Maps and Memoirs."[4] This paper was reprinted in 1920 with a separate title page and short preface, but retaining the original pagination. Now unfortunately out of print, this work is particularly valuable for its detailed descriptions of Smith's publications, with many facsimiles of his maps and sections. It contains a brief account of earlier writers on the mapping of soils and rocks, including Strachey, but not Guettard. Sheppard also gives an account of notices of Smith's work by contemporary and later authors, and describes and illustrates the various memorials to him: at Churchill, Oxfordshire, where he was born in 1769; at Tucking Mill near Bath, the cottage that he bought in 1798 and owned for about twenty years, and where he often stayed; and at St Peter's Church, Northampton, where he was buried in 1839. Sheppard's book ends with a useful bibliography, arranged chronologically, not only of Smith's own publications, but also of many later references to him.

In 1926, a memorial tablet to Smith was placed on No. 29 Great Pulteney Street, Bath, the house where in 1799, he had dictated a list of strata and their embedded organic remains, as he then knew them, to his two clerical friends, Joseph Townsend and Benjamin Richardson. An address given on the occasion of the unveiling of the tablet, by Dr. F. A. Bather, Keeper of Geology in the British Museum and then President of the Geological Society, tells us nothing new about Smith, but it does contain an excellent facsimile of the original table of strata, as it was written down, reproduced from the one in the possession of the Geological Society.[5]

A year later, in 1927, Dr. F. J. North, of the National Museum of Wales, described in the *Geological Magazine* a broadsheet by William Smith headed "Deductions from Established Facts in Geology," printed in 1835, which had escaped the notice of both Phillips and Sheppard.[6] It had probably been distributed by Smith at the meeting of the British Association in Dublin in 1835, the meeting at which he received his LL.D. from Trinity College, Dublin. This broadsheet owed its preservation to having been bound in a copy of Phillips' *Geology of Yorkshire.*

[4] T. Sheppard, "William Smith: His Maps and Memoirs," *Proc. Yorks. Geol. Soc., 19* (1917), 75. Reprinted (Hull: L. A. Brown & Sons 1920) with the addition of a title page and preface.

[5] F. A. Bather, *Address at Bath on 10th July, 1926, after unveiling a Memorial Tablet to William Smith* (Bath, 1926)

[6] F. J. North, "Deductions from Established Facts in Geology, by Wm Smith: Notes on a Recently Discovered Broadsheet," *Geological Magazine, 64* (1927), 532–540.

A few years after this, my husband and I began to collect early geological works, and in 1933 we were fortunate in finding a very good copy of Smith's large geological map of England and Wales, published in 1815. A year later we bought another copy, bound in atlas form. It was while comparing these two maps that we discovered the interesting fact that there were a number of variations between them, both in coloring and engraving. This led to our studying as many copies of the map as we could then locate in libraries, and eventually, in 1938, we published in *Annals of Science* a paper, "On the Different Issues of the First Geological Map of England and Wales."[7] From an examination of 27 different copies of the map, we concluded that Smith had added new information to the map from time to time, and that all copies of the map had not come out in 1815, as would appear from the date on them, but that they had been issued over several years. Many copies of the map have, on sheet VI, Smith's signature followed by a number, either alone or preceded by the letter *a* or *b*, and we were able to show that these were in chronological order, and that the earliest were those numbered 1 to 100, together with a few unnumbered ones, probably those first distributed, as they included a copy belonging to Sir Joseph Banks, to whom the map was dedicated. They were followed by a second series, numbered a1 to a100, and a third series numbered b1 to b100. We tabulated certain definite alterations of coloring or engraving that clearly showed how changes had been made in the map, but we emphasized that the chronological groups into which the maps are separable on the basis of geological differences do not always correspond precisely with a particular numbered series. We also described two copies, unnumbered and unsigned, differing even more from the others, which we found contained alterations and additions that had been made probably as late as 1817. Since the publication of this paper, other maps have come to light which, as far as we know, have confirmed our conclusions. In our paper we described only the general succession of the maps, with reference to a few specific areas such as the Isle of Wight and the Lake District, and there still remains ample opportunity for geologists interested in special areas to trace the changes Smith made in the geology as his own knowledge increased or as he learned of discoveries made by others.

In the year our paper on Smith's map was published, an old packing case, discovered in the Geology Department at Oxford, was opened and

7 V. A. Eyles and Joan M. Eyles, "On the Different Issues of the First Geological Map of England and Wales," *Annals of Science*, 3 (1938), 190–212.

found to contain a large number of William Smith's original papers, undoubtedly those, or most of those, used by John Phillips when he prepared his biography of his uncle.

Phillips had been appointed Keeper of the Museum of the Yorkshire Philosophical Society, in the city of York, in 1825, and a few years later he took up residence at St. Mary's Lodge, a house that had formerly been the gatehouse of St. Mary's Abbey, York. This remained his home until 1853, although from 1834 to 1840 he was Professor of Geology at King's College, London, leaving York annually to give his course of lectures there. He also worked for a time as paleontologist to the Geological Survey, and in 1844 was appointed Professor of Geology at Trinity College, Dublin. In 1853, he was appointed Deputy Reader in Geology at Oxford, and on Dean Buckland's death in 1856 he succeeded to the Readership, which he held until he died in 1874. Presumably, the packing case full of Smith's papers had lain untouched at Oxford ever since Phillips' death. Indeed, we may feel thankful for their preservation, when it is recalled that in 1877 a letter appeared in the *Geological Magazine* stating that the 16 original copper plates of Smith's 1815 map had been purchased by Mr. Edward Stanford, and suggesting that they should be bought for the Geological Society.[8] Apparently, there was no response to this appeal, and the plates were melted down!

Smith's papers, however, fortunately escaped such destruction, and after their discovery in 1938, Dr. L. R. Cox of the British Museum, who was at that time preparing a lecture on Smith, was invited by Professor J. A. Douglas of Oxford to examine them. Cox not only examined them but put all subsequent investigators greatly in his debt by sorting them into folders and preparing catalogue slips of the contents of each folder. His paper "New Light on William Smith and his Work," published in 1942, contains many valuable additions to our knowledge of Smith, derived from the Oxford manuscripts and other sources.[9] One of the plates in this paper shows the earliest surviving manuscript geological map of England and Wales, dated 1801 in Smith's hand, and there is also a facsimile of the note in his handwriting, dated January 5, 1796, concerning fossils and the "wonderful order and regularity with which Nature has disposed of these singular productions and assigned to each Class its peculiar Stratum." Extracts from Smith's diaries given

[8] G. S. Boulger, "Dr. William Smith's Geological Maps," *Geological Magazine*, *14* (1877), 378. Also see Sheppard "William Smith," p. 204, footnote.

[9] L. R. Cox, "New Light on William Smith and His Work," *Proc. Yorkshire Geological Society*, *25* (1942), 1–99.

by Dr. Cox confirm the numbering of the 1815 map that my husband and I had already noted, often indicating exactly when the copies were numbered and signed, although there is little to show in the diaries when or why Smith had actually made any alterations in the geological lines of the map. Cox provided, as an appendix to his paper, a lengthy bibliography of works by or relating to William Smith, but only as a supplement to Sheppard's bibliography. He did not include any items given by Sheppard.

Before the publication of his paper, Cox lectured to the Geological Society of London in January 1940 on Smith's life and work, and in July 1941 he described to the Society the three earliest surviving manuscript maps of the geology of England and Wales, all prepared by Smith. Brief accounts of these lectures appear in the *Proceedings* of the Society.[10]

In 1939, the Geologists' Association had intended to hold a September field meeting at Bath to commemorate the centenary of Smith's death, but with the outbreak of war this was postponed. It was actually held in March 1940, with L. R. Cox and G. A. Kellaway as principal directors. A report of the meeting, by Dr. Cox, appeared in 1941.[11] It is illustrated with excellent photographs of the different houses in which Smith had stayed, including Rugborne Farm at High Littleton, where he lodged for several years, the Swan Inn at Dunkerton, where he wrote some of his earliest recorded notes on geology, Cottage Crescent, near Bath, where he lived for a year or two, Trim Bridge, in Bath, where he had an office from 1802 to 1805, and, of course, Tucking Mill.

In 1948, the 18th International Geological Congress was held in Britain, and one of the field meetings, directed by L. R. Cox, G. A. Kellaway and F. B. A. Welch, was to the Bath district and localities associated with Smith. This excursion is described in the Congress Guide No. A.3.[12] Dr. Cox also prepared a pamphlet entitled "William Smith and the Birth of Stratigraphy," a copy of which was received by all those attending the Congress.[13] This provides a very useful concise

10 L. R. Cox, *Abstr. Proc. Geol. Soc.* Session 1939–1940, No. 1366 (29 January, 1940), 23–27; and *Abstr. Proc. Geol. Soc.*, Session 1940–1941, No. 1380 (8 August, 1941), 76–77.

11 L. R. Cox, "Easter Field Meeting, 1940, Bath," *Proc. Geologists Ass.*, *52* (1941), 16–35.

12 L. R. Cox, G. A. Kellaway, and F. B. A. Welch, *Guide to Excursion A.3. "William Smith" Excursion (Environs of Bath)*, International Geological Congress, XVIII Session (London, 1948).

13 L. R. Cox, *William Smith and the Birth of Stratigraphy*, International Geological Congress, XVIII Session (London, 1948).

account of Smith and his work, and has an appendix listing the more important of his former residences and the memorials erected to him.

A brief but important paper entitled "An Early List of Strata by William Smith," by Professor J. A. Douglas and L. R. Cox, must also be mentioned here.[14] The authors describe one of the manuscripts in Smith's handwriting found at Oxford, a list of strata dated 1797, and therefore two years earlier in date than the list dictated to Townsend and Richardson in 1799.

Cox, who died in 1965, did not publish any more on Smith, although he certainly contemplated a full-length biography. Three papers by A. G. Davis that appeared between 1943 and 1953 do require notice, however, for each contains some new information. They are "The Triumvirate: A Chapter in the Heroic Age of Geology," a presidential address describing Smith, Townsend, and Richardson;[15] "William Smith, Civil Engineer, Geologist (1769–1839)";[16] and "William Smith's Geological Atlas and the Later History of the Plates," which contains some important details concerning Smith's county maps.[17]

Some years ago, when my husband and I went to live within reach of Oxford, I obtained permission from the late Professor L. R. Wager to go through the William Smith papers there with a view to a biography. Having since then established where the principal gaps lay, I have been able to fill some of them.

That Smith was born and brought up in the village of Churchill in North Oxfordshire has always been well known, but the actual house in which he lived had never been identified. However, I found the deeds recording the purchase of the house by his widowed grandmother in 1740 in the Oxfordshire County Record Office, and after consulting these and various estate plans it has proved possible to identify the spot where the house stood, although it seems to have been demolished about a century ago. Other records have made it possible to discover more about Smith's ancestry and his family relationships.

As well as genealogical information, several reports made by Smith on such subjects as coal-finding and irrigation have turned up in County Record Offices and other repositories, and I also located a number of

14 J. A. Douglas and L. R. Cox, "An Early List of Strata by William Smith," *Geological Magazine, 86* (1949), 180–188.

15 A. G. Davis, "The Triumvirate: A Chapter in the Heroic Age of Geology," *Proc. Croydon Natural History and Scientific Society, 11* (1943), 122–146.

16 A. G. Davis, "William Smith, Civil Engineer, Geologist (1769–1839)," *Trans. Newcomen Society, 23* (1948), 93–98.

17 A. G. Davis, "William Smith's Geological Atlas and the Later History of the Plates," *Journ. Soc. for the Bibliography of Natural History, 2* (1953), 388–395.

Smith's letters to the Treasury concerning the sale of his collection of fossils to the British Museum. These are in the Public Record Office, among the Treasury papers there, and an account of them with transcripts of all the letters and relevant extracts from Smith's diaries has recently appeared in *Annals of Science*.[18]

This covers the main sources of information about Smith, although there is always some hope of more being discovered. In particular, Smith's diaries for 1794 and 1796, which were in the possession of John Phillips in 1869, according to a letter he wrote to W. S. Mitchell, were not with the other diaries among the papers at Oxford, so it is just possible they may yet be found.[19]

William Smith in Somerset, 1791–1799.

Smith received his early training as a surveyor in the office of Edward Webb of Stow-on-the-Wold, Gloucestershire, between 1787 and 1791. At the end of October 1791, he was sent by Webb to Somerset to survey an estate near High Littleton, seven or eight miles southwest of Bath, which included some coal properties. It is possible that he already had some knowledge of methods of subterranean surveys, because Webb had some connection with Monmouthshire and collieries there. In any case, young Smith (he was then 22), besides making a surface survey of the village of High Littleton, also descended underground and began making sections and plans of the collieries there. Although Smith's plan of High Littleton is now in the Somerset Record Office, I have not been able to trace any of the detailed colliery plans he must have made. Sketches for these, however, are among the Oxford manuscripts. It was this underground work that stimulated his interest in the strata, although it is also clear that surface indications of different soils, such as the Chalk escarpment seen on the way to London when he was a boy, the Cotswold limestones, the red Triassic soil of Worcestershire, and the gravels of the New Forest, all seen while he was surveying for Webb, had already been noted by his ever observant eyes.

During 1792, he was fully occupied with his surveys in the High Littleton district, but in March 1793, he was requested by members of a local committee to take preliminary levels for a proposed Coal Canal, which would link some of the north Somerset collieries with another proposed canal, the Kennet and Avon. The latter was intended to form a direct east-west link between Bristol and Bath on the Avon with New-

18 Joan M. Eyles, "William Smith: The Sale of His Geological Collection to the British Museum," *Annals of Science*, 23 (1967), 177–212.
19 Letter from W. S. Mitchell in the *Bath Chronicle*, July 29, 1869.

bury and Reading on the Kennet, a tributary of the Thames. No detailed account of the building of the Somersetshire Coal Canal has been published, but a number of contemporary references and advertisements in local newspapers have been traced and have made it possible to follow the course of events.

The Coal Canal (see Fig. 1) was planned as two branches running approximately southwest to northeast along converging valleys, each branch being a little under eight miles long. After the two branches met, near Midford (also known as Mitford), the canal was to continue northeast for another two miles to its junction with the Kennet and Avon Canal, a few miles from Bath. The producing collieries lay to the southwest, around the head of each valley and a mile or two down the valley.

It was while taking levels along these valleys in 1793 that Smith found the evidence for what he already suspected, that the dip of the strata above the coal measures was regular and not haphazard, as his friends among the miners had maintained. Moreover, the strata were dipping regularly eastwards, and he was able to make out a definite succession.

In November 1793, a map of the canal route, undoubtedly surveyed by Smith, although signed by John Rennie, the Committee's consultant engineer, was deposited with the Clerk of the Peace for Somerset; this map is now in the Somerset Record Office. In the following April, 1794, the Somersetshire Coal Canal Bill was approved by Parliament and became an Act. "Mr. William Smith, Engineer" was among those officials who gave the required evidence, as shown by records in the House of Lords Record Office. While he was in London, Smith took the opportunity to look in the bookshops to see if he could find anything to show that his discovery about the regularity of the strata was already known, but his search was unsuccessful.

Smith was then appointed engineer and surveyor to the Coal Canal Company at a fee of a guinea a day with expenses. On June 17, 1794, he began plans and levels of the whole line, and on July 14 a local Bath paper reported "from a recent survey of the Engineer, that on both lines of the Canal there is even at this very dry season, such an abundance of water as obviates every idea ever entertained of there being a probable scarcity of it in the summer months." Later on, the problem that arose was not one of lack of water, but loss of it by leakage.

In August, Smith made his well-known journey by hired chaise through England to Newcastle, accompanying two members of the

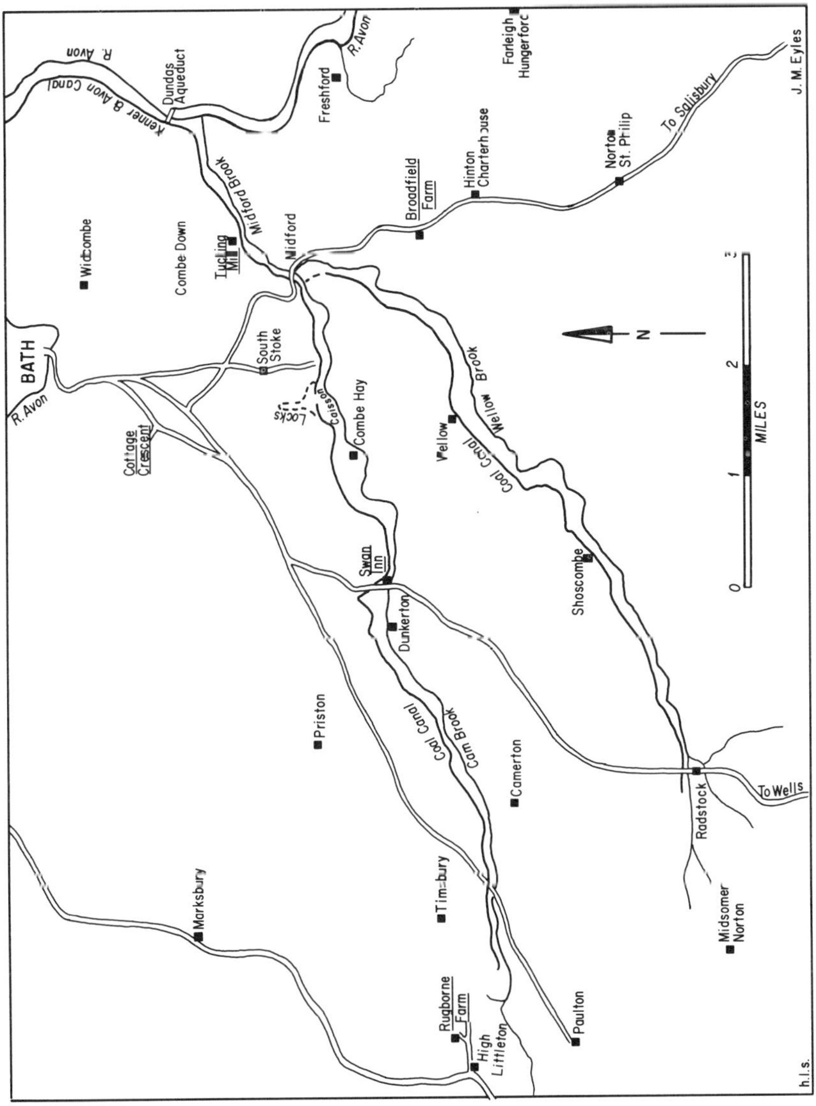

FIG. I. *A map of the district south of Bath, showing the course of the Coal Canal. Places where William Smith stayed are underlined.*

committee, and he saw much to confirm his ideas about the regular succession of the strata. In November, construction began on the Kennet and Avon Canal, between Bath and Bradford-on-Avon, and no doubt Smith took the opportunity of examining the sections exposed, though he has left no record of this.

No work had yet begun on the Coal Canal, but in the spring of 1795 the committee appointed Mr. William Bennet, "an engineer of considerable experience," as resident engineer to the Coal Canal. After this, Smith was referred to as "Surveyor." A further survey was then made and some changes decided on. For these, another Act was necessary, although this was not obtained until March 1796.

Over the full length of the canal, from the beginning to the junction with the Kennet and Avon Canal, a descent of over 100 feet had to be accommodated, and it was decided that this descent could best be made over a fairly short lateral distance, rather than by locks strung out along the valleys. For this reason, the first sections of the two branch canals were each to be cut on one level. By a clause in the 1794 Act, no preference was to be given to either line, both to be begun and completed at the same time, so that no advantage should be obtained by the proprietors of the coal-works on either line. In order to transfer the boats from the upper level of the canal to the lower one, it was decided to adopt a new invention, a caisson lock, which was a sort of gigantic lift shaft, filled with water, holding a large watertight box or caisson, into which a boat could pass; then the doors were closed, the whole contrivance was lowered, and the boat emerged from it into a lower basin. The caisson then ascended, carrying another boat. The descent in the proposed caisson was about 46 feet; hence, at least two such locks would be necessary, though it does not seem that more than one was ever built, as the first one proved structurally unsound and was abandoned. A number of ordinary locks were eventually constructed, mostly in a horseshoe sweep around a conveniently situated narrow lateral valley.

In May 1795, advertisements in local papers invited proposals from contractors for cutting the first two miles from the head of each valley, and in June it was announced that the work was to be undertaken by a contractor from Shropshire, who had agreed to complete it by the following February. In both valleys the subsoil along these sections was unfossiliferous Triassic marl, the Red Ground of the colliers, already well known to Smith. A second section of two miles along the northern line was next advertised, with estimates to be submitted by the end of September, so work here can hardly have begun before October. This section, going as far as the Swan Inn on the Bath Turnpike, would first pass over more Triassic marls, then over beds of limestone, clay and sands now called Liassic, until near Dunkerton village the eastward dipping strata brought down to canal level the hard beds of freestone now called Inferior Oolite.[20] Above the freestone were the clayey beds associated with fuller's earth, and although the canal did not reach these until it neared Combe Hay, Smith would be familiar with these beds from excavations made near Cottage Crescent, where he had moved in September or October 1795.

On January 5, 1796, Smith, at the Swan Inn, was writing his now famous memorandum on fossils, showing that he already clearly recognized their significance. His words have been quoted earlier (p. 146). From this, we may assume that the last three months of 1795 were particularly important in advancing Smith's geological knowledge. The excavation of the canal bed near Dunkerton would have given him invaluable opportunities to see and collect fossils, some of which were possibly already known to him from the north Cotswolds, particularly near Churchill.

In November and December, advertisements appeared requesting estimates for cutting a further two miles of the canal, from the Swan Inn to Combe Hay, and also for a second stretch along the southern valley as far as Wellow. This shows that work in the southern valley had not yet proceeded very far, and in fact, this section was advertised again in April 1796 and had not been finished as far as Wellow in December 1799. The section from Wellow to Midford was advertised in October 1797, but no work seems to have been done on it even at the end of 1799. The section east of Midford, to the junction with the Kennet and Avon Canal, was advertised at the end of 1797 and was in "a considerable state of forwardness" at the end of 1799. Toward the end of 1798, the shareholders began to see hopes of some return for their money, when a portion of the canal along the northern line was opened, and on October 4, five vessels carried coal from a colliery at Camerton to Dunkerton. The following year, coal was being carried quite regularly to a wharf near the Swan Inn and was unloaded there to be taken by road to Bath.

Meanwhile, public interest was centered on the caisson lock. The reservoir, nearly 90 feet long and with a maximum width of 20 feet, was begun in the summer of 1796 and completed and lined with masonry during 1797. Although the working of the caisson lock was successfully

20 The geology of the Coal Canal valleys is shown on the one-inch geological sheet 281, Frome (Geological Survey of England and Wales, 1965).

demonstrated on several occasions in 1798 and again in April 1799, there must have been some leakage, for in May 1799 the Committee announced their intention of having the caisson cistern rebuilt in order to make it watertight. There was evidently some opposition to this among the Canal Proprietors, for an advertisement appeared inviting support for a resolution to be proposed at a General Meeting on June 5, recommending the replacement of the caisson lock by an inclined plane. This resolution was evidently defeated, since advertisements for the rebuilding of the caisson cistern appeared soon after, and it is noteworthy that on June 5, 1799 (according to W. S. Mitchell, who had access to the Coal Canal minute books), Smith's employment by the Coal Canal Committee was terminated.[21] The reason for this has never been disclosed. Cox (1942, p. 73) surmised that it was due to Smith's having purchased property (Tucking Mill) adjacent to the canal while engaged in acquiring land for the Canal Company. There is no evidence for this explanation, and it seems more probable that Smith had allied himself with the anti-caisson party, recognizing full well that the caisson cistern had been excavated in a geologically unstable spot. There is much slipped ground along this side of the valley. Six or seven months later, the Committee finally abandoned the caisson lock, but dallied with other ideas, such as inclined planes, for several years; it was 1802 before contractors were invited to submit estimates for the chain of locks that were eventually built round the side valley. Remains of these locks can still be seen, although a railway took the place of the canal in 1908. That, in turn, ceased to show a profit and was closed, and now much of the excavated ground has been filled in. Smith's correspondence with his friends suggests that he was on bad terms with the Canal Company for a number of years. Bennet remained with the Company as engineer, and his name appears in the list of subscribers to Smith's map in 1815.

The sudden loss of his regular employment must have been a severe shock to Smith, particularly as it was little more than a year since he had purchased the Tucking Mill property, with about 17 acres of land, for £1,600, of which £1,200 seems to have been on mortgage. Yet the forced change of occupation was to give him opportunities for travel about the country that he would never have had while tied to the Canal Company; and naturally, this greatly enlarged his geological knowledge. Some account will now be given of the varying types of work that he undertook over the following 20 years.

21 W. S. Mitchell, "The Centenary of William Smith's Birth," *Geological Magazine 6* (1869), 356.

Smith's Work as Land Drainer, Mineral Surveyor, and Civil Engineer, 1800–1820.

William Smith was still in the employ of the Coal Canal Company when he was invited by the Company Chairman, James Stephens, to carry out some drainage work on his estate at Camerton. No doubt, Smith's knowledge of the local strata, his appreciation of pervious and impervious rocks, and his understanding of the causes of springs had already made an impression on his employers. It is not surprising, therefore, that it was in this field that he found it easy to obtain work after leaving the Canal Company. As 1799 was a particularly wet year, a number of slides were taking place on the dip slopes around Bath, often where houses had been lately built, and Smith was consulted on methods of arresting or preventing these slides. His usual method seems to have been to tunnel into the hillside at a suitable spot and drain off the water that was giving rise to the slide.

From this, he progressed to draining flatter ground for farmers in the district, and then to irrigation, recognizing how to conserve and use the water for controlled flooding. This led to his employment by two of England's most progressive landowners, Thomas Coke of Holkham in Norfolk from 1800 and Francis, fifth Duke of Bedford, from 1801. For both of these men he constructed elaborate water meadows. In 1802, he took a partner, Jeremiah Cruse, opened an office in Bath, and raised his fees to two guineas a day with expenses. In September of that year, he went to survey an estate in North Wales, and found time to ascend Snowdon and visit the slate quarries. His fame was growing, and he never seems to have been short of employment. His fossils were deposited in his Bath office, where they were examined by interested landowners. Local land surveys and valuations seem to have been carried out by his partner, with Smith available for consultation when necessary. In 1803, he was supervising a bore for coal in Yorkshire, and later in the year another one in Lancashire. The latter was successful in bringing in coal, but the former was not. In 1804, he rented a large house off The Strand in London, and his collection was moved there from Bath in the following year. During 1804, he continued to lay out water meadows, visit the two bores for coal in the north of England, and supervise the construction of an embankment in Norfolk.

In 1805 he took on an important and responsible task in Norfolk, that of closing several breaches made along the coast by the North Sea, and thereafter draining the adjacent inland areas. In this he seems to have been remarkably successful, but the work was arduous and occupied much of his time for several years, making it impossible for him to devote much time to his project of publishing a geological map of England and Wales, although he employed an amanuensis to copy out some of his papers and put them in order. The publication of his book on water meadows in 1806 led to more demands for his advice in different parts of the country. In 1808, with the work in Norfolk and Suffolk almost completed, he accepted a commission to supervise the construction of another sea-bank in South Wales, near Kidwelly; he was also appointed engineer for the Upper Ouse navigation in Sussex. During the first five months of 1809, he calculated that he had traveled 2,250 miles, mostly by coach, between London, Wales, Norfolk, and Sussex. Some weeks were spent in Bath, at the request of the Mayor and Corporation, putting right a severe leakage in one of the hot springs. For this, he received a sum of over £400.[22] Although this included the payment of his workmen's wages, it was nevertheless a considerable amount for a few weeks' work, and it reflects the esteem in which he was held.

In 1810, he was particularly busy in South Wales, although he continued to visit Bath, London, and Sussex. In 1811, in addition to the work already mentioned, which continued to occupy him throughout the year, he began making arrangements for the building of a small railway line from his Tucking Mill property to a quarry in freestone on the hill above it. This turned out to be a most unfortunate speculation, which later left him heavily in debt. Toward the end of the year, he accepted the post of engineer to the Minsmere Level drainage in Suffolk, and this occupied much of his time for the next three years. During the same period, he was consulted by several canal companies and made several visits to the Forest of Dean to advise on collieries there.

It was in 1812 that John Cary undertook to publish Smith's geological map of England and Wales, and during 1813 and early 1814 Smith somehow found time to mark the lines of the strata on the engraved maps. He was at the same time so busy with consulting work that he seems to have been quite surprised, in April 1814, at finding four sheets of the map finished and hung in the window of Cary's shop. A year later, however, he was still making corrections and additions to some of the sheets. During the second half of 1815, he devoted much of his time to scrutinizing copies of the map as they were brought him by the colorists, as well as to negotiating with Treasury officials about the sale of his collection to the British Museum. In 1816, he was again visiting Forest of Dean collieries, examining the water supply for the Wilts and Berks Canal at Swindon, visiting Suffolk to report on the River

22 Information obtained by courtesy of the Town Clerk, Guildhall, Bath.

Waveney navigation, and reporting on the Backwell colliery near Bristol. In the same year, the first two parts of his book *Strata Identified by Organized Fossils* were published. A third part, and also his *Stratigraphical System of Organized Fossils* came out in 1817, a year in which his activities are less well documented, for there is no diary until September. In 1818, the Forest of Dean, the Waveney navigation, and a proposed canal to link the rivers Don and Aire in Yorkshire all took much of his time, but he also managed to prepare geological maps of the counties of Norfolk Kent, Wiltshire, and Sussex for publication by Cary in January 1819. During the early part of 1819, he was still extremely busy, but notwithstanding all this activity, times were bad and money was extremely short. In April, his diary shows that he was seriously considering an invitation to the United States to act as advisory engineer in North Carolina, but he was persuaded by a close friend not to accept it.[23] Unfortunately, in June past debts caught up with him and he found himself confined in the King's Bench Prison, where he was forced to pass nearly ten weeks before he could obtain his release.[24] As soon as he was free, he hastened to Yorkshire to fulfill engagements there, but during his absence from London another creditor seized all the property in his London house, which of course included his valuable geological papers, most of which he did not see again for several years. This, for Smith, was the end of his work in the south of England. He turned his back on London, where he had so many friends, and set out to make a new life in the north of England. There he was to remain for the rest of his life, with only an occasional visit to his old haunts. His county geological maps continued to appear up to 1824, and he gave lectures on geology in York, Hull, Sheffield, and other northern towns, although he was handicapped by rheumatism and increasing deafness. It was 1831 before his services to geology were at last publicly acknowledged, when the Geological Society awarded him the first Wollaston Medal.

Acknowledgments

I acknowledge with gratitude the assistance afforded me in obtaining information about William Smith by County Archivists and their staff in the County Record Offices of Gloucestershire, Oxfordshire, Somerset, and Wiltshire, the staff of the Bath Municipal Library, the staff of the House of Lords Record Office and Mr. Charles Hadfield, who kindly

23 William Smith manuscripts, Oxford.

24 This information, hitherto unpublished, is based on records of the King's Bench Prison, Public Record Office, together with entries in Smith's diary.

lent me his notes on the Somersetshire Coal Canal. I am also grateful to the late Professor Wager and to Professor Vincent of the Department of Geology and Mineralogy at Oxford for the facilities given me to examine and make use of the William Smith manuscript collection.

Postscript

After submission of this manuscript for publication, the following bibliography was prepared and has since been published: Joan M. Eyles, "William Smith (1769–1839): A Bibliography of his Published Writings, Maps and Geological Sections, Printed and Lithographed," *Journal of the Society for the Bibliography of Natural History*, 5 (1969), 87.

The Extent of Geological Knowledge in the Eighteenth Century, and the Methods by Which It Was Diffused

V. A. Eyles

The great development and expansion of the geological sciences that took place in the nineteenth century resulted, it is generally accepted, from contributions made to these sciences towards the close of the eighteenth century and in the early years of the nineteenth century by a few geologists of outstanding ability. Among those who made major contributions were Nicolas Desmarest, A. G. Werner, James Hutton, William Smith, and Georges Cuvier. Concurrent advances made in other sciences, notably chemistry, also played a part.

Until comparatively recently there has been, I think, a tendency to conclude that during most of the eighteenth century much that was published on geological subjects consisted of little more than wild theorizing and fruitless speculation. Hence, in some circles, at least, it has been concluded that geology scarcely existed as a separate science worthy of a specific name before the closing years of the century. In my view, this is a conclusion that cannot be sustained upon close examination. It seems to me that if one accepted it, one might also claim, with some justification, that chemistry before the time of Black, Priestley, Cavendish, Lavoisier, and their associates was equally speculative and no more worthy of study. Yet eighteenth-century (and earlier) chemistry has attracted much more attention and has been studied in much more detail than eighteenth-century geology. While it is true to say, no doubt, that as a body, chemists have always greatly outnumbered geologists, so that the number of the former sufficiently interested in the history of their subject to study it seriously is likely always to have been greater, yet it does seem that, until comparatively lately, the serious contributions that have been made to the history of geology have been proportionately much less, and regrettably small in number.

This situation may possibly owe something to the fact that the name *geology* did not come into general use much before the close of the eighteenth century,[1] but the principal reason, it seems to me, has been the lack of serious interest in the subject and a failure to realize, particularly as far as the eighteenth century is concerned, that further research into the subject is likely to prove profitable. Recently, attention has been drawn, by R. Rappaport[2] and by the author of this paper,[3] to the scarcity of detailed monographs on eighteenth-century geology.

The object of this paper is to put forward the thesis that, undeveloped and unorganized as the earth sciences were during most of the eighteenth century, they were nevertheless the subject of widespread interest, and often of close study throughout the century, both because of their intrinsic interest and because of their economic importance; so much so that the use of a comprehensive term to include them all was long overdue when, in 1778, J. A. De Luc first put forward the name *géologie* for this purpose.[4] The corollary to my thesis is that eighteenth-century geology deserves much more detailed study than has yet been given to it.

I do not think it is open to doubt that during the whole of the eighteenth century, interest in matters that are now included under the general head of geology, or of one or other of the more specialized earth sciences, was widespread. Furthermore, it was mainly during that century that the basic physical facts and observations that led to the formation of sound geological theories were accumulated and recorded. Here again, I think the analogy with chemistry is justified. In the case of both sciences, before the close of the century, much information of an empirical nature had been accumulated. All that was lacking was an acceptable theory, or theories, capable of explaining the known facts. The enlightened intellectual atmosphere of the eighteenth century, too, was such that the time was ripe for the emergence of such theories; and, as is well known, they were propounded before the close of the century and laid the foundations on which modern chemistry and modern geology have been built.

The evidence in support of my thesis will be developed as follows. First, I shall discuss the immense body of geological literature that was

[1] F. D. Adams, "Earliest Use of the Term Geology," *Bulletin of the Geological Society of America*, *43* (1932), 121.

[2] R. Rappaport, "Problems and Sources in the History of Geology, 1749–1810," *History of Science*, *3* (1964), 60.

[3] V. A. Eyles, "The History of Geology: Suggestions for Further Research," *History of Science*, *5* (1966), 77.

[4] J. A. De Luc, *Lettres Physiques et Morals sur les Montagnes et sur l'Histoire de la Terre et de l'Homme* (La Haye, 1778), footnote p. [vii].

published in the eighteenth century, the mere volume of which leaves no room for doubt as to the general interest in the subject throughout Europe (and, I believe, in North America, too, although I have not examined this question thoroughly). Indeed, such was the extent of the literature that several bibliographies of works on geology and related sciences appeared before 1800. For example, in 1760, L. T. Gronovius published at Leiden a work of about 300 pages entitled *Bibliotheca Regni Animalis atque Lapidei*, which includes articles from journals and periodicals, as well as individual books. This was followed some years later by C. F. W. Schall's *Oryctologische Bibliothek* (Weimar, 1787; 2nd. ed., 1789). This, too, contains papers from periodicals as well as individual books. The first edition has about 1,600 entries, of which the first 384 refer to works of a general character, while the remainder are arranged geographically. In the main, the entries relate to Europe, but some extra-European papers are included, and the works cited are mostly, though not exclusively, eighteenth-century publications. Some 70 entries were added to the second edition. At about the same time there also appeared D. R. Boehmer's much more extensive *Bibliotheca Scriptorum Historiae Naturalis . . .* (Leipzig, 1785–1789), of which Part IV, in two volumes totaling over 900 pages, deals solely with mineralogy (*mineralogi*). Because the word *geology* did not come into general use much before the end of the eighteenth century, works on the subject must be sought under a variety of titles, e.g., theory of the earth, oryctology, lithology, the mineral kingdom, mineral geography, or, simply, mineralogy, which then had a much less restricted meaning than it has today. Works on mining, volcanoes, and earthquakes are important, too, for the light they throw on the development of geological knowledge and ideas.

A detailed discussion of the many individual contributions to geology made during the eighteenth century is impossible here. All that can be done is to refer to some of the more interesting and important ones that give an indication of the advances made and the state of knowledge at the time they were published. It is impossible to generalize satisfactorily about eighteenth-century geology as a whole until more detailed research into the period has been carried out. In discussing the literature, emphasis has been laid on the number of editions and translation of individual works, as an indication of the interest they aroused and the wide circulation they attained.

During the eighteenth century, very little subdivision of geology into subsidiary earth sciences took place, and such subdivisions as were used were imprecise. In this paper, modern subdivisions have been used as subheadings, as a matter of convenience, but it will often be found that individual works published before 1800 discuss more than one branch of geology. In discussing mineralogy, reference has been made to the state of crystallography and chemistry in the eighteenth century, because of the important bearing these sciences had on the development of mineralogy.

In the final section of the paper, attention has been drawn to other methods by which geological knowledge was spread; notably, by means of academic and public lectures.

Geological Literature in the Eighteenth Century

THEORIES OF THE EARTH Works of a general character that can be grouped under one head as "theories of the earth" were a feature of eighteenth century geology. Unsolved geological problems concerning the nature and origin of the rocks constituting the accessible parts of the earth's crust, and speculation about the constitution of the interior of the earth, exercised many minds in that century. The number of theories published, not always in works bearing this specific title, is an expression of the widespread interest in these problems. In the first half of the century, particularly, ideas on the subject were much influenced by the almost universal belief in the literal accuracy of the story of creation contained in the Bible, and in the reality of a Universal Flood. Authors who based their theories on these beliefs are sometimes said to belong to the physicotheological school. Not all the published theories, however, were purely speculative and imaginary, for some authors had examined rocks in the field, and modified their theories accordingly. It is only possible here to refer to some of the better known examples, but the total number published was quite large.

In discussing eighteenth-century geology, account must be taken of certain theories first published in the second half of the preceding century, for some of these circulated widely after 1700. In a few instances, these early theories were in advance of their time and did not gain general acceptance until much later on.

From the point of view of both factual geological content and theoretical conclusions, there were two seventeenth-century authors, Robert Hooke and Nicolaus Steno, whose contributions to geological thought fall into the latter category. Hooke's "Lectures and Discourses of Earthquakes," though delivered orally before the members of the Royal Society much earlier, were not published until 1705 (in his "Posthumous Works"). Steno's "Prodromus" was first published in 1669, but only one eighteenth-century reprint, a Latin edition published at Pistoia in 1763, has been recorded. Though the views of Hooke and Steno attracted much attention in the seventeenth century, the extent to which they influenced the development of geology in the eighteenth century is an open question, requiring further research.[5] It seems probable that their works were almost forgotten for a time, possibly because the naturalists of the period were more concerned with reconciling geological facts with theological teaching than with producing theories strictly in accordance with known geological phenomena.

A third seventeenth-century author who also had original ideas on geology was the philosopher G. W. Leibnitz, but his views were not published in full until the middle of the eighteenth century, and they are referred to further on in this paper.

Among the late seventeenth-century authors whose works circulated widely in the succeeding century, mention must be made of Thomas Burnet. His *Sacred Theory of the Earth* was first published in Latin in 1681, and several Latin and English editions appeared, as well as a translation into German, before 1700. During the eighteenth century, it was reprinted in English at least six times, the latest edition I have traced being dated 1759. Whatever may be thought about the geological value of Burnet's *Theory*, its influence on the development of geological thought cannot be ignored in any study of eighteenth-century geology.

The theory published by John Woodward in 1695, his *Essay Toward a Natural History of the Earth*, was considerably more important. Several eighteenth-century editions appeared in English, and it was translated into Latin, French, Italian, and German, the German version of 1744 being the latest edition known to me. Although as a theory of the earth, Woodward's *Essay* is not now thought to be of much importance, it must be remembered that he was among the earliest to accept unequivocally the marine origin of fossils, and he also recorded the important fact that different strata contained different fossils, an observation based on his own field work. Furthermore, Woodward's influence on the development of geological thought in the eighteenth century was important. It is known, for example, that the Swiss naturalist J. J. Scheuchzer was converted to a belief in the marine origin of fossils through reading Woodward's *Essay*, and this and other works by Woodward were quoted throughout much of the eighteenth century.

[5] See, however, V. A. Eyles, "The Influence of Nicolaus Steno on the Development of Geological Science in Britain," *Acta Historica Scientiarum Naturalium et Medicinalium*, *15* (1958), 167.

Another theory, first published in 1696, was William Whiston's *New Theory of the Earth*, of which at least six editions were published in English, the last of which I am aware appearing in 1755; and a translation into German, dated 1713, has been recorded. Whiston was primarily a mathematician, and the publication of his *Theory* shows that interest in the subject was not confined solely to naturalists, such as, for example, Woodward and John Ray.

Ray's intelligent interest in geology was expressed in two works. The first, *The Wisdom of God manifested in the works of creation,* originally published in 1691, was reprinted many times throughout the eighteenth century. It was also translated into French (1714) and German (1717). Ray's second work, *Miscellaneous discourses concerning the dissolution and changes of the world* (entitled *Three physico-theological discourses* in later editions) first appeared in 1692, and was reprinted several times, there being three eighteenth-century editions in English, two translations into Dutch, and at least three German translations, the latest dated 1756.

The large number of editions and translations of these geological works by Burnet, Woodward, Whiston, and Ray indicate that they were widely read, not only in Great Britain, but throughout much of Europe. Although, for the most part, the theoretical views they expounded were unsound, they did record a number of geological facts and observations of value. When first published, they attracted much comment, such criticisms mostly appearing in the form of contemporary pamphlets. One commentary of some length, John Keill's *Examination of Dr. Burnet's Theory of the earth, together with some remarks on Mr. Whiston's New theory of the earth* (1698), was republished as late as 1734.

As the eighteenth century advanced, other authors published works in which they attempted to advance geological theory. From Italy, for example, came several works of considerable importance, such as A. Vallisnieri's, *Lezione accademica intorno all'origine delle fontane* (Venice, 1715; 2nd. ed., 1726) and *De' corpi marini che su' monti si trovano* (Venice, 1721; 2nd. ed., 1728), the latter containing much more than simply a discussion of fossils, though the title does not indicate this; and L. Moro's *De crostacei e degli altri marini corpi che si truovono su'Monti* (Venice, 1740). A translation of the latter work into German appeared in 1751.

About 1750 serious attempts were made to break away from the physico-theological school of thought, as exemplified in the works of Burnet, Ray, and others. In 1748, De Maillet's *Telliamed* was published posthumously at Amsterdam, and later editions, also in French, appeared at Basle in 1749, and La Haye in 1755. There was also a translation into English published at London in 1750. Manuscript versions of De Maillet's book had been in circulation for some twenty years before it was published (see Professor Carozzi's exposition of De Maillet's geological ideas, included in this volume).

Another theory of the earth published at this time, also posthumously, was that of G. W. Leibnitz. His *Protogaea* was first printed in full, in Latin, at Göttingen in 1749, and a translation into German appeared in the same year at Leipzig. It had been published originally in *Acta eruditorum* (Leipzig, 1693, pp. 40–42), but in such an abbreviated form that it is improbable that it had any effect on the development of geological thought. His theory is of considerable interest, but it was developed in such a manner as not to run counter to biblical teaching about the deluge. Leibnitz' interest in geology was probably stimulated by his contacts with Steno, whom he had met at the court of the Duke of Hannover, for he thought highly of Steno's *Prodromus*, about which he wrote to a friend, as follows (a translation):

> One must admit that all of Monsieur Steno's scientific works are excellent; but the one most worthy of esteem is his treatise *De Solido intra Solidum*. I have often urged him to enlarge it, and to draw from it conclusions regarding the origin of the human race, the universal flood, and some other great truths that confirm what is told us about these matters in the Holy Scriptures.[6]

In the same year, 1749, Buffon published, in the first volume of his *Histoire naturelle*, a theory of the earth that attracted much attention at the time and was frequently reprinted and translated into several languages. In it Buffon denounced the idea of a universal deluge. The wide circulation it received no doubt did much to stimulate interest in geology; but it was of less importance than his *Epoques de la nature* published thirty years later, also in a volume of his *Histoire naturelle*.

A book written by Elie Bertrand, *Memoires sur la structure intérieure de la terre* (Zurich, 1752; 2nd. ed., 1760) is perhaps worth a mention here because of the well-documented summary of the views of many earlier writers on geology that it contains.

As interest in geology grew in the second half of the eighteenth century, so the volume of literature on the subject increased. A theory published in 1763 by the German geologist and mine adventurer R. E. Raspe, *Specimen historiae naturalis globi terraquei*, deserves attention. In this work, which has been largely ignored by historians of geology,

6 G. Scherz (Ed.), *Nicolai Stenonis Epistolae et Epistolae ad Eum Datae* (2 vols.; Copenhagen: Nyt Nordisk Vorlag, 1952), Vol. 1, p. 66.

the author devotes considerable space to a discussion of the geological views of Hooke. Other works by Raspe show that he held enlightened views on geological theory. A translation of his *Specimen Historiae naturalis*, written in Latin, is being prepared for publication by Professor A. Carozzi, of the University of Illinois.

In 1778, the British author John Whitehurst published an *Inquiry into the original state and formation of the earth.* Two later editions appeared in 1786 and 1792, and a translation into German in 1788. Althought this work contains much that is unsound, Whitehurst was evidently well informed about the stratigraphical succession in parts of England, recognizing that sediments occur in a regular succession; and he was familiar with the fossil contents of the rocks he had examined. He also recognized the volcanic origin of certain rocks associated with the sediments in Derbyshire.

In 1779, Buffon published his important *Epoques de la nature.* This appeared first as a part of the author's *Histoire naturelle,* in a volume dated 1778 but not actually published until 1779. It was also issued in two volumes, as a separate book, in 1780. The *Epoques* has recently formed the subject of a detailed monograph,[7] and need not be discussed here, except to point out that it contains one of the earliest attempts to compute the age of the earth by an experimental method, ignoring calculations based on biblical history.

In the same year, also, J. A. De Luc published his five-volume *Lettres physiques et morales sur l'histoire de la terre,* which is of less importance than its length might suggest, but this author requires some notice in relation to the critical attacks he made later on Hutton's *Theory of the earth.*

In 1780, an obscure, newly graduated medical practitioner, G. H. Toulmin, published still another theory of the earth. The first edition (1780) bore the title *The antiquity and duration of the world.* Later versions appeared with changed titles: *The antiquity of the world* (1783), *The eternity of the world* (1785), and *The eternity of the universe* (1789). Until Professor Tomkeieff drew attention to this work in 1948,[8] its existence seems to have been completely overlooked by historians of geology. Tomkeieff pointed out the general similarity of some of Toulmin's conclusions to those published soon after by James Hutton. As far as I can ascertain, Toulmin based his theoretical views entirely on his reading of earlier literature, and there is virtually

7 J. Roger, "Buffon: Les Epoques de la Nature," *Mémoires du Muséum National d'Histoire Naturelle* (Sciences de la Terre), *10* (1962).
8 S. I. Tomkeieff, "James Hutton and the Philosophy of Geology," *Transactions of the Edinburgh Geological Society, 14* (1948), 272.

no evidence suggesting that he had ever studied rocks in the field, as had Hutton for many years before he produced his *Theory*. A further remarkable fact is that, assuming that the book was widely read, as must be supposed from the number of editions, it received almost no notice from contemporary geologists.

Next in the list comes the theory of the earth propounded by James Hutton. This, as is well known, was first announced publicly in 1785, but was not printed, except in the form of a privately circulated short abstract, until 1788; and the final version "with Proofs and Illustrations" was not published until 1795. Contemporary reception of Hutton's *Theory* was lukewarm, and general acceptance of those of his views that were sound did not become universal until a number of years after his death. However, his *Theory* received some notice outside Great Britain not very long after publication. The first detailed exposition of the *Theory*, published in 1788 in the *Transactions of the Royal Society of Edinburgh*, was noticed at some length in the *Magazin für das neuste aus der Physik und Naturgeschichte* (Gotha) in 1790; and what is apparently a complete translation into German of the 1788 paper appeared in the *Sammlungen zur Physik und Naturgeschichte* (Leipzig) in 1792. In France, a translation of the *abstract* was published in the *Journal de Physique* in 1793.

Still other theories of the earth appeared before 1800, but this is a convenient point at which to conclude this account of the eighteenth-century theories.

STRATIGRAPHY It would be surprising if no progress had been made in stratigraphy during the eighteenth century, when quarrying and mining were active. Miners and quarrymen must long have known that some rocks were stratified and that individual strata varied greatly in character. Such knowledge is likely to have spread gradually among interested persons, though it may not always have found its way into print. It would have been used in the search for economic minerals, such as coal. In fact, important contributions to stratigraphic knowledge were made known by eighteenth-century mining engineers; and the broad subdivision of rocks into three main groups was suggested long before Werner began his teaching on the subject. The earlier observers, however, did not maintain that these subdivisions were universal, as Werner did, dogmatically.

The names of authors who made important contributions to stratigraphy during the eighteenth century are, for the most part, well known. Some of the lesser known contributions to the subject are mentioned here.

In 1723, Benjamin Holloway, a contemporary of John Woodward (who himself must have been well informed about the details of British stratigraphy) published in the *Transactions of the Royal Society* a letter to Woodward,[9] in which he referred to the outcrop of Cretaceous rocks (Greensand under Chalk) that stretches northeastward from Oxford by Bedford to the region of Cambridge. Holloway mentioned the pits of "fullers' earth" in the ridge of sandhills (Lower Greensand) in Bedfordshire and neighboring counties, and he stated that these occurrences confirmed what Woodward had said about the regular disposition of strata over large tracts of country. He suggested, on the basis of these stratigraphical observations, that other occurrences of fullers' earth ought to be found elsewhere along this tract of country. This is a very early instance of the application of stratigraphy for economic purposes.

Another early pioneer in stratigraphy was the country squire John Strachey, who lived in Somerset, in the district later to become well known through the observations of William Smith. There were a number of working coal pits near Strachey's house, and his interest in these mining activities led him to take a wider interest in geology. In 1719 and 1725 he published, in the *Transactions of the Royal Society*[10] two papers on the stratigraphy of this area, both accompanied by a remarkable diagram showing steeply inclined Coal Measures overlain by horizontal strata, including "Red Earth" (Trias), "Lyas or Limestone" (Jurassic), and, in the later paper, "Chalk hills." This section, which is substantially accurate, was based partly on mining information and partly on direct surface observations made in the country surrounding the mines. His section shows the Coal Measures broken by a fault that he says makes the seams "trap" up or down, and he shows seven named coal seams. He records the occurrence of shells in the shale above one particular vein of coal, and states their presence is a "certain indication that they are coming to the vein." While Strachey certainly did not anticipate William Smith's grasp of the general significance of fossils, this is the earliest instance known to me of recognition that they had some stratigraphic value.

In Italy, some years later, G. Arduino, an inspector of mines and later professor of mineralogy at Venice, made a detailed study of the rocks in northern Italy, and as a result proposed a broad threefold

[9] B. Holloway, "An Account of the Pits of Fullers-Earth in Bedfordshire," *Philosophical Transactions of the Royal Society*, 32 (1723), 419.

[10] J. Strachey, "A Curious Description of the Strata Observ'd in the Coal-Mines of Mendip in Somersetshire," *Philosophical Transactions of the Royal Society*, 30 (1719), 968; and "An Account of the Strata in Coal-Mines,&c.", *ibid.*, 33 (1725), 395.

division of these rocks. He also prepared geological maps of the areas he examined. One of these, constructed in 1740, was published in 1744. Others, including a strikingly modern-looking horizontal section, have since been reproduced from his original manuscripts.[11] Judging from these maps, Arduino's mapping technique was considerably in advance of that of Guettard, and in my view he has a much better claim than the latter to rank as the first geologist to construct geological maps, but unfortunately his work seems to be little known outside Italy. Incidentally, Guettard was one of a number of naturalists with whom Arduino corresponded. His work requires further study in relation to that of his contemporaries.

Other European geologists who made important contributions to stratigraphical geology were J. G. Lehmann and G. C. Füchsel, and their work is better known. Lehmann's *Versuch einer Geschichte von Flötzgebürgen,* first published in 1756, also appeared in French translation in 1759. Füchsel's stratigraphical work appeared in 1762 in Volume 2 of the little known periodical *Acta Acad. Elect. Moguntinae Scientiarum* (Erfurt), and in this paper he recorded that the beds he described were characterized by their fossil content.

Another eighteenth-century stratigrapher was John Michell, F.R.S., astronomer, physicist, and geologist, who held the Woodwardian lectureship in geology at Cambridge from 1760 to 1762. As a geologist, he is chiefly known as the author of *Conjectures Concerning . . . the Cause of Earthquakes* (1760). This work indicates that he had a considerable knowledge of stratigraphy in general and of the fact that strata change little when traced over large tracts of country, probably based on personal observations in Britain. In this book, too, he suggested that continuity of strata was a phenomenon that would be met in other parts of the world, making, in this connection, a specific reference to Lewis Evans' map and account of the geology of North America. This is one among several instances illustrating the exchange of geological information between Britain and America during the eighteenth century. Michell also compiled a table of the English strata from the Coal Measures up to the Chalk, with estimates of thickness, but this was not published during his lifetime.[12]

11 The 1740 map was published in J. J. Spada, "Corporum Lapidefactorum Agri Veronensis Catalogus" (Verona, 1744), without attribution to Arduino. The attribution is given in G. Stegagno," Il Veronese Giovanni Arduino e il suo Contributo al Progresso della Scienza Geologica," (Verona, 1929), in which this and the other maps are reproduced.

12 J. Michell, *Conjectures Concerning the Cause, and Observations upon the Phaenomena, of Earthquakes* (London, 1760). Reprint of a paper published in the *Philosophical Transactions of the Royal Society,* 51 (1760), 566.

An important geological work of a general character by the Swedish mineralogist and chemist Torbern Bergman was his *Physisk Beskrifning öfver Jord-klotet*, published in 1766. This was evidently widely read in Europe for a second (two-volume) edition in Swedish appeared in 1773–1774, a Danish translation in 1771, and translations into German in 1769, 1780 and 1791, and into Russian in 1791–1794. Until recently, the work has been neglected by English-speaking authors, but an assessment of Bergman's contributions to stratigraphy forms the subject of a paper by Professor Hollis D. Hedberg in this volume.

Mention must be made, too, of the important stratigraphical studies carried out by the Abbé Giraud-Soulavie in the Vivarais, in the south of France, which were published in 1780–1784 in his seven-volume *Histoire Naturelle de la France Meridionale*. Soulavie divided the sedimentary succession in this area, including the associated volcanic rocks, into five epochs, or ages, by means of the fossil assemblages they contained. He noticed that the fossils in the lowest beds had no living analogues, whereas his third limestone contained fossils similar to those inhabiting existing seas. One limestone contained worn pebbles of basalt, and from that fact he inferred that volcanic eruptions had preceded the formation of this limestone and that its products had been destroyed by erosion. From this fact he concluded that an enormous period of time must have elapsed since the volcanoes were active.

The suggestion that the knowledge of stratigraphy in the eighteenth century may have been more general among interested geologists than is indicated solely by their publications is exemplified in some correspondence of James Hutton. This shows that at least as early as 1770, Hutton had acquired an extensive knowledge of the stratigraphical formations in both England and Scotland.[13] This knowledge was used by him in formulating his *Theory of the Earth*, but he does not seem to have been interested in stratigraphy as a restricted subject for study, only for this wider purpose. It must not be overlooked, either, that William Smith's great discoveries were made in the last few years of the eighteenth century, although they were not published immediately, and their effect on the development of the subject was delayed until the early years of the nineteenth century.

This brief account of eighteenth-century stratigraphy indicates, however, that it was much studied throughout the period, and that several contributions of definite scientific importance were published during the century.

13 V. A. and J. M. Eyles, "Some Geological Correspondence of James Hutton," *Annals of Science*, 7 (1951), 323–325.

PALEONTOLOGY The observation that many sediments contain organic remains goes back very far in time, but the old arguments about the origin of these fossils were largely over by the opening of the eighteenth century. On the other hand, the convenient belief, arising from biblical teaching, that all fossils, as well as the sediments enclosing them, were deposited by a universal flood persisted longer.

Paleontology today is divisible into two main divisions, stratigraphic paleontology and biological or evolutionary paleontology. While no such subdivisions had been established at the opening of the eighteenth century, it was during this period that many of the facts that led to the establishment of these subdivisions were accumulated.

So far as stratigraphic paleontology was concerned, the observation that fossils found in one particular bed or group of beds might differ markedly from those found in other beds must, it seems to me, have gradually become common knowledge as the century advanced. Yet the full significance of this fact was not realized until the close of the century, in spite of the tentative suggestion by Robert Hooke that fossils might be used as a chronological index. Such an acute observer and thinker as James Hutton, for example, seems to have been aware at least as early as 1770 that different beds contained different fossils, but, he states, "being neither botanist nor zoologist in particular, I never considered the different kinds of figured bodies, found in strata, further than to distinguish betwixt animal & vegetable, sea & land objects; the mineralisation of these objects being more the subject of my pursuit. . . ."[14]

So far as biological paleontology was concerned, the eighteenth century saw the formation of many museums and private collections, and fossils as such were much studied. The fact that not all fossil remains could be matched by analogous living species had been observed at quite an early date by, for example, Hooke and Ray, and the suggestion had been made that living forms to correspond with such fossils might ultimately be found in distant oceans. Before the century had advanced very far, however, the idea that there might be extinct species among fossil remains was put forward; and before the century was out the evolution theory had begun to take form.

The number of works of paleontological interest published during the eighteenth century was very large, and only a few typical examples can be mentioned here. They can be divided into three groups: descriptive works of a general character, regional works describing the fossils found in a particular area, and treatises on particular groups of fossils.

[14] *Ibid.*, p. 323.

Their scientific value varied widely, of course, but they should not be judged by modern criteria. They served a useful purpose by figuring and describing fossils, more or less accurately, by drawing attention to their widespread occurrence and great variety, and occasionally by indicating that certain beds were characterized by particular fossils or groups of fossils.

In the first category, a book published by the Italian author, A. Scilla, *La vana speculazione disingannata . . . circa i corpi marini, che petrificati si trovano in varii luoghi terrestri*, is of interest, because it is an argument for the marine origin of fossils. Though first published in 1670, it was reprinted three times in the eighteenth century (1752, 1754, and 1759) with a new title, *De corporibus marinis lapidescentibus*. Works of a general character were also published in Switzerland, by J. J. Scheuchzer, and in France by Dezallier d'Argenville. Special mention must also be made of Knorr and Walch's beautifully illustrated folio volumes (1755–1778), whose colored plates figure many fossils. This work was issued with German, French, and Dutch texts.[15]

Among regional works, E. Lhuyd's work on British fossils, *Lithophylacii Britannici Ichnographia*, first published in 1699, and reprinted in 1760, illustrates about 250 fossils. The localities from which they were collected are recorded, and an attempt is made to classify them. Swedish fossils were illustrated by Magnus von Bromell in his *Lithographiae Svecanae*, published at Upsala in Volumes II and III of *Acta Literaria et Scientarum* in 1727–1730, and by other Swedish authors. Spanish fossils were illustrated by J. Torrubia in his *Aparato para la Historia Natural Espanola*, published at Madrid in 1754. A German work of a similar character was G. A. Volkmann's *Silesia Subterranea* (1720). A specialized work, *Fossilia Hantoniensia* (1766), in which many fossils from the Eocene (Tertiary) of Hampshire, England, were accurately described and figured by Dr. Solander, is of interest as the first catalogue of fossils issued by the British Museum.

Among treatises on fossils of particular groups, mention may be made of J. S. Buttner's *Coralliographia Subterranea* (1714) and J. T. Klein's work on living and fossil echinoderms, *J. T. Klein . . . Naturalia Dispositio Echinodermatum* (1734), of which a French translation appeared in 1754 and a revised edition edited by N. G. Leske in 1778.

The works cited above form only a more or less random sample,

[15] G. W. Knorr and J. E. I. Walch, *Sammlung von Merckwürdigkeiten der Natur und Alterthümern des Erdbodens, welche Petrificirte Körper enthält &c.*, (Nürnberg, 1755), continued as *Die Naturgeschichte der Versteinerungen* (3 vols.; Nürnberg, 1768, 1773, 1774).

and many others might be added. The knowledge of fossils that had been accumulated in the eighteenth century, taken in conjunction with the considerable knowledge of stratigraphy that existed towards the close of the century, had laid a sound foundation for the more scientific approach to stratigraphic paleontology that was developed in the nineteenth century; it also provided basic material for later studies in scientific paleontology and biological evolution.

MINERALOGY AND PETROLOGY The third division of geology that was extensively studied during the eighteenth century was mineralogy. In fact, during much of the century the term *fossil* was in general use to refer to anything dug out of the earth, whether mineral, rock, or petrified organic remains, and very few attempts seem to have been made to establish a clear-cut distinction between rocks and minerals, in the sense in which these terms are used today. In consequence, eighteenth-century works on mineralogy almost invariably include descriptions of rocks as well as of minerals, *sensu strictu*, and frequently of fossil organic remains, too. Various adjectives came into use to distinguish organic from other fossils — "extraneous" was used by John Woodward in 1729, in a work of his referred to later, and also by later authors. Other adjectives used were "accidental," "adventitious," and "foreign," or their equivalents in other languages.

One consequence of this view of the scope of mineralogy was that the study of rocks, as such (that is, petrology or petrography), was not developed as an independent subject during the eighteenth century, and what there is to say about petrology can most conveniently be included under the general head "mineralogy."

Two basic problems confronted eighteenth-century mineralogists, the identification of individual minerals, and the erection of a satisfactory system for classifying them. Two methods of classification came into use, one based on external form and such other physical characters as were then easily determinable, e.g., color, hardness, specific gravity, solubility, and fusibility. The second method was based on the view that chemical composition would be a more desirable basis for classification. Some authors attempted to combine both methods into one system. It might be thought, using the criteria available in the eighteenth century, that there were no insuperable difficulties in devising a workable system but, in fact, there were many difficulties to contend with.

The chief handicap was the state of chemistry. The smelting of metallic ores, based on centuries of metallurgical practice, had provided much information about ore minerals, but this was of strictly limited value in relation to mineralogy as a whole. However, considerable advances in other methods of chemical analysis were made during the eighteenth century. For example, mineral analysis by the blowpipe was developed at a fairly early date by the Swedish chemist A. F. Cronstedt. By one means or another, a considerable body of empirical knowledge of the chemical composition of minerals and rocks was accumulated. Such knowledge was limited, however, by the fact that before 1800, relatively few elements had been discovered, either by isolating them or through the reactions of their compounds. Nor had it yet been realized how extremely complex is the chemical composition of many minerals and some rocks. The principal handicap, however, was that the phlogiston theory dominated chemical thinking until the last decade or so of the eighteenth century, and until that was overthrown no major advances could have been made in the determination of minerals by chemical methods.

External characteristics, too, though certainly not without value, were on the whole an unreliable and ineffective means of identifying either minerals or rocks, and it would have been difficult to have devised a satisfactory system of classification based on such characteristics alone. A. G. Werner's classic work on the external characters of minerals, *Von den äusserlichen Kennzeichen der Fossilien*, was received with acclaim when published in 1774 because it was the most practical handbook of *determinative* mineralogy published up to that time—but it did not contain a system of classification.

The possibility of identifying minerals with precision by their crystal form was only suggested toward the close of the eighteenth century. At the opening of the century, crystallography was virtually nonexistent as a science. It had been recognized that certain minerals occurred naturally in the form of crystals, and it was known to chemists that substances precipitated from solution might be deposited in the form of crystals. A number of eighteenth-century authors attempted, unsuccessfully, to explain this phenomenon, but it was not until near the close of the century, after the basic laws of crystallography had been discovered by Romé de l'Isle and the Abbé Haüy, that it became possible to use crystal form as an important aid in the identification of minerals.

The optical methods commonly used today for identifying minerals had not been discovered. A few records exist of the use of the microscope before 1800 to examine minerals, but it is not always clear whether authors refer to the microscope proper or to a pocket lens, and the technique of examining minerals and rocks in thin transparent slices under a microscope was not devised until some time after 1800.

Thus, the precise identification of minerals, except for the simplest forms, and the determination of the mineral composition of most rocks, was a matter of great difficulty during most of the eighteenth century. Although the progress made in mineralogy was less rapid than in some other branches of science, there was a continuous and growing interest in the subject, which was undoubtedly stimulated by the growth of mining activity and industry generally. This interest was shared by the mineralogists with the chemists, for whom minerals provided the raw materials for their researches. It resulted in the publication of a large number of works on the subject, and the statement made by A. G. Werner in 1774, that the "number of mineral systems are annually increased by at least one,"[16] emphasizes the extent of interest in the subject. His remark was close to the truth, if the number of editions of any one work and the numerous translations of such works are taken into account. Nor is it surprising, in view of the difficulties mentioned, that he goes on to say "that oryctognosy has not yet made the progress that is much to be desired." The term *oryctognosy*, as used by Werner, signified the determination of minerals.

There is, in fact, much confusion and little precision in the eighteenth-century textbooks of mineralogy. However, it gradually came to be realized that of the two methods in use for classifying minerals, the one based on chemical composition was likely to prove more satisfactory. Ultimately, external form and physical properties passed out of use as a method of classification, though of course these properties continued in use by mineral prospectors and geologists as a rough and ready means of identifying minerals.

Mention is made here of a few of the more important, interesting, and also lesser known contributions to mineralogy and petrology made during the eighteenth century.

John Woodward, in 1728, published a book with the curious title *Fossils of All Kinds Digested into a Method Suitable to Their Relation and Affinity*. This work is essentially a textbook of mineralogy and contains a tabular statement of Woodward's mineral system, which was principally based on external characters. His last work, published posthumously in 1729, was entitled *An attempt towards a natural history of the fossils of England*. In this work, he lists a very large number of specimens of British (and some foreign) minerals, which are carefully described and localized; and occasionally he records having examined particular specimens, such as fine sands, under the microscope. Translations of Woodward's 1728 work into French (1735), Italian (1739), and

[16] A. G. Werner, *On the External Characters of Minerals*, trans. A. V. Carozzi (Urbana: University of Illinois Press, 1962), p. xxiv.

German (1748) have been recorded. As a pioneer in the careful and systematic study of minerals, Woodward deserves an important place in any study of eighteenth-century mineralogy.

Of some historical interest is the fact that Linnaeus, in his *Systema Naturae*, first published in 1735, places the mineral kingdom first among the three kingdoms of nature he describes. He divided minerals, i.e., fossils in the eighteenth-century sense, into three broad groups corresponding roughly to simple minerals, metallic ores, and aggregate minerals (which included organic remains).

In 1748, John Hill published *A history of fossils* (forming part of his three-volume *General Natural History, 1748, 1751, 1752*), in which he divides "fossils" into series, classes, orders, and genera. His systematic classification has little merit, but his detailed descriptions of individual minerals are often interesting; he records crystal form wherever applicable; and he frequently records the use of the microscope for examining specimens.

In 1757, E. Mendes da Costa published Volume I, Part i, of a work entitled *A natural history of fossils*, which is of interest because it is possibly the only work published in the eighteenth century that can be regarded as strictly a textbook of petrology. Da Costa intended to deal with the rest of the "Fossil kingdom" in later volumes, but these were not published. In this volume, he divided rocks into two principal classes, "Earths" and "Stones." The former included soft, uncompacted rocks and soils; the latter class has several subdivisions, each headed "Stones forming continuous strata." While the rocks described in these subdivisions are mainly sediments, two subsections described granites and porphyries, rocks whose igneous origin was as yet unsuspected. Da Costa's general description of granite as a rock of "irregular texture, concreted of separate large grains . . . distinct from each other . . . of *quartz* and *micae*," and of porphyry as a stone of "an exceeding fine compact uniform texture like flint, in which detached pieces . . . of *quartz* are imbedded," shows that he was thoroughly familiar with the appearance of the rocks he described, if not with their origin.

E. Bertrand's *Dictionnaire universal des roches propres et des fossiles accidentels*, published in two volumes at La Haye in 1763, and reprinted at Avignon in the same year, is a useful reference work for students of eighteenth-century mineralogy.

Among the most valuable contributions to mineralogical science made in the middle of the century were those of the Swedish mineralogical chemists A. F. Cronstedt and Tobern Bergman. The importance of their contributions lay in the fact that they were the first to stress the value of chemical composition for classifying minerals. Contemporaneous recognition of the value of their work is indicated by the number of editions of their mineralogies that were published, and the fact that translations into English, French, German, Italian, Russian, and Portuguese were made.

Another, but little known, mineralogical textbook, J. R. Forster's *Introduction to Mineralogy*, published in 1768, is of interest because the author's approach to his subject was rather more scientific than that of some of his contemporaries. The text includes, in addition to the descriptive matter usual in similar works, a section recording the action of various acids on minerals and rocks, being a translation of tables supplied to him by J. G. Lehmann when he was in St. Petersburgh. There is also a list of the specific gravities of many minerals and rocks, and an account of the electrical conductivity of mineral bodies (taken from Priestley's *History and Present State of Electricity*).

A work worth passing notice is M. F. Ledermüller's *Physicalisch-mikroskopische Abhandlung vom Asbest, Amiant, Stein- oder Erdflachs, und einiger anderer mit demselben verwandten Fossilien* (Nurnberg, 1775). A translation into French appeared in the same year). This author, in preparing his treatise on varieties of the mineral asbestos, used a microscope to examine his specimens. His book has six colored plates showing specimens viewed in this manner, one specimen being enlarged approximately 25 times. As far as I know, Ledermüller was the first, and for a very long time the only mineralogist to publish illustrations of minerals as enlarged under the microscope.

One of the most important of the late eighteenth-century mineralogies was Richard Kirwan's *Elements of Mineralogy*, first published in 1784, followed by a second edition in two volumes (1794, 1796). Kirwan's approach to the subject was largely chemical, but the main interest of these two editions is that in the earlier one he still adheres to the phlogiston theory, which detracts from the value of the analyses he gives, whereas in the second edition he has abandoned this theory and we find that oxygen has a place in his mineral analyses. Even so, as late as 1794 he still defines minerals, in the eighteenth-century manner, as "All bodies found in or on the earth, and not of vegetable or animal origin,"[17] rather than in the restricted sense now in use.

If the mineralogists and chemists of the eighteenth century failed to establish mineralogy on a completely sound scientific basis by the end of the century, it was not for lack of interest or want of trying. They laid the foundation for the great advances that were soon to be made in the succeeding century.

[17] R. Kirwan, *Elements of Mineralogy* (2 vols.; London, 1794, 1796), Vol. 1, p. 1.

The Teaching of Geology and Mineralogy in the Eighteenth Century

The evidence provided in the preceding sections of this paper seems to me to establish beyond doubt that geology and mineralogy were subjects of widespread interest, seriously studied throughout the eighteenth century. The question naturally arises as to whether these subjects were taught before 1800, and if so, by what means and to what extent. While I have not investigated this question exhaustively, the available evidence indicates that they were taught more frequently than has hitherto been realized, especially in the second half of the century. Instruction was given formally in universities and mining academies and, occasionally, in courses of public subscription lectures.

The earliest public lectures on geology delivered in the eighteenth century were, as far as I know, those delivered in Paris by the chemist G. F. Rouelle, as part of the course in chemistry he gave at the Jardin du Roi from 1742 to 1768. The detailed account of Rouelle's work recently published by R. Rappaport[18] records that, in fact, he taught geology as early as 1740; and his views on geology have been summarized by one of his pupils, the French geologist Nicholas Desmarest.[19] Although Rouelle quotes other authors — John Woodward, for example — it seems that the views he expounded were his own, founded in part on his own investigations, and that he kept his lectures up to date by incorporating new information. Several manuscript copies of his lectures have been preserved. The only one I have examined personally was dated 1755–1756. In this, there were 18 pages dealing with general geological topics, and a further 440 pages on "Mineralogy." Judging from the contents of contemporary works on mineralogy, the latter section probably covered a wider field than its title suggests, but I was unable to examine it in detail. Among the more distinguished pupils who attended Rouelle's lectures was the chemist Lavoisier, and his known interest in geology may well have been stimulated by attending these lectures.

Evidence of another course of public lectures came my way by chance. It was contained in a four-page pamphlet entitled *Syllabus of a Course of Lectures on Fossils by Emanuel Mendes da Costa,* dated London, October 9, 1778, which I found bound into my own copy of da Costa's *Natural history of fossils* (1757), to which I have referred above. The *Syllabus* summarizes the contents of 27 lectures, of which

18 R. Rappaport, "G-F. Rouelle: An Eighteenth Century Chemist and Teacher," *Chymia,* 6 (1960), 68.

19 N. Desmarest, "Rouelle. Notice sur sa Doctrine Relative à Plusieurs Points Importans de l'Histoire Naturelle de la Terre," *in Encyclopédie Méthodique,* [Section *Géographie Physique*], Vol. I (Paris, 1794), pp. 409–431.

the first 21 deal with rocks and minerals, and the last six were on "extraneous" fossils.

Lectures on geology and mineralogy were also delivered in several British universities during the eighteenth century. The earliest formal lecture courses on these subjects were those given in the Scottish universities. They were included in courses on natural history or chemistry, for no chair or lectureship in geology was established in Scotland until much later, in the nineteenth century.

The first of these courses was that given in the University of Aberdeen, in Marischal College. Here, a lectureship in natural history was established in 1753, the first holder being Francis Skene. He included as a major part of his course a series of lectures on the "*Regnum lapideum*, or Kingdom of Unorganized Bodies." The earliest extant manuscript of these lectures, preserved in the University of Aberdeen library, is dated 1768, so the actual year in which this part of the course was first delivered is uncertain. As might be anticipated, the subjects dealt with were mainly mineralogy and petrology, but there were also references to the occurrence of "extraneous" fossils. The course included details of the nature, mode of occurrence, distribution, and uses of the substances described. If the information falls very far short of what is now expected in accuracy and detail, it must have served to stimulate interest in the subjects dealt with. One other manuscript of the natural history course, dated 1775, is also preserved in Aberdeen. It, too, contains a section on the "Mineral Kingdom," generally similar to that in the earlier manuscript but rather more detailed. Francis Skene died in 1775 and was succeeded by his son George Skene, so it is at least possible that the lectures on mineralogy were continued after that date, as part of the course in natural history.

In 1776, Joseph Black was appointed professor of chemistry in Edinburgh. In a recently published book, *Notes from Doctor Black's lectures on chemistry, 1767/8*,[20] the notes on these lectures taken by a student, Thomas Cochrane, have been printed. These show that Black included some instruction in geology and mineralogy in his course in the session 1767–1768. Other manuscript copies of Black's lectures, preserved in the library of the University of Edinburgh and elsewhere, indicate that later on his treatment of these subjects was rather fuller than in Cochrane's notes. Under the subject heading "Earths," Black had something to say about geology in general, dealing with such matters as stratigraphy, faulting, erosion, and organic fossils, and then dis-

20 Edited by Douglas McKie (Imperial Chemical Industries, Limited, Pharmaceuticals Division, Cheshire, England, 1966).

cussed various "earths," or stony substances, and minerals proper. Other notes of general geological interest can be found in these manuscripts under the heading "Mineral Waters." So far as is known, Black continued to include some instruction in geology and mineralogy in his chemical lectures for nearly thirty years, until a few years before he died in 1799. When Thomas Hope took over the chemistry course from Black, in 1797, he continued to include geology in the course, and in his lectures added a detailed comparison of the Huttonian and Wernerian systems of geology.

Lectures much wider in scope than those of Skene in Aberdeen and Black in Edinburgh were delivered in Edinburgh from 1782 until 1800 by John Walker, professor of natural history. In these lectures, geology was named separately and treated as a separate subdivision of natural history, mineralogy forming an additional subdivision. Some other geological topics were treated in a third subdivision of the lecture course, hydrography. These lectures are of great interest, not only because the treatment given to the earth sciences was much fuller than in any other eighteenth-century course in Britain, but because they bridge the gap between eighteenth- and nineteenth-century thinking. They were the work of a man who had not only read the earlier literature thoroughly, but had traveled extensively in Britain, studying rocks and minerals in the field. Walker was interested in facts rather than theories. A manuscript version of his lectures, with an extensive commentary by Professor H. W. Scott of the University of Illinois, has recently been published.[21]

In the English universities, less attention was paid to the subjects of geology and mineralogy during the eighteenth century. It is well known that John Woodward, who died in 1728, left a sum of money to found a course of lectures at Cambridge, and that the Woodwardian chair in geology at this university can be traced back to this foundation, but it is possibly less well known that the subjects of this lectureship were to be, at discretion, those treated in Woodward's *Natural History of the Earth*, his "Discourse on Vegetation,"[22] or his book *The State of Physick* (1718). In fact, none of the eighteenth-century holders of this lectureship are known to have delivered any formal lectures on geology, and only two made any attempt to advance geological science. The only one who made an important contribution to geology was John Michell,

[21] *Lectures on Geology by John Walker*, ed. H. W. Scott (Chicago and London: University of Chicago Press, 1966).

[22] J. Woodward, "Some Thoughts and Experiments Concerning Vegetation," *Philosophical Transactions of the Royal Society*, *21* (1699), 193.

to whom I have already referred. The last appointment made in the eighteenth century was that of John Hailstone, in 1788. He at least made an attempt to familiarize himself with the subject by attending Werner's lectures in Freiberg; and he subsequently published, in 1792, *A Plan of a Course of Lectures on Mineralogy*, based on Werner's lectures. He added materially to the geological collections in the university and is believed to have demonstrated them to interested persons, but as far as is known, he gave no formal lectures.[23]

There is, however, a record that some instruction in geology and mineralogy was given in Cambridge before the close of the century by Thomas Martyn, who had been appointed professor of botany in 1762. He published in 1782 a pamphlet entitled *Heads of a Course of Lectures in Natural History Read at the Botanic Garden.* His course comprised 34 lectures, of which 8 were on the "Fossil Kingdom." The subjects dealt with under this head included earths, stones, minerals, and organic fossils, here called "petrifactions." It seems unlikely that Martyn contributed anything original to the subject, for he names the authors on whose works the lectures were based. It has been stated that, having failed to make botany a popular subject, he was forced to include geology and mineralogy in his course to secure an adequate attendance at his lectures.[24]

Apart from the universities, mineralogy was taught for a time in an English school, famous in its time, the dissenters' Academy at Warrington, in Lancashire. Here, Johann Reinhold Forster, a naturalist who later accompanied Cook on his second voyage around the world, taught natural history from about 1767 to 1770[25] and he included mineralogy in his course. In 1768, he published the textbook of mineralogy to which reference has been made earlier, in which he states that it was specifically prepared for use by his students at Warrington. The contents of this book suggests that his students must have received a sound grounding in the subject.

So far, with the exception of Rouelle's lectures, I have mentioned the teaching of mineralogy and geology only in Great Britain. While I have made no serious attempt to investigate the extent to which these subjects were taught in other European countries, some examples have

23 *Life and Letters of the Reverend Adam Sedgwick*, ed. J. W. Clark and T. McK. Hughes (2 vols.; Cambridge, England: Cambridge University Press, 1890), Vol. i, pp. 166–198.

24 R. T. Gunther, *Early Science in Cambridge* (Oxford: Oxford University Press, 1937), p. 395.

25 J. W. A. Smith, *The Birth of Modern Education: The Contribution of the Dissenting Academies* (London: Independent Press Ltd., 1954), p. 167.

come to my notice. It is probable that further investigation would disclose additional examples, particularly because the "mineral kingdom" was often included in courses on chemistry or natural history.

A. F. de Fourcroy, the French chemist, taught mineralogy from about 1778 to 1791 in his own laboratory in Paris, as part of a general course in chemistry and natural history, giving 37 lectures on the "mineral kingdom." The substance of these lectures was incorporated in his well-known textbook, *Eléméns d'Histoire Naturelle et de Chimie* (1782, with later editions and translations). These appear to have been private lectures, but he also included 32 lectures on mineralogy in a course at a Lycée in Paris from 1787 onwards.[26]

Mineralogy and geology were also taught in several European mining schools that were established in the eighteenth century. The Freiberg academy was opened in 1766. In fact, one authority states that students traveled to Freiberg at an earlier date for instruction by the mining engineer J. F. Henckel, and that after his death they continued to visit the town for instruction even before the mining school was opened. Schools of mining were also opened at Schemnitz in Slovakia in 1763, in St. Petersburg in 1773, and in France in 1783. Mineralogy must also have been taught in Italy in the eighteenth century, for the Italian geologist Arduino held the post of professor of mineralogy and metallurgy at Venice at one time during his career.

It also seems probable that geology was taught in Sweden in the eighteenth century, probably at Upsala and possibly elsewhere, but I have not been able to obtain details.

In the New World, a mining school was opened in Mexico City in 1792, with the Spanish mining geologist Fausto de Elhuyar as director. He was one of Werner's pupils and had also studied under Bergman. He was assisted by A. M. del Rio, another of Werner's pupils and author of the first textbook of mineralogy published in the Americas, *Elementos de Orictognosia* (2 vols., Mexico City, 1795, 1805).

Other Channels Through Which Geological Knowledge Was Diffused in the Eighteenth Century

Among the additional channels by which geological knowledge was diffused during the eighteenth century, the most obvious and most important was by means of the publications of learned societies, such as, for example, the Royal Society and the Académie des Sciences, and other national academies. There were, too, a number of important scien-

[26] W. A. Smeaton, *Fourcroy: Chemist and Revolutionary; 1755–1809* (Cambridge, England: W. Heffer & Sons Ltd., 1962), pp. 177–179 and 186.

tific journals not specifically connected with societies, such as the *Journal de Physique*, which published original articles of geological interest. Of less importance, but still worth consulting, were periodicals or magazines of a popular or semipopular character in which scientific information and criticism was published. As examples, I may mention the *Hamburgisches Magazin*, first published in 1747, and continued until 1780; the *British Critic*, first published in 1783, and the *Monthly Review*, first published in 1749 (in which J. A. De Luc published his criticisms of Hutton's *Theory of the Earth*. Nor should it be overlooked that in the eighteenth century educated persons had the leisure to write and that they corresponded freely. Much correspondence of interest to students of eighteenth-century geology has been preserved in libraries and among the archives in public record offices, both in Europe and in the United States. Little of this has been published, but one or two examples are worth a mention: *A Calendar of the Manuscript Correspondence of Sir Joseph Banks in the British Museum; British Museum (Natural History) and other Collections in Great Britain* (British Museum (Natural History), 1958); and *Tobern Bergman's Foreign Correspondence*, Vol. 1 (Stockholm, Almquist & Wiksell, 1965). The Banks correspondence is voluminous and covers almost all branches of natural history, while that of Bergman is of particular interest to mineralogists. Some unpublished correspondence still awaits careful examination and some, I suspect, still awaits discovery.

Finally, the eighteenth-century naturalists often had the leisure to travel, so that eighteenth-century travel books and works on topography not infrequently contain information of geological interest.

The "other channels" for the diffusion of geological information that I have discussed briefly here are, of course, of varying importance, looked at simply as sources of geological history. Taken as a whole, however, in the context of this paper, they lend further support, if such is needed, to my main thesis that throughout the eighteenth century, geology certainly existed as a live branch of science that aroused widespread interest.

The Development of Mineralogy in Philadelphia, 1780–1820: A Summary

John C. Greene

Medical schools provided one avenue by which mineralogy, as an adjunct of chemistry, entered the academic curriculum in the United States, but mineralogy failed to take a firm hold in this way at the School of Medicine of the University of Pennsylvania. James Woodhouse gave some attention to the analysis of minerals in his writings and lectures, but his successor, John Redman Coxe, did very little in this line, although his *Philadelphia Medical Museum* provided an outlet for the mineralogical articles of Woodhouse and Adam Seybert. Seybert, who had studied chemistry and mineralogy at Paris and Goettingen and had brought back a cabinet of minerals, was the leading mineralogist in Philadelphia until 1809, when, discouraged at not securing Woodhouse's chair in chemistry, he went into politics and sold his collection. His son, Henry Seybert, who was also trained in Paris, published some good papers on mineralogy before turning to philanthropy after his father's death. Thomas Cooper taught mineralogy and geology in the undergraduate college of the University of Pennsylvania from 1816 to 1820, stressing geology, but he abandoned the post because of inadequate financial reward.

The American Philosophical Society gradually accumulated a sizable collection of minerals, and these were eventually arranged in scientific order by Silvain Godon, a Parisian mineralogist who had emigrated to the United States in 1807. Godon was very active in the Society and in other scientific groups in Philadelphia until the end of 1812, when his imprisonment for debt discouraged him from further scientific endeavors. On the whole, however, the American Philosoph-

The full text of this paper is in *Proceedings of the American Philosophical Society* *113* (1969), 283–295.

ical Society was not a major center of mineralogical research, although it published a few valuable mineralogical articles by Godon, Joseph Cloud, a practical chemist in charge of melting and refining operations at the United States Mint, and one or two others.

Mineralogy was studied as a branch of natural history, as well as in connection with chemistry. The Linnaean Society of Philadelphia gave mineralogy a prominent place in its activities, but it was short lived. The Academy of Natural Sciences of Philadelphia, organized in 1812, was more successful. The study of mineralogy flourished there from the beginning, owing primarily to the mineralogical and geological interests of its first two presidents, Gerard Troost and William Maclure. Trained in the French school of crystallography under the Abbé Haüy, Troost introduced French methods and concepts at the Academy, both in his lectures and in his contributions to the *Journal* of the Academy. Through his influence, the Academy acquired the collections of Seybert and Godon, to which were added extensive donations from William Maclure and numerous specimens collected by other members of the Academy. The *Journal* of the Academy became an important avenue for mineralogical publication, and the contacts with the mineralogists of Paris initiated by Troost and Maclure were developed still further by a younger generation of Philadelphians, including Isaac Lea, Lardner Vanuxem, William Keating, and Henry Seybert, who went to Paris to study. Thus, although mineralogy failed to gain a firm footing in the academic institutions of Philadelphia before 1820, it became securely established at the Academy of Natural Sciences.

The Influence of Torbern Bergman (1735–1784) On Stratigraphy: A Résumé

Hollis D. Hedberg

Torbern Olof Bergman was born in 1735 in the province of Västergötland, Sweden. He was a student of Linnaeus at Upsala University. Bergman did scientific work in entomology, mathematics, physics, geology, and astronomy, but attained greatest fame as a chemist and mineralogist. He became professor of chemistry at Upsala in 1767 and made many lasting contributions to chemistry and mineralogy. He is known as a founder of quantitative analysis and was also particularly celebrated as a teacher. He died in 1784 at the age of 49 years.

In contrast to his fame as a chemist and mineralogist, Bergman's contributions to stratigraphic geology are surprisingly little known. His influence in this field stemmed largely from his monumental geographical-geological work entitled "Physisk Beskrifning Öfver Jordklotet," or "Physical Description of the Earth," which was first published in Swedish in Upsala in 1766. An "improved and enlarged" edition was published in two volumes in Upsala in 1773–1774. The work was widely read at the time and was translated into Danish, German, and Russian; however, no English translation has ever been published.

It is of interest that Werner's famous stratigraphic classification was drawn in large part from Bergman, and Werner is lavish in his praise of "the cool-blooded, sharp-minded, great Bergman — the philosopher, chemist, and mineralogist — the idol of his colleagues." Voigt, and later Kant, borrowed heavily from Bergman. Finally, Playfair lists Bergman as one of the five principal writers on geology on whom Hutton chiefly relied.

In the paper given at the New Hampshire Inter-Disciplinary Conference on the History of Geology, I discussed the development of stratigraphy in Europe up to Bergman's time and then presented a paraphrase of the three chapters of Bergman's work dealing particularly with stratigraphy, entitled, respectively, "On the Earth's Strata," "On Fossils," and "Opinions on the Origin of the Earth." A final section of the paper dealt with the sources of Bergman's information on stratigraphy and his influence on subsequent workers. Because of the length of the paper only a brief résumé is given here. The complete paper has been published by the Swedish Academy of Sciences.*

The importance of Bergman's geological work lies not so much in any new ideas that he may have presented as in the fact that it constituted a review and appraisal of the status of stratigraphic knowledge and thought, by a keen and discriminating mind, just at the time when geology was at the threshold of emergence as an independent science. Bergman, in his geological writing, was persistently concerned with the stratification of rocks and the layering of the earth, and he may therefore rightfully be considered one of the important early contributors to stratigraphy.

From a review of stratigraphic sequences in various parts of Europe, Bergman reached the opinion that these had clearly been deposited in water; hence, he concluded that most layers had an aqueous origin, that the whole earth had at one time been covered with water, and that the different rock strata of the earth had resulted from episodes of settling from waters that were locally and from time to time varied in composition. With regard to character, age, and origin of the rock strata, he recognized four major classes or layers of rocks: Uråldrige, Flolägrige, Hopvräkte, and Vulkaner.

The *Uråldrige*, or Primitive, rocks make up the innermost layer and are probably essentially as old as the earth itself. They appear in the cores of major mountain ranges and probably widen out downwards under the mountains to make a continuous solid mass or shell, which, from the behavior of the compass needle, must be magnetic. The Uråldrige rocks are less clearly bedded than younger strata and are so commonly cut by veins that they might as a class be called "Gångberg" They consist of granites, crystalline limestones, quartzites, and so forth. Nowhere do they contain fossils. These rocks were formed by deposition, either from solution or very fine suspension — probably both — as water was lost from the earth, presumably through evaporation. Because of local differences in water composition and mixing by water movements, the resulting sediments were not everywhere of the same character.

* Hollis D. Hedberg, "Influence of Torbern Bergman (1735–1784) on Stratigraphy, "*Acta Universitatis Stockholmiensis, Stockholm Contributions in Geology,* 20 (1969), 19–47.

The second class of rocks — *Flolägrige*, or Bedded — constitute the next oldest division. These rocks consist principally of sandstone, limestone, clay-marl, alum-shale, coal, traprock, and so forth — all of which were deposited dominantly from suspension and show little indication of crystallization out of solution. These rocks are commonly found flanking the Uråldrige rocks or covering them. They are usually well stratified and only moderately inclined. Fossils are often abundant, proving that the beds were formed as deposits on the sea bottom, covering the animals living there. The Flolägrige rocks were largely derived from submarine erosion of the Uråldrige crust.

After parts of the earth became dry land, and vegetation and land animals appeared, the youngest layers, consisting of *Hopvräkta* (swept together) and *Vulkaner* rocks, were formed. In this period, a great many drastic changes in the earth's surface took place, and these are still taking place. The Hopvräkta rocks, although deposited in water like the Flolägrige, differ in that they originated in very shallow and restricted water bodies and streams and thus their strata are less regular and continuous. The Vulkaner rocks, on the other hand, have been formed by the piling up of ashes, stones, and lava, which have been erupted through the activity of the earth's interior fire.

Bergman clearly recognized the importance of stratigraphic division and correlation and emphasized the continuity of individual beds and sequences throughout southern Sweden, to which Linné had previously called attention. He enumerated individually Lehmann's 30 units from the Harz Mountain sequence near Mansfeld, totaling some 1,500 feet of strata and including such famous stratigraphic divisions as the Zechstein, Kupferschiefer, and Rothliegendes. Also, he writes:

> Often one finds on both sides of a valley the same bed at the same height and position in sequence, as if it had formerly been connected and through some powerful force afterwards broken apart. The study of beds on both sides of sea-straits should very much clarify the question whether all lands now separated from each other were originally connected and only with time have become separated.

That Bergman also recognized the importance of fossils is evident from his chapter "On Fossils," which begins: "By Stenvandlingar (petrifacts and fossils) I understand here not only stones which have a likeness to organic bodies or their parts, but also imprints of the same, and all kinds of remains of animals and plants, which in unusual places are found. I have mentioned these in the foregoing, and in general hold them to be very important." Then follows his famous statement: "They are actually medallions of a sort, which were laid down on the originating earth surface, whose layers are archives older than all annals, and which, appropriately investigated, give much light on the natural history of this our dwelling place."

With regard to the biblical deluge, Bergman says:

> It is not believable that such a huge and unnatural action could take place without great changes in the earth. However, since it originated and passed away rapidly, the results probably consisted principally in disturbing loose earth beds, separating different rocks from each other in some places and bringing them together in others, opening new seaways and separating lands from each other, displacing huge rocks, flinging around forests and imbedding them, carrying animals and plants far from their birthplaces, and so on. But from this flood no regular rocks, neither Uråldrige nor Flolägrige, originated. The results of the Deluge also are seldom to be differentiated from earlier and later accumulations, although some special circumstances such as the laying down of plants from India in France and the burial of elephants in Siberia may be specifically attributed to it.

Bergman was remarkably well read, and obviously many of his views came from impressions he had received from writings of his predecessors, which he reworked, updated, and developed. This bringing together and digesting of the views of earlier contributors to geology from all over Europe was in itself an important contribution to stratigraphy. With respect to theories of earth origin, Bergman precedes his own conclusions with concise but fairly comprehensive reviews of the thoughts of Burnet, Woodward, Linné, Whiston, Descartes, Leibnitz, de Maillet, Rajus, Hooke, Moro, Kruger, and Buffon. Among early contributors to geology and paleontology to whose work he makes reference are such names as Bayer, Bertrand, Bourget, Boyle, Delius, Donati, Erhart, Gesner, Gmelin, Guettard, Jussieu, von Justi, Knorr, Lang, De Luc, Lehmann, Mylius, Palissy, Pallas, Pennant, Plancus, Plot, Pontopiddan, Rosinus, Rouelle, Scheuchzer, Spada, and Vallisnieri. (Strangely enough, he makes no mention of Steno, Strachey, Arduino, or Füchsel). Among his own countrymen, he refers frequently to the work of Cronstedt, Gyllenhal, Hermelin, Kalm, Linné, Rinman, Scheele, Swedenborg, Tilas, Triewald, and Wallerius.

Bergman undoubtedly owed much to Linné and Lehmann, both of whom he quotes frequently, but it is strange that nowhere does he mention Lehmann's quite similar, and earlier published, recognition of an Uranfängliche (or Ganggebirge) and a Flötzgebirge class of rocks, which so closely correspond to his own Uråldrige and Flolägrige classes. (Perhaps these terms were in such common use that no reference seemed necessary.) On the other hand, while Bergman may have drawn from Lehmann, he also added much to the concepts of Lehmann. Thus, while both postulated an original watery sphere, Bergman gives a much more reasoned account of gradual sedimentation, decrease of waters, and emergence of land. Also, Bergman largely rejects the origin of any substantial body of rocks from the deluge, whereas Lehmann had postulated the deluge as the source of his Flötzgebirge.

There seems no question but that Werner's famous classification was drawn from Bergman. As first published in 1787, Werner's classification consisted of Uranfängliche Gebirge (Bergman's Uråldrige), Flötzgebirge (Bergman's Flolägrige), Vulkanische Gebirge (the Vulkaner part of Bergman's Hopvräkte), and Aufgeschwemmte Gebirge (the sedimentary part of Bergman's Hopvräkte). Later, in 1796, Werner changed Uranfängliche Gebirge to Urgebirge, and also introduced his Transition or Ubergangs-Gebirge for rocks that seem to be equivalent to the lowest part of Bergman's Flolägrige, formed largely by reworking of Uråldrige rocks.

With regard to the geological theories expressed in his book, Bergman concludes:

> I recognize the quality of the here-expressed thoughts only too well to hope for general approval. Even essential truths do not escape refutation, and what can one expect then from suppositions on the origin of the earth which in great part cannot be verified? Every subject has many sides and can be placed under different viewpoints from which it may be observed — often very differently, especially when prejudices and such miscarriages enter in. For this reason I excluded entirely my thoughts on this subject from the first edition of my book. Only because a number of friends found it extraordinary, after such a considerable quantity of experiences had been collected, to avoid conclusions, have I here given the guides which I believe to have found to the origin of the present arrangement of the earth. These thoughts I present, not necessarily to maintain, but rather to correct, or wholly give up, as new observations may require.

There seems no doubt that, though little recognized today, Bergman's "Physisk Beskrifning," which was well known and widely distributed at the time, constituted, through its reasoned and scientific approach, even though still burdened with faulty inheritances, an important step toward the great flowering of stratigraphy that was to take place near the close of the 18th century. Its moderation of tone and the modest, inquiring, and truth-seeking attitude of the author are in pleasant contrast to the many dogmatic works characteristic of the period. Many of the theories to which Bergman gave support, such as the persistent increase of land relative to water-covered areas, the sedimentary origin of granitic rocks, and the carving of mountains by water action at the bottom of the sea, came to seem ridiculous to geologists of the immediately succeeding century; but some of these may appear not quite so bizarre today in the current atmosphere of continental accretion, granitization of sediments, and the cutting of deep sea canyons by turbidity currents.

References

Adams, Frank Dawson (1954). *The Birth and Development of the Geological Sciences.* New York: Dover Publications, 506 pp. (reprinted from original 1938 edition)

Bergman, Torbern (1766). *Physisk beskrifning öfver jord-klotet, pa Cosmographiska Sällskapets vägnar författad.* Upsala: Werlds-beskrifning, 431 pp.

German translation by Lampert Hinrich Röhl (with additions by Bergman), (1769). *Physicalische Beschreibung der Erdkugel, auf Veranlassung der Cosmographischen Gesellschaft Verfasset.* Greifswald, 487 pp.

Danish translation by Johan Ditlev Bredkling Brandt (1771). *Physisk Beskrivelse over Jordkloden, paa det Cosmographiske Saelskabs Vegne Forfattet.* Oversat, og med nogle nye Tillaeg of Anmaerkninger, af Hr. Forfatteren selv tilsendte, forbedret og forøget af Johan Ditlev Bredkling Brandt. Copenhagen, 648 pp.

——— (1773–1774). *Physisk Beskrifning Ofver Jordklotet, pa Cosmographiska Sällskapets Vägnar Författad.* (2nd. ed.; 2 vols.), Upsala: Werlds-beskrifning, Vol. 1, 470 pp.; Vol. 2, 535 pp.

German translation by Lampert Hinrich Röhl (1780). *Physikalische Beschreibung der Erdkugel, auf Veranlassung der Cosmographischen Gesellschaft Verfasset* (2 vols.; Zwote vermehrte und verbesserte Auflage). Greifswald: Welt-Beschreibung, Vol. 1, 387 pp.; Vol. 2, 426 pp.

Russian translation (1791–1794).

German translation by Lambert Hinrich Röhl (1791). *Physicalische Beschreibung der Erdkugel, auf Veranlassung der Cosmographischen Gesellschaft Verfasset* (2 vols., bound together; Dritte vermehrte und verbesserte Auflage). Greifswald: Welt-Beschreibung, Vol. 1, 387 pp.; Vol. 2, 428 pp.

Carlid, Göte, and Nordström, Johan (1965). *Torbern Bergman's Correspondence: Vol 1, Letters from Foreigners to Torbern Bergman.* Prepared by Göte Carlid; revised and published by Johan Nordström, Stockholm: Almquist & Wiksell, 466 pp.

Lindquist, John (1900). Framställning af Torbern Bergman's Fysiska Geografi. Stockholm: Gustaf Lindströms Boktryckeri, 110 pp.

Moström, Birgitta (1957). Torbern Bergman, a Bibliography of His Works. Stockholm: Almquist & Wiksell, 120 pp.

Nathorst, A. G. (1909). "Carl von Linné als Geolog. In *Carl von Linné's Bedeutung als Naturforscher und Arzt.* Jena: Gustav Fischer, 86 pp.

Regnell, Gerhard (1949). "On the Position of Paleontology and Historical Geology in Sweden Before 1800." *Arkiv för Mineralogi och Geologi, bd. 1, no. 1* (1949), 1–64. (Reference received by the author after submission of paper.)

Wolf, A. (1961). A History of Science, Technology and Philosophy in the 18th Century. New York: Harper & Brothers, Vol. 2, 404 pp.

Zittel, K. von (1901). History of Geology and Paleontology. Trans. by Maria Ogilvie-Gordon, London, 1901. Reprint Weinheim: J. Cramer, 1962. Original German edition, 1899.

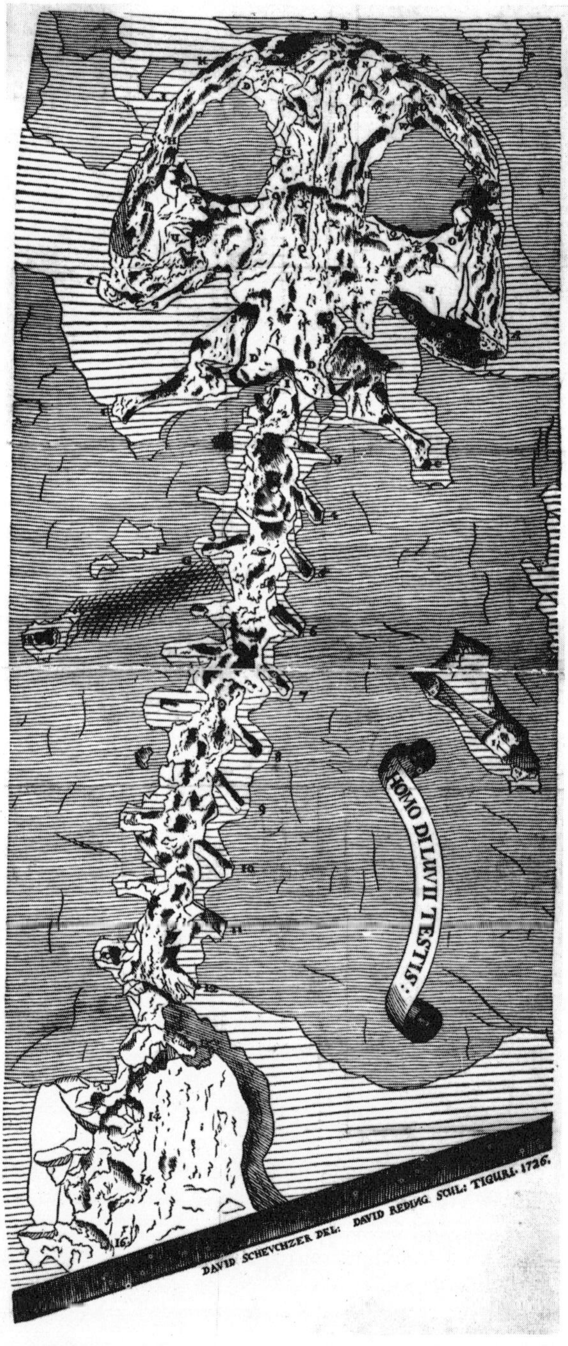

FIG. I. Homo diluvii testis *from Scheuchzer's treatise:* Homo Diluvii Testis.

Some Notes on Dr. Scheuchzer and on *Homo diluvii testis*

Melvin E. Jahn

During the late seventeenth and early eighteenth centuries, the burgeoning science of paleontology was undergoing great change. In a seemingly undirected manner, it was being refined into a secure academic discipline. From the hands of the early speculators it devolved into the stronger custody of the virtuosi — those intrepid *amateurs* whose curiosity led them into so many areas of study. It was during this middle period of development (i.e., before the advent of the professional scholar) that paleontology began to drift in many directions — seemingly without care or purpose. This interval, which continued through the mid-eighteenth century, provided an opportunity for the scrutiny of the developing science and an evaluation of its goals, its purpose, and, most of all, its theory of causes — an interval so necessary to the progress of learning. Indeed, the science had great need of the many speculations and aberrant notions to insure its very survival.

No science arrives at its final state unaided, and it sometimes seems more realistic to pay tribute to the many and few unguided speculators who dared to theorize and set their thoughts to paper than to the great synthesizers who advertently or inadvertently brought about revolutions. A revolution can only be measured in terms of its effect on contemporary thought, and few, if any, are the revolutions in paleontology. Steno, whom many consider to have brought about a revolution, did cause a great deal of stir. But was this revolution or controversy? John Woodward was highly indignant when he was accused of borrowing ideas from Steno, and on one occasion wrote heatedly to Scheuchzer on the subject:

> . . . M. Valkenier wrote to me to get Dr. Langius Elected Fellow of ye Royal Society; but I refused him, & was surprized he should make yt Request for

so unworthy a Man. I have read over his Book *De Origine Lap. fig.* & admire at his great Ignorance & Stupidity. All men here pity & laugh at him. It would not be amiss if you made some brief Reply to him. I admire at his Impudence in Suggesting p.3. yt I took any thing from N. Steno, or Ag. Scilla. We hardly agree in any Thing. If you please I will send you a Collation of our 3 Hypotheses, that you may ye better judg of that Matter.[1]

If one admits that a "revolution" is little more than a temporary "final" synthesis of the positive *and* negative sides of earlier theorizers, the mistake in employing this term of convenience untempered becomes apparent. As such, any "scientific revolution" exists only in retrospect — our retrospect.

To the diluvial school of paleontology in the seventeenth and eighteenth centuries is often assigned the honor of first recognizing fossils as organic remains. This assignment, while seemingly meaningful, is misleading. Any attempt to determine which theories supported the organic origin of fossils is basically thankless — it is like arguing whether Christ changed water into red or white wine at the dinner at Cana. Strange as it may seem, there are no firm criteria on which to base such a judgment. The preformationists, for instance, most definitely recognized fossils as organic remains — being productions from viable seed happily deposited by any one of several methods in a nurturing (albeit an especial) earth. If such an argument be recognized, how, then, do we distinguish (and the distinction is a fine one) among the theoretical assertions of various schools? Where, then, does one set the boundaries of the organic and the inorganic hypotheses?

While these thoughts may appear to be unnecessarily contentious, they are important to the final establishment of an organic theory even remotely similar to that accepted today. It is through the complex interaction of preformation and the so-called inorganic theories, buffeted by the often logically arbitrary diluvialists, that our present notions on the origin of fossils came into being.

Biographical Notes

Johann Jacob Scheuchzer (1672–1733), famed defender of the diluvial hypothesis and Woodward's theory of the earth, is remembered today primarily as the discoverer of *Homo diluvii testis* — the Man who Witnessed the Flood (Fig. 1). He is known to the literary public through Karel Căpek's novel, *The War with the Newts* — the central figure of which is *Andrias scheuchzeri.* As a figure in the history of science, his contributions and worth have been vastly underrated. Scheuchzer possessed the restless intellect and pervasive sort of genius that char-

[1] Letter from John Woodward to Scheuchzer, dated July 3, 1710. Zentralbibliothek Zurich, MsH 294, pp. 212–213.

acterized many of the virtuosi. His work in many areas is that of an innovator, but an innovator with a degree of perception and a sense of intuitive method not often found.

Scheuchzer was born in Zurich, August 2, 1672,[2] the son of Johann Jacob Scheuchzer, the Resident Physician, and his wife, Barbara Fäsi. Although he attended the local gymnasium, it was to his father that Johann Jacob owed his interest in natural history and medicine. Following his graduation from the gymnasium and preparatory medical studies with Johann Jacob Wagner and Johannes von Muralt, he entered the University of Altorf (1692), where he came under the influence of Johann Christoph Sturm, his teacher in mathematics and physics. A year later he moved to Utrecht, where he took the degree of Doctor in Medicine in 1694. The next two years were occupied with travel and study; most especially, he visited museums and private collections to widen his knowledge of natural history. On his return to Zurich in 1696, he secured the position of Second Resident Physician (*Poliater*),[3] and was promised that when next vacant he would receive the professorship of mathematics — a position he obtained in 1710. The year following his appointment as *Poliater* (1697) he married Susanna Vogel; this union produced, through the years, eight children, seven sons and one daughter.

Virtuoso endeavor began more and more to occupy Scheuchzer's thoughts and time in the years following 1694. In addition to holding private classes in the natural sciences for students at the university, he began his work in geography and geology. He was also taken into (or perhaps helped to found) one of the hebdomadal welfare clubs so dear to the seventeenth and eighteenth centuries. Very little is known of the history of the Society of the Benevolent (*Gesellschaft der Wohlgesinnten*), even though records of many of its meetings survive at the Bibliothek Basel. These records, kept by Scheuchzer as secretary, contain lists of the questions aired and discussed after each lecture. Regrettably, no mention is made of the results of these debates — of which the subjects alone give so excellent an insight into contemporary culture:

Whether comets are omens of divine punishment or world catastrophies. Whether a creature born with two heads has two souls or should be regarded as two individuals. Why the Moors are black. Whither the storks and swal-

2 Rudolf Wolf, *Biographien zur Kulturgeschichte der Schweiz* (Zurich: Druck und Berlag von Dress, Futzsi & Comp., 1858–1862). (Scheuchzer: Vol. I, pp. 181–228). Excepting correspondence, this work was the principal source of biographical notes.

3 The senior Resident Physician was the *Archiater*, the second, *Poliater*. Both were paid by the *Chorherrenstift* and were required to perform services for the city and foundation.

lows migrate for the winter. Whether the magistrate should allow in good conscience public theatres, rope acts, etc. Whether the devil could seduce a female physically. Whether Judas committed a greater sin in betraying Christ or in killing himself. Why seedlings are damaged so much by rain falling on top of dew and especially May's dew. Whether Atheism necessarily causes corruption of morals. Where Christ found garments after His Ascension. Whether the mountains had been created in the beginning of the world or had developed during the deluge or through other influences.

His efforts for the Society of the Benevolent, if nothing else, convinced him of the necessity of cooperation with other scholars and societies in order to achieve constructive results in chosen areas of study. To this end, he issued in 1697 the first of his many sets of queries.[4] In the invitational letter, he requested answers to some 200 questions, particularly concerning geographical, climatological, and physical phenomena in Switzerland and their effects in various parts of the country.[5] The response to so lengthy a set of queries was discouraging, and it was three years before he undertook the publication of two shorter sets of instructions for making observations in natural history. One of these contained an almost definitive international directory of scientists who concerned themselves with natural history.[6] Finally, in February 1705, he began to publish, at his own expense, a weekly paper: *Seltsamer Naturgeschichter des Schweizerlandes wochentliche Erzehlung* (*Weekly Tales of the Curious Natural Phenomena of Switzerland*).[7] This small journal does not seem to have had any formal connection with the Society of the Benevolent (whose primary concern was with the development of a sociopolitical consciousness), but rather to be an outgrowth of Scheuchzer's desire to make the results of his researches known to his countrymen. There was also the hope that this means of communication might call forth data denied by the necessarily formal nature of the earlier queries, thereby further widening his knowledge of the nature, culture, and history of Switzerland.

4 Scheuchzer, *Charta Invitatoria, Quaestionibus, quae Historiam Helvetiae Naturalem concernunt, praefixa* ([Tiguri: Davidis Gessneri, 1699.])

5 Also included were inquiries concerning phenomena of a more general nature: longitude and latitude; the prevalent winds and diseases; atmospheric temperatures and their daily and annual variations; thunderstorms and earthquakes; magnetic declination and its variations; unusual minerals, plants, and animals; about fossils, glaciers, and springs; about the length, width, and depth of rivers and lakes; and so forth.

6 Scheuchzer, *Historiae Helveticae Naturalis Prolegomena* (Tiguri: Davidis Gessneri, 1700); *Stoicheiologia ad Helvetiam Applicatam Placidae Eruditorum Censurae submittunt Praeses* (Tiguri: Davidis Gessneri, 1700).

7 The journal included stories of all varieties, without any particular system or order, taken from all fields of natural science. Among the myriad of subjects treated are mountains, animals, plants, physics and meteorology, geography, and alpine agriculture.

Scheuchzer's interest in the past and present state of Switzerland led him to undertake many walking tours, usually accompanied by several of his students. In addition to geological and mineralogical observations, particular attention was paid to climatological phenomena, meteorological observations, and altitude measurement. Scheuchzer was, in fact, the first to carry physical instruments to the Alps and to record barometrical altitude measurements. His travels led to the publication, in 1712, of a detailed map of Switzerland,[8] a map that remained authoritative until the beginning of the nineteenth century. This was the first map to determine the exact geographical location of Zurich. He became, through the years, the principal Swiss correspondent of many learned societies and was chiefly responsible for the Swiss meteorological data published in scientific journals.[9]

Although Scheuchzer's scientific activities embraced many fields of inquiry, "fossilology" became his primary area of interest. His first paper on figured stones was published in 1697.[10] In it, he expressed the belief that their origin could be traced to the laws of physics and chemistry — that a fossil (or, more properly, figured stone) was, in fact, nothing more than a *lusus naturae*. In a larger work published in 1702,[11] he returned to the subject of fossils but submitted no theory regarding their origin — a fact that did not escape the notice of John Woodward:

> . . . I may hint that to you by the by, several Gentlemen of ye Royal Society much admire that at this time of day you should haesitate as yw there do (in your Dedication & Praeface, & in the Book also pag. 30.) whether the Bodyes yw treat of be Shells brought out by the Deluge, or lusus of Nature. They think that that is now no longer a Dispute, but set in a clear Light so that any reasonable man must acquiesce & be satisfyd.[12]

It is generally accepted that Woodward's *Essay toward a Natural History of the Earth* (1695) had begun to influence Scheuchzer's thought, and that he was, at the time he wrote *Specimen lithographiae Helveticae*

[8] Scheuchzer, *Nova Helvetiae Tabula Geographica, Illustrissimis et Potentissimis Cantonibus Et Rebuspublicis Reformatae Religionis* ([Zurich, 1712–1713].)

[9] Among the journals to which he regularly contributed such data were *Sammlung von Natur- und Medicin-Geschichten* and the *Academie des Sciences*, Paris, *Histoire*.

[10] Scheuchzer, "De Generatione Conchitarum . . ."*Miscellanea Curiosa Medico-Physica Academiae Naturae Curiosorum*. Ann. IV, Dec. III (1697): Appendix, pp. 151–166.

[11] Scheuchzer, *Specimen Lithographiae Helveticae Curiosae, Quo Lapides ex Figuratis Helveticis Selectissimi Aeri incisi sistuntur & describuntur* (Tiguri: Davidis Gessneri, 1702).

[12] Letter from John Woodward to Scheuchzer, dated October 2, 1703. Zentralbibliothek Zurich, MsH 294, p. 101.

curiosae, grappling with the question of diluvialism. As Woodward himself remarked many years later:

> . . . Of the many I could name I shall mention only one, whose Authority is equal to that of Many, I mean Dr. Scheuchzer, a Person of distinguish'd Parts and Judgment, consummate Learning, and who is deservedly ranked among the first Naturalists of Europe. He publish'd in the Year 1695 [1697] a Dissertation De Generatione Conchitarum, wherein he endeavours to prove that these Bodyes ought to be reputed native and genuine Fossils. But afterwards, upon a carefull Perusal of my Book, he publickly acknowledged his Mistake; confessing he had too hastily embraced that Opinion. Thereupon, as became a sincere Labourer in the Cause of Truth, he gave up his own, and came over to my Sentiments: and the many learned Works, wherein he has from that Time asserted and demonstrated his other Writings, abundantly shew the great Progress he has made in these Studies.[13]

In 1704, Scheuchzer published a Latin translation of the *Essay*. The dedication (to Woodward) is a confession of his indebtedness and a modest panegyric in its own right:

> The writing which is yours, quite properly returns to you. But, with your gracious permission, it has laid aside its English garb for the Latin Toga, bequeathed to it by Rome, and indeed by the whole Society of Letters. No longer is your work restricted to the shores of England; the scope of its services is expanded to the whole world. Your gentle spirit, as well as the bond of our intimate friendship, restrains me from heaping praises upon this work. I prize the role of admirer more highly than that of trumpeter. I leave it to your readers to render judgment when they have weighed your cogent arguments on the honest scales of their minds. This only will my candor not suffer me to cloak in silence: that I hold my place, though it be the lowest, in the ranks of those in whom the accuracy of your experiments and their sound superstructure of reasoning, have engendered so strong a conviction, that they elect to lay the palm of victory at the feet of Truth, and would rather abjure and forsake the hypothesis — however widely favored and publicy defended — which holds the earth to be the place of origin of many figured stones, than to cling tenaciously to prejudices once firmly rooted. Whatever effort I have expended in the task of interpreting this admirable specimen of your major treatise, which honorable men anticipate in all earnestness, I tender to you as a public acknowledgement of your good-will toward me, and a pledge of my devotion to you, begging you to receive it as affectionately as you do myself.[14]

Four years later, he presented to the learned world his great polemic, *Piscium Querelae et Vindiciae*.[15] In the introduction to the German version of the work (published as *Bildnissen verschiedener Fischen . . .*) he wrote:

[13] John Woodward, *Natural History of the Earth, Illustrated, Inlarged, and Defended* (London: Tho. Edlin, 1726), pp. 5–6.

[14] Scheuchzer (translator), *Specimen Geographiae Physicae* (Tiguri: Davidis Gessneri, 1704).

[15] Scheuchzer, *Piscium Querelae et Vindiciae* (Tiguri: Gessnerianis, Sumtibus Authoris, 1708).

The learned and unlearned World is full of opinion about the petrified shells, snails, fish, crabs and plants which are dug up here and there and exhibited in cabinets of curiosities. One says that these stones are but a whim of nature, the next has different ideas, as the origin should be a sort of brood oven in the earth's insides. I thought at one time that they were brought forth by nature according to her own laws, without any male-female help. However, after collecting a great number of such fossils, my eyes were opened and now I see their origin in the Biblical Deluge. . . . [The fish], which are found here and there in quarries and mines, are not bastards of nature but belong to the animal kingdom.[16]

Scheuchzer's fossilizing in the Alps led to remarkable discoveries — discoveries which he duly recorded in his many publications. A volume on fossil plants appeared in 1709[17] and a more general treatment of petrifactions in 1716.[18] As his writings on the deluge and its productions proliferated through the years, so did his correspondence with naturalists throughout the world. Among his paleontological correspondents, few were as important to Scheuchzer as John Woodward. Their exchange of letters began in 1701 and lasted until shortly before Woodward's death in 1728. Woodward, himself an ardent diluvialist, encouraged Scheuchzer's growing preoccupation with the deluge. The cause for this progressive reliance on the Noachian flood is not, perhaps, as obvious as it may at first seem. Scheuchzer was a thorough naturalist and beyond any doubt recognized fossils as organic remains. Nevertheless, he was by no means free from the foibles of the eighteenth century. For the most part, his work on fossils was methodical and perceptive; his descriptions were accurate and, for the time, succinct. Still there remains his *bête noire*: the deluge. Cuvier himself observed (apropos *Homo diluvii testis*): "Nothing less than total blindness on the scientific level can explain how a man of Scheuchzer's rank, a man who was a physician and must have seen human skeletons, could embrace such a gross self-deception."[19] But Scheuchzer's quest for *Homo diluvii testis* must not be allowed of itself to obscure his many contributions to learning—neither those to science nor those to letters.

For several years preceding 1710, Scheuchzer was occupied with research into the history of Switzerland and Swiss literature. His studies of historical documents and deeds (especially those regarding the Toggenburg War) led to his being called to the Assembly during the poli-

16 Scheuchzer, *Bildnissen verschiedener Fischen* (Zurich: In Verlegung des Authoris, 1708).
17 Scheuchzer, *Herbarium Diluvianum* (Tiguri: Davidis Gessneri, 1709).
18 Scheuchzer, *Museum Diluvianum* (Tiguri: Henrici Bodmeri, 1716).
19 Cuvier, *Recherches sur Ossemens Fossiles . . . Troisième Édition* (Paris: Chez G. Dufour et E. D'Ocagne, 1825), p. 431.

tical reform movement of 1713.[20] They also led to a *Chronological History of Switzerland* (30 folio volumes in manuscript), a work that was never published due to a chronic shortage of time and funds, as well as to the threat of censorship. His various enterprises had begun to be remarked upon, and the situation between Scheuchzer and his colleagues, most notably the theologians, became very disharmonious. In his first lecture as Professor of Mathematics, he endeavored to show his theology students the importance of mathematics and natural science in relation to the Bible.[21] This principle he reiterated in the Introduction to the *Physica Sacra*: "Since the early days in the Mathematics-Physics lectures in the Gymnasium, I have sought to explain the text of the Holy Bible according to the principles of the newer philosophy and the natural sciences."[22] The study of natural science had been discredited, and the Ptolemaic System defended from the pulpit. Scheuchzer's position, as a determined Copernican, became progressively less tenable, and the antagonism against him more open, so open that a Zurich official could recount:

> . . . Dr. Scheuchzer had a white crow, which flew away and landed on the Baptisten's roof. Dr. Scheuchzer climbed barefoot on the roof, talked to the crow and caught her. Suddenly he slipped and came to rest against the drainpipe. He caught himself and climbed back into his house, but did not let go

[20] Scheuchzer was called by the citizenry to the Assembly in 1713, to aid in the reform of the government. He served as leader of the Citizen's Committee (*Bürgerausschüsse*), a body which communicated directly with the deputies of the Assembly. Wolf, Biographien, Vol. I, pp. 220–221.

[21] The text of this lecture (delivered on either November 26 or 28, 1710) was published in 1711 as *Praelectio De Matheseos Usu In Theologia* (Tiguri: Joh. Finsleri, 1711). Scheuchzer sent a copy of the publication to Woodward with some mention of the negative reception it had received, for in a letter dated September 28, 1711, Woodward remarks:

> . . . I read your Praelectio de Math. usu in Theol. as I do every Thing of yours, with great Satisfaction. But I do not wonder that the Divines should not care to have some of Their Notions brought to the Test of Mathematics. T'would be greatly to their Honour & Advantage that they Studyed Nature and Mathematicks. But tis not an easy thing to put them out of their Way. (Zentralbibliothek Zurich, MsH 294, pp. 225–226.)

[22] Scheuchzer, *Physica Sacra*. (Augustae Vindelicorum & Ulmae, 1731). Praelectio, leaf a. Wolf, in his biography of Scheuchzer (*Biographie*, Vol. I, pp. 208–209) records the following note by one of Scheuchzer's students (Joh. Jakob Zimmerman):

> . . . Mathematics was taught by the famous Joh. Jakob Scheuchzer, a man of great diligence. He read in his public lectures not about algebra or mathematics, but *Loca scripturae physica*, or about the Copernican System, or Spheres in general. In mathematics he did less. He knew the opinions of the learned physicists and often read from the writings of the learned men of Paris, London and Prussia. Thus his philosophy of physics was more historically inclined than it was concerned with principles.

of the crow in his hand. It is known that, had he killed himself, the canons would have given a feast for the crow.[23]

In December 1712, Scheuchzer received a letter from Leibnitz, who wrote on behalf of the Czar of Russia, Peter the Great. The letter invited him, on the recommendation of Leibnitz, to accept the position of Court Physician.

> . . . Your enlightened knowledge of Nature will doubtless render you most acceptable to this eminent Prince, should you be willing to enter his service, and will compensate you beyond what has been promised you. Hence, if it is compatible with your family affairs, I am sure that in accepting this position, you will advance both your best interests and your satisfaction. Besides being a very honorable engagement, it will offer you opportunities to perform signal services for the world of letters through the discovery of countless magnificent products of nature and art in a sort of virgin territory which has not so far been explored by scholars.[24]

A counteroffer was soon made by the Zurich Municipal Government. The increased emolument offered by the government and Woodward's vehemence against the "uncertain Capriche" of the "Czar of Muscovie"[25] persuaded Scheuchzer to remain in Zurich. The promises made him in 1714 were not fulfilled until 1733. He was made Professor of Mathematics and Physics and promoted to the highest rank of Resident Physician. These new circumstances he enjoyed only for a short time. He died on June 23, 1733, at the age of 61. Bourguet, his friend and defender, wrote in his obituary:

> Dr. Scheuchzer was of medium height, slight of build, but of sharp and prominent features. His disposition was buoyant, his conversation pleasant and witty. He had mastered several languages, some of which he spoke quite fluently. His erudition was universal in scope, his memory tenacious and accurate, his judgment clear and solid. His mind was very penetrating. He loved truth above all else, and his judgment of the various opinions and divergent persuasions of men in general, but of scholars in particular, was sound and impartial; a character trait which is the more estimable that it is extremely rare. His manners were impeccable, and he was honest and gen-

23 Letter from Landschreiber Gwerb to Landvogt Füssli, dated July 6, 1714. Wolf, *Biographien*, Vol. I, p. 225.

24 Wolf, *Biographien*, Vol. I, p. 226.

25 Letter from John Woodward to Scheuchzer, dated January 20, 1712 [1713]. Zentralbibliothek Zurich, MsH 294, p. 229:

> . . . Your Letter of Decemb. 16. 1712 came Safe to my Hands. The Czar of Muscovie is a Prince very capricious & unsteady: & I am averse to Hazard ye Fortunes of a Man of your Character & Worth, by Subjecting them to his uncertain Capriche. Besides ye Pension he offers is not near sufficient: & I take his Affairs to be in an unsetled Condition; so yt you can have no Security of any Thing there. His inveterate Enemy, ye K. of Sueden, will never leave him without a Blow.

erous, which did much to endear him to those who knew him. These and other enviable qualities of mind and heart appeared to the best advantage in him and in a manner becoming to a Christian philosopher whose piety is the equal of his intellectual stature. His practice of medecine was performed with a high degree of wisdom and circumspection. As a result, it was most frequently successful. His correspondence with men of letters was unusually extensive, and there were very many who received his letters with unfeigned pleasure. Dr. Scheuchzer has bequeathed his family a selective library, an impressive collection of medals, and a bounteous cabinet of curiosities of nature — especially fossils, and more specifically such as are related to the Flood.[26]

That Scheuchzer was a man of "indefatigable industry and extensive knowledge" must by now be apparent. That he was also beset by an adamant preconception is equally apparent. While it is necessary to transcend this preconception for proper evaluation of his work in a larger sense, it is just as necessary to understand the preconception itself — to know what was written and what was ventured. Fossilizing was very popular in the seventeenth and eighteenth centuries, and this led to the inevitable speculation — often (to us) with unhappy results. These theoretical peregrinations are vital to the progress of any branch of science. To the modern scientist, Scheuchzer's failing lay not only in his acceptance of the Noachian deluge as being responsible for fossils, but in allowing this acceptance to prejudice his observations regarding a fossil salamander. Certainly, as Cuvier remarked, this was a grievous error. But is it reasonable on our part to negate the value of Scheuchzer's many studies (including the writings on fossils) because of *Homo diluvii testis*, any more than it would be reasonable to set aside his studies merely because they issued from the mind and pen of a diluvialist? To do so would itself be tantamount to owning to a preconception meriting note. Very little has been written about Scheuchzer, and the little that has is largely confined to his anthropolith. In justice to Scheuchzer, it must be said that very few twentieth-century writers have scrupulously examined his two most remarked-upon works: *Piscium Querelae et Vindiciae* and *Homo Diluvii Testis*. It is hoped that the very brief biographical notes presented here will revive interest in a worthy scholar, and that the inclusion of primary source material will answer some of the flood-man questions raised in secondary sources.

Homo diluvii testis

Scheuchzer announced the discovery of parts of the skeletons of two petrified humans in a letter to Sir Hans Sloane (dated at Tiguri, December 25, 1725). The letter was printed in number 392 (January–February,

[26] Wolf, *Biographien*, Vol. I, p. 228.

1726) of the *Philosophical Transactions* (*XXXIV*, pp. 38–39). Similar descriptions appeared in the *Journal des Sçavans* (*LXXXI* [January 1726], pp. 24–28), *New Memoirs of Literature*, London (*III* [June 1726], pp. 440–442), and the *Sammlung von Natur- und Medicin-Geschichten* (*XXXII* [April 1726], pp. 406–408).

Late in 1726, Scheuchzer enlarged upon these papers and issued a formal declaration of the discovery of the Noachian flood-man in a short treatise (24 pages), *Homo Diluvii Testis*, which allegedly[27] concludes with a couplet that has outlasted the work:

> *Betrübtes Beingerüst von einem altem Sünder*
> *Erweiche, Steine, das Herz der neuen Bosheitskinder!*
>
> Afflicted skeleton of old, doomed to damnation
> Soften, thou stone, the hearts of this wicked generation!

The treatise itself was anticipated by a broadsheet containing a single column of text and a plate of the fossil: *Homo Diluvii Testis. Beingerüst, eines in der Sündflut ertrunkenen Menschen.*[28] Scheuchzer returned once again to the subject of his "ancient sinner" in his *Physica Sacra* (1731), in which he largely repeats the earlier description published in the *Philosophical Transactions*, but with the addition of a plate of the famous fossil (Vol. I, plate XLIX; see Fig. 2).

Homo diluvii testis remained firmly ensconced in paleontological literature until, in 1787, the famed anatomist Petrus Camper visited Oeningen, where he studied many of the fish fossils in Scheuchzer's museum (described in *Piscium Querelae et Vindiciae*), as well as his Noachian flood-man. Camper at once detected the error into which Scheuchzer had fallen. He communicated his findings to François Xavier Burtin, who included an abstract of Camper's letter in his "Response à la Question Physique, Proposée par La Société de Teyler,

27 I use "allegedly" here because two well-known sources of the Scheuchzer-*Homo diluvii testis* story — Louis Figuier and Herbert Wendt — have included the couplet in their books. Herbert Wendt says, for example:

> . . . Even in this critical hour Scheuchzer could not resist moralizing. He added a couplet that has outlasted the centuries. It was often quoted at student's gatherings, and has won a place for the father of prehistoric research in the history of unconscious humor. (Wendt, *In Search of Adam*, p. 16.)

My copy of *Homo Diluvii Testis*, however, does not contain the couplet. Nor have I been able to locate it in either German or Latin in any of Scheuchzer's works.

28 I have seen two reproductions of the broadsheet, one in Wendt's *In Search of Adam* (plate 1) and another in William Elgin Swinton's *The Wonderful World of Prehistoric Animals* (p. 18). There is no copy of the broadsheet in the Zentralbibliothek Zurich, nor is it cited in Rudolf Steiger's bibliography of Scheuchzer's works (*Naturforschende Gesellschaft*, Zurich. *Vierteljahrsschrift*, *LXXVIII* [1933], *21*: *Beiblatt*, pp. 1–75). The plate is the same cut as that accompanying the treatise, and the text (which is in German) is of a very general nature, concluding with a brief anatomical description of the fossil.

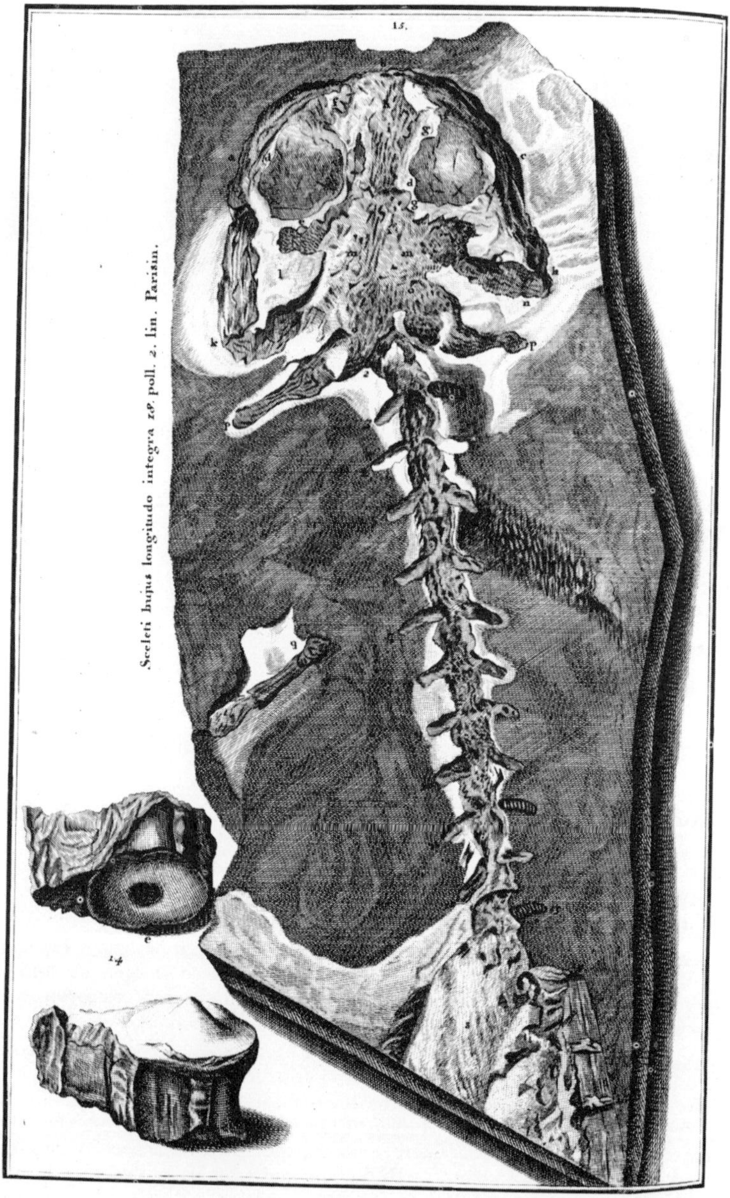

FIG. 2 Homo diluvii testis *from the first volume of Scheuchzer's* Physica Sacra (1731).

Sur les Revolutions Generales."[29] Camper himself erred in his conclusions on the nature of the fossil, for he concluded that the remains were those of a lizard: ". . . mais comme un lézard pétrifié a pu passer pour un anthropolithe. . . . " [pp. 35–36]

It remained to Georges Cuvier to place the fossil in its proper class. In his *Recherches sur Ossemens Fossiles*, he devoted an entire article to *Homo diluvii testis:* "Sur le prétendu Homme Fossile des carrières d'Oeningen, décrit par Scheuchzer, que d'autres naturalistes ont regardé comme un silure, et qui n'est qu'une Salamandre Aquatique [*Cryptobranchus primaevus* or *Andrias scheuchzeri*] de taille gigantesque et d'espèce inconnue."

To trace effectively the history of the famous "anthropolith," it is necessary to go back to 1708, in which year Scheuchzer first came upon the *reliquiae diluvianae* of man. This initial discovery consisted of "a pair of petrified, human dorsal vertebrae" found at the foot of the gallows in Altorf:

> . . . Last to appear are a pair of petrified, human dorsal vertebrae from the Altorf Field in the Territory of Noremberg. Their substance is almost wholly stone, shiny black, and the structure of the bony fibres is beautifully delineated. In their entire form and in the arrangement of their processes, they have the exact specification of vertebrae of the human back. I found not only this pair, but six others joined in series with these, within the walls of the gibbet of Altorf when, on one occasion I journeyed there in the zeal of my devotion to stones, accompanied by that Palatine Myrtelus and Faithful Achates of my excursions, Meister Langhans. . . .[30]

The story of the finding of the vertebrae at the foot of the gallows has been told many times; Cuvier himself gave a very brief version of it in his *Recherches sur Ossemens Fossiles* (Vol. V, p. 451). Unfortunately I have been unable to trace the origin of this story — even the identity of Scheuchzer's "Faithful Achates," Langhans, remains a mystery. From several very fragmentary remarks by Scheuchzer and the account given by Cuvier (evidently the source of succeeding versions), the following emerges.

Scheuchzer and Langhans were on a walking tour through the environs of Altorf, searching, it is supposed, for diluvial relics. One of the pair — it is not certain which, for Cuvier says Langhans, while Scheuchzer says it was himself — found near the foot of the gallows a string of eight vertebrae in a piece of ash-colored marble. Whether the

[29] *Teyler's Tweede Genootschap*, Haarlem. *Verhandelingen, VIII* (1790), pp. 1–242, 243–389. Camper's notes appear on pp. 35–36 (French) and pp. 264–265 (Dutch).

[30] Scheuchzer, *Piscium Querelae et Vindiciae*, pp. 22–23.

vertebrae were found by Langhans or by Scheuchzer, it was Langhans who, as Cuvier says, "in a seizure of panic," hurled the stone over a wall, where it shattered. Scheuchzer retrieved all that he could — two single vertebrae — and had them engraved. These he figured in plate III of his *Piscium Querelae et Vindiciae*. The two vertebrae were also figured with the plate of *Homo diluvii testis* in the *Physica Sacra*. There is an obvious discrepancy here, for in *Piscium Querelae et Vindiciae*, as well as the paper contributed to the *Sammlung von Natur- und Medicin-Geschichten*, Scheuchzer tells us that of the eight vertebrae he retained two for his museum and presented the six remaining to a friend, whom he does not name.

Very little is known of the finding of the more complete skeleton of *Homo diluvii testis*. Scheuchzer tells us only that the skeleton was discovered in an Oeningen quarry and was presented to his museum. Herbert Wendt, in his book *In Search of Adam*, gives a fanciful account of the discovery, for which, while possibly accurate in some particulars, he cites no documentary evidence.

Following, then, are two of Scheuchzer's four papers heralding the discovery of *Homo diluvii testis* — the Man who Witnessed the Flood. Both have been translated from the Latin. The communications to the *Journal des Sçavans* and *New Memoirs of Literature* have been omitted, as they are nearly identical to the paper published in the *Sammlung von Natur- und Medicin-Geschichten*. These are followed by an abridged translation of Cuvier's article on *Homo diluvii testis* from the third edition of his *Recherches sur Ossemens Fossiles*. Scheuchzer's papers contain cursory references to a number of authors and works which require more complete citation. Rather than clutter the text with insertions in brackets, a short bibliographical list has been appended at the end of this paper.

"A Portion of two petrified human skeletons. From a letter of Johann Jacob Scheuchzer, M.D., F.R.S., to Sir Hans Sloane, Baronet, S.R.V.Pr." Royal Society of London, *Philosophical Transactions, XXXIV*: 392 (January–February, 1726), pp. 38–39.

You are aware, most learned Sir, that few remains of the race of men submerged in the Deluge are extant. Nor had I hitherto, in my quite extensive collection, more than a pair of petrified dorsal vertebrae, of shiny dark quality. Now, however, by a stroke of good fortune, there has been added to my modest museum a fragment set in fissile stone from Oeningen wholly deserving of thorough scrutiny, in which one can plainly discern, not the figments of capricious imagination, but numerous parts of the human head — genuine residuaries distinguishable from any other class of animals: the brain case, the frontal bone, the sincipital and occipital bones, the orbits, the base of the brain and the remains of the medulla oblongata, the interior prominence of the occipital bone which separates the lobes of the cerebellum, neck vertebrae (seven in all) partly bared, partly covered by a stony hull, all of this being, as it were, a frontal section of the posterior area of the head.

But no sooner had I acquired this monument than by a remarkable favor of Providence, a new guest from the aforementioned quarry arrived — the dearer that it is the rarest, dearer still that it has no need of clothing or provisions, coming well supplied with all such, and surpassing the former specimen in size, age and intriguing features. Here, immersed in the same sort of fissile stone, we have from an adult, human skeleton, or more exactly from its anterior structure, the periphery of the frontal bone, the malar bone, the orbits, the overlapping panels [*cranii tabulae cum diplöe*], vestiges of the infraorbital foramen (serving as the passage for the fifth pair of nerves), the remnants either of the brain itself or of the dura mater, cribrose and spongey bones, the vomer (which divides the nose), a portion of the fourth maxillary bone (which is the cheek bone), bits of the nose, a section of the masseter, (the frontal section passing through the condyloid apophysis of the lower jaw to the angle of this jaw), vertebrae in continuous series (sixteen in all, several of them complete with transverse processes), the extremity of the right clavicle (which is connected to the scapula), and the central part of the left clavicle enclosed in the stone. From the proportions of this skeleton, I estimate that the total stature of the man approximated my own: 58 and one-half Parisian thumbs.

"The Remains of a Man Drowned in the Flood," *Sammlung von Natur- und Medicin-Geschichten, XXXII* (Fruhlings Quartal, April 1726), pp. 406–408, 1 plate.

The truth of the Universal Deluge, acknowledged for many centuries, was never more manifest than in our present age, with its many remains of silted and petrified animals and plants — preserved from that age to our own time. Nevertheless, and surprisingly so, one chances upon nothing, among all these remains, so rarely as those of man himself. It is as though the fiery wrath of God had willed to consume the principal among the many causes in the corruption of sin-sodden men, so that little or nothing of their bodies should remain. Consequently it will be taken for a very great rarity if, of such human remains, something really worthwhile should be drawn out of the depths of the earth. We therefore receive most favorably this communication, one of many which the industrious and accomplished Swiss Physicist, Herr Professor Johann Jacob Scheuchzer, has authored. The Latin account, which has been sent to us, reads as follows:

In our day we have as many witnesses to the Deluge, wherein the primaeval world and its inhabitants perished, as there are provinces, cities, hamlets, mountains, valleys and quarries of the modern world. On another occasion in my *Herbarium Diluvianum*, I afforded Academicians a systematic conspectus, based on Tournefort's classification, of the numerous plant groups. In their "Grievances and Claims" [reference is made to his book *Piscium Querelae et Vindiciae*, Tiguri, 1708], the Fishes, notwithstanding their native muteness, championed their own cause. The celebrated Woodward has epitomized the entire system in his *Geographia Physica*, which I translated from the English. I myself have compiled a *Lexicon of Diluvial Antiquities* [never published], which not merely nominally but objectively assigns these findings to their proper genera and species — along with all the synonyms I have been able to discover. This work, still nestled among my manuscripts, is the companion and key to my exceptionally complete and trustworthy *Museum Diluvianum.* Increasingly these relics are being amassed into collections as numberless and boundless as swarms of ants. The testacea are very numerous, while the insects are only sparsely represented. To the present, very few remains of men drowned in the Flood have been found. Perhaps the reminders of such innocent creatures as plants, testacea, fishes and even lowly insects are more abundant because these are more deserving of remembrance than men — all of whom, except some (though not all) of Noah's family, had taken to paths of infernal, as well as physical, corruption, and had richly deserved to be consigned to eternal oblivion. These human corpses floated on the waves and their flesh rotted. As a consequence we cannot always, or even easily, judge whether the bones that have been submerged are human. Many persons have been astonished at the two human dorsal vertebrae which I found immersed in extremely hard rock in the field of Altorf — vertebrae which were lustrous black and petrified, and which measured up exactly to the specifications of dorsal vertebrae in the structure of their osseous fibers, the position of their processes and in their overall form. They were joined to six others which I presented to a friend. I am not so convinced of the genuinity of *The Granite Man* (Worm *Mus.*, p. 82), of *The Stony Human Foot, The Greaved Foot, The Stone Knee,* all in the Calceolarian Museum, pp. 412 and 417; or of *The Two Bones of a Human Leg* found in a vein of iron and now exhibited in the Grew Museum [actually the Royal Society Museum], or finally, of the *Cheir* or *Chir:* an ashen stone bearing the imprint of a man's hand, in Besler's Museum, p. 96, Tab. 34, and another very similar in form, dug up near the Blankenstein Arch, as described in Sachs' *Gammarologia,* p. 159. It would be well to point out here that in this area of study, a cloud is not infrequently mistaken for Juno. An example in point is *The Stone Teeth of Giants,* commonly alluded to by naturalists and displayed in most museums — though actually they belong to some quadruped, possibly an elephant, rather than to man. Returning to my vertebrae, should someone contend that these are Ichthyospondyls, I would decline to join issue with him, were he first to show me that they are more congruously associated with the back of a fish than with that of a man. Quite otherwise is the case of the stone presented to my museum a short time ago — a stone which incontrovertibly pertains to the family of man. I here offer an illustration of this exceedingly rare relic, and a brief description of its features.

We have here not merely some vague, general outlines which a fertile imagination could sketch into a head — much as we fancy castles, walls, ruins, crosses, mountains, meadows, groves, moss, letters, etc. in Florentine marble, in agate, jasper, topaz, dendrite, granite, especially if these are inscribed — but rather parts which are as similar to parts of a head as one egg is to another, or as the concave diluvial cochlea is to its marble counterpart. What is more, the residual parts visible in the stone are not at all of the same substance as the stone itself, but altogether different in color and solidity — indeed, though it is almost beyond belief, they are the hardened, genuine remains of the brain. Besides these, there are, in this same Oeningen quarry, remains of plants, leaves, conches, cochleae, insects, crabs and fishes — along with the true relics of vertebrae, bones and flesh, all of which I have treated at some length in my *Herbarium* and in the *Grievances and Claims of the Fishes.* I venture to say that no-one who would direct even a passing glance, not to say an attentive study, upon this stone would fail to recognize it as a real and authentic relic of the Flood, for this most unusual handiwork of Nature will meet even the exactions of the Anatomist. Here, then, are the particular and specific features which distinguish this head from that of any other living thing (Fig. 3):

ABCDEFG. The circumference of a child's cranium.
CDE. The frontal bone.
CB. The left bone of the syncipit.
EF. The right bone of the syncipit.
ARGF. The occipital bone.
HIK. Intropressed portion of the frontal bone.
LMN. Another portion of the frontal bone (on the right side) with the orbit.
OP. Remains of the base of the brain and of the medulla oblongata.
PQRS. Interior prominence of the occipital bone, separating the lobes of the cerebellum.

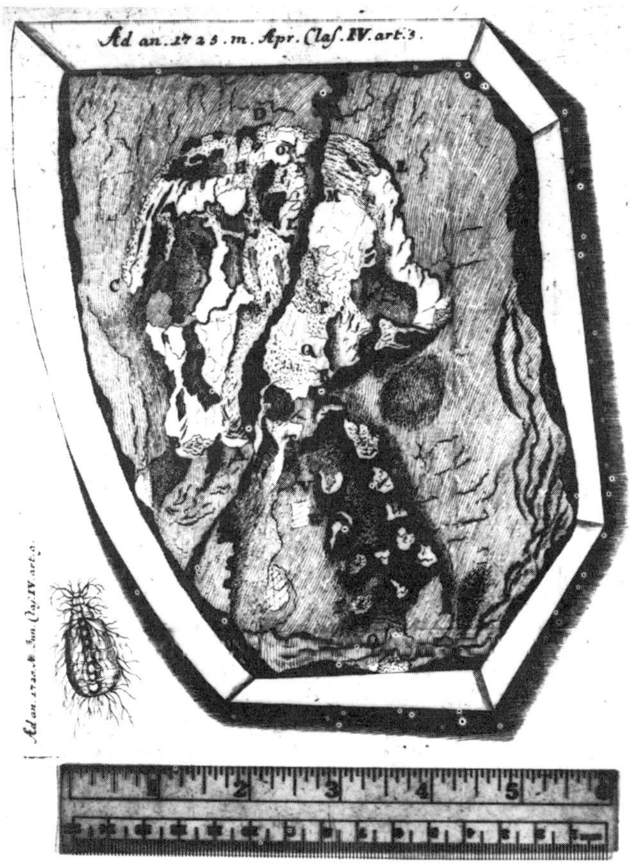

FIG. 3. Homo diluvii testis *from Scheuchzer's article "Hominis in Diluvio submersi Reliquiae,"* Sammlung von Natur- und Medicin-Geschichten, XXXII (1725), *406–408.*

T. First vertebra of the neck.
U. Second vertebra of the neck.
W. Third vertebra of the neck.
X. Fourth vertebra of the neck.
Y. Fifth vertebra of the neck.
Z. Sixth vertebra of the neck.
♀. Seventh vertebra of the neck.

It should be noted that these are true bones and flesh, displaying that yellow hue which Nature herself bestowed upon them. The stone is fissile, and gives off the odor of burnt horn when rubbed, as do all those in this quarry. It is grey in color, and six lines thick.

Georges Cuvier. *Recherches sur Ossemens Fossiles . . . Troisième Édition.* Paris: Chez G. Dufour et E. D'Ocagne, 1825. Vol. V, pp. 431–440.

> On the alleged Fossil Man from the Oeningen quarry described by Scheuchzer, which other naturalists have taken to be a *Silurus*, and which is none other than an aquatic salamander of enormous dimensions and unknown species.

Not without reason are those who attribute all petrifications to the Flood baffled by their failure to find any identifiable human bones among the remains of all the other animal species.

Scheuchzer, who championed this theory more vigorously and persistently than anyone else, was naturally most eager to find relics of our species. With something akin to ecstasy, he acquired a schist from Oeningen which appeared to him to exhibit the imprint of a human skeleton. He published a short description of this fragment in the *Philosophical Transactions* in 1726 . . . and later devoted a dissertation, entitled *Homo Diluvii Testis*, to it. He included a sketch of it in his *Physica Sacra* (I, plate XLIX), with the assertion (p. 66) that "it is immune to all dubiety that it contains half, or only slightly less, of a man's skeleton — the very substance of the bones and, more significantly, that of the flesh and of even softer parts are embodied in the rock. In a word, this is one of the rarest relics we possess of that accursed race buried under the waters."

Nothing less than total blindness on the scientific level can explain how a man of Scheuchzer's rank, a man who was a physician and must have seen human skeletons, could embrace such a gross self-deception. For this fragment, which he propagated so sententiously, and which has been sustained for so long on the prestige of his word, cannot withstand the most cursory examination.

Johann Gesner still cites this fragment as an anthropolith in his *Tractatus Physicus de Petrificatis* (Leyden, 1758). It would seem, however, that this naturalist, who came into possession of a similar fragment, was the first to raise doubts concerning the species, and to conjecture that it might well be nothing more nor less than a *Silurus glanis* Linn. — an opinion to which the naturalists rallied with all the confidence they had previously accorded to Scheuchzer.

No engraving was made of Gesner's fragment, nor of the other which was reportedly kept in the Augustinian Convent at Oeningen. It has been learned, however, that a more complete specimen than that found by Scheuchzer was formerly in the collection of Dr. Amman of Zurich, and was subsequently placed in the British Museum. An engraving of it was published by Karg in the *Mémoires de la Société de Souabe*, plate II, figure 3. We inserted a copy of it in our first edition. As seen there, it is one-sixth its natural size (plate XXV, figure 3). Next to it we placed Scheuchzer's fragment (figure 2), also reduced to one-sixth, and a silurid skeleton (figure 4) in the same proportions. At the top of this series is a skeleton of a land salamander in natural size.

This was the most convenient method of aiding our readers to follow the lines of reasoning which have at last been wholly confirmed.

A comparison of the first of these fragments with a human skeleton would in itself suffice to disabuse one of the idea that this was an anthropolith.

The proportions of the parts present very perceptible differences. The size of the head is indeed that of an average man, but the length of the sixteen vertebrae is several inches more than it should be. Also, each vertebra taken separately is disproportionately long for that of a man.

Other differences found in the shape of the parts are no less striking. The roundness of the head, a feature which was primarily responsible for the illusion, is only related to the human head by broad analogy. What has become of the upper portion which should constitute the forehead? If we assume that the forehead has been deleted, then the roundness of the whole is a mere effect of chance and proves nothing.

How have the orbits become so large? Neither a comparison of the head from front to back nor a vertical section can explain this disproportion of the orbits. The more this section is deepened, the smaller the orbits will become.

The interstices between the orbits is filled with whole bones partitioned by a longitudinal suture. Does the human anatomy have anything similar? Why do we perceive neither the bones nor the cavity of the nose? And if we have no more than the posterior portion, how do we account for the formation of this structure?

Whether this head be compressed or cut, why do we find no trace of teeth, the more so that teeth are always the best preserved part of fossils? Scheuchzer surmises that the bones on either side of the first vertebra are the remains of the lower jaw. But why is there here no appreciable resemblance, and again, how does he explain the absence of teeth?

These considerations and many others undoubtedly motivated the effort to reclassify this fossil as something other than human. . . .

When I was in Haarlem in May of 1811, the learned H. Van Marum, Director of the Teyler Museum, authorized me to make some incisions in the stone containing Scheuchzer's alleged anthropolith, with the view to exposing such bones as might be concealed in it. The operation was conducted in the presence of Van Marum and H. Van den Ende, Inspector-General of Studies — so deserving of recognition for his contributions to the development of primary instruction in the United Provinces. We placed before us a sketch of a salamander skeleton, and, with no little pleasure, noted that with each bit of stone ejected by the chisel, there appeared a bone to match one that was already in the sketch. Thus it came about that the illustration of this schist, which in a hundred years has been reproduced a score of times exactly as it appears in plate XXV, figure 2, was finally altered to the form seen in plate XXVI, figure 2 [of *Recherches sur Ossemens Fossiles*].

First we noted on the circumference of the rotundity, both right and left, a double row of small teeth, which indicated that the rotundity was the effect, not of the skull, but of the jaws.

We also discovered some small ribs at each of the transverse apophyses, just as in Amman's relic and in the salamander.

We established that the head articulated on the first vertebra by means of a double condyle (a,a),[31] as is common to all batracians.

Proceeding to the anterior extremities, which were revealed only by a small surface of the left humerus, we uncovered both of them.

On each side there was an omoplate (b), considerably dilated at its spinal edge, semicircular in contour, closely resembling that of an aquatic salamander. The clavicle and the coracoid were evidently lost.

Near the omoplates are the two humeri (c,c), twice as long as the omoplates, slightly enlarged at top and bottom, with a furrow separating the condyles — again, similar in all respects to the aquatic salamander.

At the end of the humeri are the bones of the forearm (d,d), half as long as the humeri, one of them slightly longer than the other. Lastly, the bones of the fingers, incomplete on the left side but complete on the right, though somewhat in disorder. These bring the number of parts to exactly the same as that of the aquatic salamander: four fingers, a metacarpal and two phalanges for each, except the third, which has three phalanges. . . .

Thus there is no doubt that the supposed anthropolith of Oeningen was an aquatic salamander of gigantic dimensions for its genus.

All these dimensions can be given as follows:

Length of head	0,120
Width of back of head	0,176
Diameter of orbits	0,045
Overall length	1,100
Distance between posterior and anterior extremities	0,340
Length of tail	0,500
Length of arm	0,052
Length of forearm	0,033
Length of hand through longest finger	0,040
Length of femur	0,055
Length of leg	0,032

Works Cited in Scheuchzer's Papers

Besler's *Museum:* Lochner von Hummerstein, Johanne Henricus. *Rariora Musei Besleriani quae olim Basilius et Michael Rupertus Besleri collegerunt* ([Nuremberg], 1716).

Calceolarian *Museum:* Cerutus, Benedictus. *Musaeum Franc. Calceolari. Ivn. Veronensis . . . Luculenter Descriptum, et Perfectum, In quo multa ad naturalem* (Veronae: Angelum Tamum, 1622).

Grew *Museum:* Grew, Nehemiah. *Musaeum Regalis Societatis. Or a Catalogue and Description of the Natural and Artificial Rarities Belonging to the Royal Society And preserved at Gresham Colledge* (London: W. Rawlins, printed for the Author, 1681).

Sachs' *Gammarologia:* Sachs, Philip Jacob. *Gammarologia, sive Gammarorum, vulgo Cancrorum consideratio Physico-Philologica-Historico-Medica-Chymica, in qua Praeter Gammarorum Singularem Naturam* (Francofurti et Leipzig: Esaiae Fellgibelii, 1665).

[31] Letters in parentheses refer to plate XXVI of Cuvier's *Recherches sur Ossemens Fossiles* (3rd edition).

Scheuchzer, "Lexicon of Diluvial Antiquities": This work was never published. The manuscript, however, is in the Zentralbibliothek Zurich (Ms Z VIII 21 and a.). Rudolf Steiger, in his bibliography of Scheuchzer's works (p. 37, number 127), cites the preceeding as: "Lexicon diluvianum". 2 Bde (360 Bl.) 38/24 cm. Ms Z VIII 21 & a. (See Steiger, "Verzeichnis des wissenschaftlichen Nachlasses von Johann Jakob Scheuchzer." *Naturforschende Gesellschaft*, Zurich. *Vierteljahrsschrift*. LXXVIII:21 (1933), *Beiblatt*, pp. 1–75.)

Worm *Museum:* Worm, Olao. *Museum Wormianum. Seu Historia Rerum Rariorum, tam Naturalium, quam Artificialium, tam Domesticarum, quam Exoticarum, quae Hafniae Danorum in Aedibus Authoris servantur* (Lugduni Batavorum: Elseviriorum, 1655).

Theories of the Earth in Renaissance Cosmologies

Sister Suzanne Kelly

A theory of the earth can include anything about the earth from its origin to its projected destruction. No one theory ever covers all topics, and while one aspect or approach may have intrigued some writers at a given time, other writers, if we judge by their silence on the topic, have found it uninteresting. Yet within a given period, some ideas have gained prominence and have received more general attention than others. Such is the case in dealing with theories of the earth in the fifteenth, sixteenth, and early seventeenth centuries. Few, if any, writers discussed all of the topics that are included in this paper. Some dealt only incidentally with one or more of them in works primarily concerned with other topics. Theories of the heavens or hypotheses about the origins of specific earthly phenomena (metals, springs, winds, etc.) were more popular topics than general theories of the earth. For most men of these times, the earth was as God had made it. Still, variant theories did appear in some of the writings of the philosophers, theologians, scientists, and literary men of the period. Their ideas are a mixture of the traditional and the original, and at times they are a reflection of the change in thought patterns taking place in other fields.

In the literature I have surveyed, several topics appeared more frequently than others. This paper will be limited to three: the origin and future destruction of the earth; the structure of the earth; and changes in the surface of the earth. I have not included a consideration of the many discussions on the origins of metals or rivers and springs written during this period for two reasons. These topics, which are of importance to an over-all theory of the earth, would expand the paper too much, and they have been well covered in existing literature. I have also not mentioned the many writers who repeated the traditional Christian modification of the Aristotelian teaching in regard to these topics. These men were by far the most numerous, especially in the early part of the period under discussion, but they are also, to me, the least interesting.

The theories of the earth included in the cosmologies of the thinkers of Renaissance-Reformation period are generally neither detailed nor complete. Often they are buried in discussions concerning other topics or immersed in nonrelevant philosophic digressions. Frequently, they appear naive to the modern reader.[1] However, taken in the context of their time, they form an integral part of that complex phenomenon called the Scientific Revolution.

Origin and Destruction of the Earth

There is comparatively little original discussion on the origin and destruction of the earth during the Renaissance-Reformation period. Most of the writers who mentioned either of these topics willingly accepted the scriptural account of creation and the prophecies of the consummation of all things at the end of time. "In the beginning God created heaven and earth and all things" was interpreted by some as an instantaneous creation of all matter in a state of chaos and darkness. This was then followed by a six-day sorting process comparable to the standard six-day creation, without the start from chaos.[2] The writers of many of these accounts seemed more interested in expounding Scripture, than in accounting for the origin of the earth. Others appeared to consider this a self-evident truth that needed no discussion.

Nicolas of Cusa (1401–1464) at the beginning of the period, Paracelsus (1494–1541) in the middle of it, and Giordano Bruno (1548–1600), Nicolas Rymers (d. 1600), and Nathanael Carpenter (1589–

1 For example: "These early fables of geological science should be read by all who are in need of mental recreation and who possess the required leisure and a certain sense of humor, although many of them make a further demand upon the seeker after amusement and recreation in this fairyland of science, namely that he shall seek this relexation in the somewhat unaccustomed field of medieval Latin." Frank Dawson Adams, *The Birth and Development of the Geological Sciences* (New York: Dover, 1954), p. 210.

2 The following accounts are typical: Martin Luther, *Luther's Commentary on Genesis,* trans. J. Theodore Mueller (Grand Rapids, Mich.: (Eerdmans, 1958), Vol. I, pp. 9–37; John Calvin, *Commentaries on the First Book of Moses Called Genesis,* trans. John King (Grand Rapids, Mich.: Eerdmans, 1948), Vol. I, pp. 67–100; Guillaume de Salluste, Sieur du Bartas, *Bartas: His Divine Weekes and Works,* trans. Joshua Sylvester (Gainesville, Fla.: Scholars' Facsimiles & Reprints, 1965), pp. 2–3, 9–13 16–19 *passim;* Loys le Roy, *Of the Interchangeable Course or Variety of Things in the Whole World,* trans. R. A. (London: Charles Yetsweirt, 1594), ff. 1r–17v.

1628) are notable exceptions. All of these men rejected either the Aristotelian cosmology or the literal interpretation of Genesis or both. The worlds of Cusa and Bruno defied limits of both space and time. For Cusa, creation was one continuous act of God,[3] while Bruno's monads had eternally been moving, forming, and reforming.[4] Neither postulated an end to the process he had described. Neither was specifically interested in the earth.

A similar near-infinite, possible eternal world was the one described by Nicolas Rymers. His earth-centered universe had no outer limit, so the fixed stars extended indefinitely.[5] Rymers stated that this world was a self-existing being, but he did not develop the thought further.[6] His work dealt primarily with a new astronomical system, and his remarks on the terrestrial region were only passing ones, which imply a non-changing earth that will exist eternally.

Paracelsus' and Carpenter's theories are more traditional and are good examples of attempts to combine the scriptural account of creation with the use of natural causes. Both of these men postulated an instantaneous creation by God, but then they differed as to what happened next. Paracelsus had water as the first thing made, and from it the heavens and earth. He then adopted the usual six days of creation, at the end of which everything was in its assigned place.[7]

When the earth was first created, according to Carpenter, its surface was round and uniform and completely covered by water. He speculated that God could have separated the land from the water either "by giving supernaturall bounds and limits" to the waters or by redistributing portions of the earth and thus making cavities and mountains. The second possibility was more reasonable to Carpenter because it was more natural and gave the earth a more pleasing shape.[8]

[3] Nicolas Cusanus, *Of Learned Ignorance*, trans. Germain Heron (New Haven: Yale University Press, 1954), pp. 71–75. The second book of *Of Learned Ignorance* treats of the universe, pp. 65–122.

[4] Bruno's main cosmological works are *La cena de la ceneri; De la cause, principio et uno;* and *De l'infinito universo et mondi.* All three were first printed in London in 1584. Two are available in English translations: Dorothea Waley Singer, *Giordano Bruno, His Life and Thought. With Annotated Translation of His Work on The Infinite Universe and Worlds* (New York: Henry Schuman, 1950); and Sidney Greenberg, *The Infinite in Giordano Bruno. With a Translation of His Dialogue Concerning the Cause, Principle and One* (New York: King's Crown Press, 1950). See, especially, Singer, pp. 50–61, 71–92, 93–115.

[5] Nicolas Rymers, *De astronomicis hypothesibus, seu systemate mundano, tractatus astronomicus & cosmographicus* (Pragae Bohemorum: Apud Avtorem, 1597), f. 38v.

[6] *Ibid.*, f. 37r.

[7] Paracelsus, *Selected Writings*, trans. Norbert Guterman, Bollinger Series XXVIII (New York: Pantheon Books, 1951), pp. 87–92, 300.

[8] Nathanael Carpenter, *Geography Delineated Forth in Two Books. Containing the Sphaericall and Tropicall Parts Thereof.* (Oxford: Henry Cripps, 1625), p. 10.

The rest of the creative process was also the result of God's action on the newly formed earth.

There is little that is of special importance in any of these theories. They stand out only as exceptions among statements that were more conventional for that period. They may represent a beginning of a break with tradition, but it is more probable that they are simply isolated instances of independent thought that had little effect on contemporary or subsequent ideas.

The Structure of the Earth

Considerations about this part of the theory of the earth in the period under discussion involved two things: the material or materials of which the earth was made and the arrangement of these materials. Since the time of Aristotle, the earth-water-air-fire system, although consistently if not effectively challenged, had prevailed, and it continued to do so well into the seventeenth century. For most men before, during, and for some time after the Renaissance, the earth was a solid, spherical body composed of assorted metals, rocks, and earth (all of which were predominantly formed of the element earth, with varying proportions of the other elements). Within this solid body were underground regions of water, air, and fire. Bodies of water covered some areas of the earth, a sphere of air surrounded the earth, and a sphere of fire surrounded the air.

The exceptions were frequently only minor variations on the theme. The element fire was most frequently questioned, and it was so commonly eliminated from the list of the four elements by the end of the sixteenth century that Pierre de La Primaudaye (b. 1545?) devoted an entire chapter of his *French Academie* to refuting the arguments of those who denied fire its rightful place in the universe.[9] Air and water, although occasionally considered exhalations from the earth rather than distinct elements, usually went unquestioned. The earth survived, if no other element did, but sometimes had its properties changed. William Gilbert's (1544–1603) earth was magnetic;[10] Franciscus Maurolycus's (1494–1575) was rocky.[11] Such considerations allowed for no discussion on the structure of the earth.

9 Pierre de la Primaudaye, *The Second Part of the French Academie*, trans. B[ishop] R. N[ewberry] a. R. B[arker] (n.p.: 1594), pp. 180–185.

10 William Gilbert, *Gvilielmi Gilberti Colcestrensis, medici Londinensis, de magnete, magneticisqve corporibus, et de magno magnete tellure; physiologia nova, plurimis & argumentis, & experimentis demonstrata* (London: Peter Short, 1600), pp. 211–214.

11 Franciscus Maurolycus, *Cosmographia* (Paris: Gulielmum Cauellat, 1558), f. 21v.

The Italian Platonists offered different elemental forms of matter. Bernardine Telesio (1509–1588) had a basic substance, which, when acted upon by his two primary agents, heat and cold, produced the heavens and earth, respectively. When the two agents acted together on the same matter, other beings were formed, which varied according to the relative effectiveness of the agents involved.[12] Francesco Patrizi (1529–1597) substituted space, light, heat—three incorporeal elements —and *fluor*, a material substance, for the primary elements. When light and heat acted on the *fluor* in space, the three secondary elements—earth, water, and air—were formed. From these, all terrestrial things were made.[13]

Nicholas Hill (1570–1610) and Giordano Bruno each offered a form of an atomic theory. Bruno's basic units were monads in constant motion. The kinds of monads were infinite, their number was infinite, and they occupied infinite space.[14] The basic units of Hill were similar to Bruno's but had more order and regularity than his. This order had been given to them by God at the time of creation, and their motion, which varied, was somewhat directed by the Creator.[15]

The writers debating the nature and number of the elements were generally unconcerned about the physical structure of the earth as a whole. The spherical nature of the earth was defended repeatedly, either in terms of the water and earth together forming a sphere or by considering the heights of the mountains insignificant when compared with the diameter of the sphere.[16] The possible proportions of whatever elements were being used were estimated, and a general description of the interior and surface of the globe was given. That the interior was hard and more or less homogeneous throughout was taken for granted; that the surface was irregular and the irregularities changing was more debatable.

Telesio and Patrizi, those writers having agents working on some basic substance, and Hill and Bruno, the two employing basic units, did not write on the distribution of their elements or on the structure

[12] Bernardine Telesio, *Bernardini Telesii de rerum natura, a cura di Vincenzo Stampanto* (3 vols; Modena: A. F. Formittini, 1910), Vol. I, pp. 12–17.

[13] Francesco Patrizi, "Panarchias," *Nova de universis philosophia* (Ferrariae: Apud Benedictum Mammarellum, 1591), ff. 61r–79v.

[14] Singer, *Giordano Bruno*, pp. 71–79.

[15] Nicholas Hill, *Philosophia Epicurea, Democritiana, Theophrastica proposita simpliciter, non edocta* (Coloniae Allobrocum: Prostant in officina Fabriana, 1619), pp. 30, 101.

[16] For example: Du Bartas, *Bartas*, pp. 88–89; George Buchanan, "Sphera," trans. James R. Naiden, in *The Sphera of George Buchanan (1506–1582). A Literary Opponent of Copernicus and Tycho Brahe* (Philadelphia: Allen; Turin: Botlega d'Erasme, 1952), pp. 96–97.

of the earth other than in general terms. None of them seems to have asked the question: "What is at the center of the earth?"

Giovanni Paolo Gallucio (fl. 1590) offered a novel structure for the interior of the earth in an otherwise conservative world system. Underneath Gallucio's region of earth and water, the part of the earth in which metals, earthquakes, and other such surface or near-surface events occurred, was Limbo, followed by Purgatory, and then, in succession, the regions for the fighters, the vainglorious, the gluttonous, the oath takers, the angry, the covetous, the proud, the traditors, and finally, at the center, Lucifer. The explanation given for this arrangement was that the inhabitants of these regions were to be as far from God as possible, and the center of the earth was the place the greatest distance from the empyrean.[17]

Such speculations, like those regarding the origin and destruction of the earth, had little direct effect in shaping contemporary thought regarding the theory of the earth. However, they may have been slightly more important than the other speculations because they introduced the possibility of diversity within the earth in a manner that had not been possible in the previous systems.

Changes in the Surface of the Earth

Discussions on the changes on the surface of the earth can also be put into two categories: the minor changes that were proper to the sublunar region, which consisted of the alteration, generation and corruption, growth, and decay, as described by Aristotle;[18] and the greater changes that caused the features of the surface of the earth to vary from one age to another. The former included the statements on origins of metals, rivers, and springs and varied from simple declarations such as "that successive alteration which is in the inferior part of the world, consisteth principally in the fower Elements"[19] to complete works on these subjects.

The larger changes, however, were not easily dismissed. The writers who had incorporated an element of change in their world systems could account for variations in the surface of the earth in terms of these built-in factors. Those who had followed the Genesis account and finished creation in six days had to find some other explanations.

Aristotle had explained the interchange of land and sea by the dry-

17 Giovanni Paolo Gallucio, *Theatrum mundi, et temporis* (Venetiis, 1558), pp. 38–40.

18 Aristotle, *Physica* iii. 1–2. 200b12–202a11.

19 Le Roy, *Of the Interchangeable Course*, f. 3r.

ing action of the sun.[20] This applied to the coastal regions, where the shorelines would vary over a considerable period of time, rather than to major changes on the surface inland. This thought is repeated by many without further explication. The most frequently expressed opinion I have found is that at the time of creation the seas and mountains were placed where they now are and that there have been only minor changes in them since that time. However, there was enough dissenting opinion to keep the discussion open.

In the last half of the sixteenth and the very early part of the seventeenth centuries, several discussions on the origins of mountains appeared that go beyond the earlier statements that earthquakes occasionally produced mountains. The first of these, *De montium origine*, by Valerio Faenzi, lists ten possible natural causes of mountains and, in a passing remark, attributes the origin of some mountains to an immediate action of God. The work is a dialogue in which Rudolphus gives various theories as to the origin of mountains while Camillus comments on these or asks the proper questions.[21]

Earthquakes and the expansion of air or fire under the earth are the mentioned causes that bring mountains into existence quickly. Other causes are less dramatic. When the dry elemental earth in some part of the globe absorbs more water than the land surrounding it, this part of the globe might "expand and rise up in the form of certain protuberances which, after they have hardened, take the shape of mountains."[22] This process was a slow one, possibly taking many years, and was reversible. Plains could develop into mountains and mountains could sink into plains.

The action of water in its course over the earth was another slow producer of mountains. The increase and decrease in the speed of the water in rivers and streams could sharpen or smooth the irregularities of the surface over which the water traveled. Thus, a fast-moving stream could, in time, change a plain into a valley between two mountains. Likewise, the action of wind was considered a plausible explanation for the gradual growth of small hills. As the wind picked up earth from one part of the surface and deposited it elsewhere, the deposits would accumulate in some spots, form barriers, and thus collect more dust from the wind. In time, these accumulations became hills.

20 Aristotle, *Meteorologica* 14. 351^a 19–353^a 26.

21 I have not been able to get a copy of this book and so have used the partial English translation in Frank Dawson Adams, *The Birth and Development of the Geological Sciences,* pp. 348–357.

22 *Ibid.*, p. 350.

The action of water due to the effect of the moon and other celestial bodies on it could account for some minor changes in the location of hills and valleys. If, at the same time, all the celestial bodies caused the water to move in the same directions, the movement could result in a permanent change in the location of that body of water.

One explanation for the origin of mountains is questioned in the dialogue, and another is rejected. After Rudolphus explains how some believed that the action of the sun in attracting the vapors within the earth to itself can form mountains, Camillus replies that this is a "view as easy to refute as it is to support,"[23] and then proceeds to another topic. The theory that the soul of the earth, when it meets opposition to its motion within the earth, becomes violent and pushes portions of the earth above the natural level is rejected.

The last reason given as the cause for the formation of mountains is the action of man in piling up earth, rocks, or other materials. While this is possible in the normal manner, Rudolphus scorns the legend that some of the mountains were made by giants.[24]

Joannis Velcurionis published a commentary on Aristotle's physics in 1588 that suggested possibilities for the origins of mountains. Along with the acceptable earthquakes, Velcurionis included wind, water (floods), man, and the giants who supposedly formed the mountains in Thessaly. A passing gesture was made to God as the maker of mountains, but this follows a statement on the necessity of physics' searching for the causes of these inequalities in the surface of the earth and precedes another on the action of water in varying shorelines, islands, etc.[25] The statements seem to have been inserted as a formality. The whole discussion is a minor topic in Velcurionis' commentary and received only a short treatment. However, in an otherwise traditional work, he does allow for the building as well as the destruction of mountains by natural means. There is nothing said about the time allowed for the processes and no indication that he was thinking in terms of extensive ages.

Another late sixteenth-century writer who permitted continual changes in the earth's surface was Loys le Roy (c. 1590). Besides the usual exchange between land and sea, and the drying up of rivers and fountains, le Roy stated that countries were changed from their normal

23 *Ibid.*, p. 355.
24 *Ibid.*, p. 356.
25 John Velcurionis, *Ioannis Velcurionis commentariorum Libri IIII. In universam Aristotelis physicem nunc recens summa fide exactaque diligentia castigati & excusi* (London, 1588), p. 200.

condition into marshes, pools, deserts, or woods, that mountains degenerated into plains and the plains raised up into mountains. While he is quite definite about these changes occurring, he seems less sure about the causes of them. Earthquakes, heat, winds, water, air, fire, the sun, and the other heavenly bodies are all mentioned as possible causes, but no detailed description of any action is given.[26]

In the second book of geography,[27] Simon Stevin (1548–1620) discussed three types of changes in the surface of the earth: the change in the soil itself; the interchange in location between land and sea; and the change in the height of the land, that is, the building and leveling of mountains. The principal agents in all three of these processes were the actions of wind and water.

The starting basis for Stevin's distinction among the soils of the earth was their location. The soils on the plains, which could be heavy or clayey, sandy heaths or a sulphurous kind, differed from the soil on the mountains, which was always sandy. The sandy soil of the heaths changed into the clayey variety whenever there was sufficient moisture over long periods of time for the fruits and leaves of the plants growing in this soil to rot and thus enrich the soil. The plants themselves were a help in holding the soil so that the wind could not move it and also in assisting the soil to hold moisture.[28] The change to the sulphurous soil, or peat, was somewhat more complicated.

As the sandy soil in the heaths changed to the heavy or clayey kind, the two variations were mixed. Where water carried a mixture of this black, greasy earth and the sand to a wider bed, the decreased speed of the water caused the sand to settle from the water first; the black earth was thus separated from the sand, carried further, and deposited by itself. These deposits, in time, became peat marshes. If the beds of the mixture were not eventually under water, they could, by themselves, become peat marshes if there were sufficient moisture in the locality to separate the soils so that the sand layered out.[29]

The interchange in location between land and water, Stevin thought, was due primarily to the action of the water. As rivers ran their courses into the sea, they deposited sandy or clayey soil, which in time raised the surface of the land where the river empties into the sea. At the same time, elsewhere, the action of the water beating against the

26 Le Roy, *Of the Interchangeable Course*, f. 5 r–v.
27 Simon Stevin, *Les Œuvres Mathematiques de Simon Stevin*, Augmentées par Albert Girard, A. Leyde: B. & A. Elsevier, 1634). The second book of geography is found on pages 128–137.
28 *Ibid.*, pp. 130–132.
29 *Ibid.*, p. 132.

shore would wash the land away or break up the rocks on the shore and thus lower the land level at these places. Here, the sea replaced the land on the surface of the globe. Finally, as the sedimentation and changes in the rate of flow of the river flattened the banks on one or both sides, new plains were formed where the rivers had been.[30]

The third change in the surface of the earth that Stevin discussed was the rising and leveling of mountains. The rivers and sea and the wind all had parts in this process. The rivers, by their rapid courses, could cut deep gorges in the plains of the earth's surface and thus form mountains and valleys. The rivers could also build up areas by sedimentation and raise land areas, but Stevin never let this process accumulate to the mountain-building stage.

The sea had its part in the formation of mountains in the following manner: the waves of the sea breaking upon the shore scattered the sand along the shore. After the heat of the sun dried this sand, the wind whipped it into small mounds.[31] Stevin compared this process to the drifting of snow, and was obviously using the formation of sand dunes as his model for the formation of mountains. This was consistent within his system, since, for him, all mountains were predominantly sand except the ones made by man or by some extraordinary event such as an earthquake. Stevin mentioned these only in passing. He was interested in the natural causes of the irregularities of the earth's surface, not in the unusual ones.

Thomas Lydiat (1575–1646) used the action of subterranean fire and the motion of the earth's water to account for the ordinary exchange of land and sea on the surface of the globe. He thought that simple floods, as man commonly knew them, were the result of excessive rainfall, but the great flood described in Genesis had been caused by the waters above the heavens. At the command of God, these supercelestial waters had left their assigned place beyond the ninth sphere and had descended to the terrestrial region. After the flood, they had returned to their place and apparently had not made any significant change on the face of the earth.[32]

Nathanael Carpenter was the only one of the writers I found who had a theory concerning the changes in the surface of the earth that postulated a gradual and continual decline in the proportions of surface irregularities. For him, the existing mountains were always decreasing

30 *Ibid.*, pp. 136–137.
31 *Ibid.*, pp. 131–132.
32 Thomas Lydiat, *Praelectio astronomica de natura coeli & conditionibus elementorum; tum autem de causis praecipuorum motuum coeli & stellarum* (London: Ioannes Bill, 1605), pp. 125, 149, 180–200.

in size, and no new ones were being formed. He divided the earth's history into two periods: the age of Paradise, including the period up to the universal deluge, and the modern era, which included the time from the deluge to the end of the world. Throughout this span, three factors had been at work changing the surface of the earth. The first of these, man's industry, was evaluated as helpful. In the processes of providing homes and villages, farming, and caring for the land around him, man had protected the surface of the earth from the assaults of wind and water and had often even improved the soil. The second factor, water, was both a good and a bad influence. The water spreading over arid land was helpful; water in the form of too much hard rain or raging floods was harmful. The third influence on the features of the earth's surface was the distribution and inclination of the population. Where there were no inhabitants or where the people were lazy, the surface was not protected, and the degenerative process that Carpenter believed was natural to the earth continued unchecked.[33]

All these changes were surface ones. While Carpenter was willing to admit that not all of the present rivers and fountains had existed from the beginning, he would not permit major changes in the earth's surface. Even the great flood had not changed the earth in an extraordinary manner. Carpenter gave several reasons to support his statement. The first of these was Scripture; the book of Genesis declared that the waters covered the mountains by fifteen cubits. This was coupled with his own logic. If the surface of the earth had been radically changed by the flood, the mountains and rivers existing after this event would not be the same as those before it. If they were not the same, they would not have the same names. But they did have the same names, and therefore they were the same. Furthermore, he added that the waters were not strong enough to move large quantities of earth. His concluding reason was that the earth was better before the flood than after it, and so "we may conclude, that it is farre more probable that the great *Mountains* were so created in the beginning and not made by the flood."[34] The only change the flood might have made would have been small hills, and some of these might also have been produced by the industry of man since that time.[35]

The most significant feature of these discussions is the acceptance of the possibility, where there is such acceptance, that the surface of the earth was not the same as it had been at the time of creation and

33 Carpenter, *Geography* pp. 7–11.
34 *Ibid.* p. 169.
35 *Ibid.*, pp. 115, 167–169, 174–178.

the opposition to this possibility where it is not accepted. Since there is very little discussion of a great age of the earth, these changes would have had to have happened rather quickly. The means that are suggested for them to have come about are the ones with which the writers were familiar — wind, rain, sun, earthquakes.

Conclusion

In many ways, this is a work-in-progress paper. When I started reading the astronomical works, the commentaries on Aristotle and the cosmological and theological works of the period, I was simply searching for ideas about the earth. Somewhat later, I began to wonder what the correlation was between a new theory of the earth and a writer's attitude toward the Copernican theory and/or the religious controversies of the time. Unfortunately, I have not been able to make an extensive enough survey of the literature of the period to support my hunches along these lines.

At this point in my study, it seems that a break with some other part of the Aristotelian cosmology is a necessary accompaniment to a new idea about the structure or changes in the earth. The innovations cut across religious lines a little more freely, although, on occasion, the innovators in the religious world were the conservatives in the scientific one, and the reverse. I had hoped to find some relation between the acceptance of the Copernican theory and the development of new theories about the earth. I may still be able to find this, but it will probably be only a slight one. Very few of the men interested in the Copernican theory diverted their attention from the heavens to discuss the earth. It may well be that it was the next generation that did so.

Bernard Palissy

Aurèle La Rocque

If ever a man was born to trouble, it was Bernard Palissy. Protestant in a Catholic country, militant and outspoken during troubled times when heretical opinions were punishable by prison and death, intellectually near genius but born into a poor family with no hope of achieving a good education, a natural scientist in disagreement with tradition and the official science of his day, we could expect him to knuckle under to the overwhelming handicaps that beset him. Instead, we find him educating himself and developing various talents that brought him the protection of high personages in spite of his outspoken views on religion and his challenging of the mighty Sorbonne itself by teaching natural history, especially geology, almost in the shadow of its walls. Such nonconformism could not last forever, and we lament his end in prison at 90 or more years of age, not by formal execution, not by the sufferings of torture, but by plain starvation and the attacks of vermin on an enfeebled old man.

His exact birthplace is unknown. According to a contemporary, François Grudé de La Croix du Maine, he was born in the diocese of Agen in Aquitaine. There is no supporting evidence from other sources, but La Croix du Maine seems to be the most believable of Palissy's contemporaries who have written about him. Likewise, the date of his birth is unknown. According to La Croix du Maine, he was born between 1514 and 1520; Pierre de l'Estoile, *Grand Audiencier* of the Chancellery of France gives 1510; and d'Aubigné, the chronicler, gives 1499.

Again, we are in the dark concerning his parents. Some French essayists have thought that they were poor artisans, but Leroux (1927, p. 15) believes that internal evidence of Palissy's works strongly suggests that they were peasant farmers.

Whether he was born a peasant or not, his life from adolescence on was certainly not typical of French peasant ways. For one thing, he learned to read, write, and count, an unusual accomplishment for the times except for children who were early admitted to monastery schools. Palissy himself tells us that he knew neither Latin nor Greek, which he would certainly have begun to learn (Latin, at least) if he had studied in a monastery school. We shall return to this point later, for even if he could not read these two languages, he certainly knew something of the classics, which he may have been able to read in translations. He also learned something about drawing, modeling, and sculpture, again rather unusual accomplishments for the lower classes at the time. Who supported and encouraged these studies? There is no evidence on this point, only speculation; it is possible that the Gontaut family was responsible (Leroux, 1927, p. 15), but the idea is far from certain.

Palissy was apprenticed to a master stained-glass worker early in his life. Did he give up his studies at this point? Or is it possible that he studied on his own, with much trouble and effort, as he himself tells us, learning "avec les dents," which is exactly what he would have to do in Saintes, not a large or important town, cut off from most learned circles, if any existed at all in the region of Saintes at the time. In this part of his life, as in so many others, more unanswered questions beset an understanding of Palissy's life and background, and it is not likely that even manuscript sources can resolve all of them. What did an apprentice stained-glass worker learn besides the techniques of his work? He certainly had to know how to stain the glass itself, which meant that he had to learn how to handle ovens, regulate their temperature, and mix pigments with the glass. Perhaps he even had to learn to make glass itself from raw materials. He had to learn to handle lead and to make it into ingenious joints between the polygons of colored glass that made up a window. Perhaps, in the last stages of his apprenticeship, he had something to do with the designing of windows, the drawing of figures in the time-honored scenes most in demand for churches, such as the Nativity, the Crucifixion, Resurrection, Ascension, and Last Judgment. At any rate, he finished his apprenticeship and must therefore have acquired not only the skills of the glass worker but also the standing and prestige attached to the artist, who would belong to a guild. Palissy wrote that the profession was "noble and the men who work at it are noble," a passage that has led to some misunderstanding, for the word *noble* here has nothing to do with nobility. Some writers have gratuitously given Palissy the nobiliary particle *de*, which, as Leroux (1927, p. 17) observes, he may have deserved but never had.

The custom of the time demanded that once a tradesman had finished his apprenticeship he should travel, undertaking the *tour de France*, which still has such appeal and importance for a Frenchman. Where did Palissy go? His books tell us occasionally — and sometimes in a rather offhand way — of things he observed in one part of France or another, but we do not have his diary (if he kept one), and he has not told us whether he started out in the North or the South or even whether he wandered more or less aimlessly as fancy dictated. Audiat (cited by Leroux, 1927, p. 20) thought that he went southward first and returned by the North, but the idea is mere speculation based on Palissy's love for southern France and his affectionate description of nature in that part of his country. On the accompanying map (Fig. 1), I have marked the areas he is known to have visited. One conspicuous omission stands out — Paris — but perhaps the capital had less attraction

FIG. 1. *Map of France about 1550, showing provinces and towns (underlined) visited by Palissy.*

for Frenchmen in those days than it has now. Versailles was still very much in the future, the Louvre was a royal palace, not a museum, and so many things that make Paris attractive today — not only to Frenchmen — did not exist at the beginning of the sixteenth century.

Wherever he went, his observant eye and eager mind noted things that he would remember later and use to advantage. For example, his ideas on sedimentation were aided by his observations of salt pans and the methods of obtaining salt from subsurface brines; his visits to many mines gave him sound knowledge concerning crystals and their occurrence in sedimentary as well as igneous rocks; and everywhere he observed living nature in its most inconspicuous representatives, which he remembered later when he decorated his ceramics with insects and lizards, snakes and snails.

Some time after 1535 he returned to his original home, Saintes, and married. We know neither the name of his wife nor those of his children, but we know that she understood little of his work and aspirations, that the children were numerous (more than six), and that possibly some of them followed in their father's footsteps when he had become a ceramist.

His large family required more income than his trade could supply. Stained glass was in diminishing demand, and Palissy had to use other talents. His early studies of geometry caused him to turn to surveying, and he tells us in a delightful digression in the *Recepte Véritable* (1563) that he was often employed as a surveyor in court cases and felt that he was well paid for his work.

His ability in this work was recognized a little later (1544), when he received a commission to survey the salt flats of Saintonge newly subject to the *gabelle*, or salt tax, by royal edict. He drew double profit from this work, first a respectable — though intermittent — income, and second, precious observations on the soil and geology of the coast and islands of Saintonge.

His restless mind was not content with two trades. About 1539, the sight of an enameled cup made such an impression on him that he began seeking energetically to reproduce it. The details of his quest are given firsthand in his own book (1580, pp. 266–295; La Rocque translation, 1957, pp. 188–203), and they have been retold at length, sometimes with unnecessary embellishments by Morley (1852 and later editions) and his devout paraphrasers, who were more concerned with the martyr in Palissy than the scientist or artist. The facts as Palissy tells them are spectacular enough and equal to Don Quixote's love affair with chivalry in the single-minded struggle against apparently impossible odds. It is true that he had never mixed clay for a pot, nor learned to fire one; that he had no knowledge of enamels and still less of the clays and other materials used in making fine ceramics. Nothing stopped him in his quest. He trudged back and forth to the village of Vieilles-Verreries a few miles out of Saintes, where he persuaded the glassmakers to bake his pots and where he observed the design of their ovens. Later, he built his own oven and used up much of his income buying wood to fire it. When he could no longer buy wood, he burned his furniture and the floor of his house (was he perhaps exaggerating a little here?) to the sound of sarcastic remarks from his wife, cries of famine from his children, and the jokes and insults of his neighbors. But unlike Don Quixote, he succeeded at last and managed to build up an income from his increasingly popular *rustiques figulines*. His patrons were the nobles of Saintonge, whose respect and admiration for him were to serve Palissy in other ways before long.

Palissy's conversion to Protestantism around 1546 had a profound influence on his life and deserves to be examined in greater detail. It was a turning point in his life in that he had to leave Saintes and go to Paris; from a provincial artist with a local reputation he became a member of the royal household. He had been arrested in 1559 or shortly afterwards and sent to Bordeaux to await trial and probably the death penalty. He was saved by the intervening of Anne de Montmorency, his patron, who prevailed upon Catherine de Médicis that he be made *inventeur des rustiques figulines du roi* and attached to the royal household. When Palissy dedicated his *Recepte véritable* "A ma très chère et honorée dame, Madame la Royne Mère," he was doing more than customary homage to royalty. She, in turn, understood and appreciated his genius and called him to Paris to decorate the new palace of the Tuileries, next to the Louvre.

Just over 400 years ago, in 1567, Palissy left Saintes forever and went to Paris to live, to work at the embellishment of the gardens of the Tuileries and to enlarge his intellectual horizons at the court of the last of the Valois, dominated by the Queen Mother, Catherine de Médicis. Here and elsewhere in Paris he had many friends, among them Ambroise Paré, surgeon to three French kings, Ducerceau, an architect, Jean Viret, a mathematician, and many others who may not have been intimate friends but who were sufficiently aware of his genius to pay to attend his lectures in 1575. He names 34 of them in his *Discours Admirables* (1580, pp., 209–211) and Leroux (1927, pp. 66–88) has given enlightening biographical details concerning most of them. We shall return later to these fascinating public lectures of Palissy's.

In spite of the fact that he now lived in an environment in which famous people were writing history in official documents, in diaries, letters, memoirs, and books, the mystery that shrouds Palissy's life continues to pique our curiosity. For example, he escaped the St. Bartholomew massacre of the Huguenots in August 1572. We know that Catherine de Médicis saved the life of Ambroise Paré; perhaps she did likewise for Palissy. If so, he is strangely silent on the subject. The massacre and the troubles that preceded it are explanation enough for his flight to Sedan, at the time ruled by a Huguenot prince, but he was back in Paris by the middle of 1573. Soon afterward, he added another string to his already well-strung bow by becoming an independent teacher on what we would now call geology and other things. His classes were given during Lent of 1575 and attracted men whom Leroux (1927, p. 79) has called the most learned minds of the capital. The substance of his lectures, perhaps augmented by other topics that occurred to Palissy later, was incorporated in his *Discours Admirables* of 1580.

The lectures won Palissy renown, but Paris society soon had more pressing topics of conversation than the origin of springs and the various kinds of earths, metals, and rocks. The ineptitude of King Henry III before the Catholic League was quite evident. Paris was ruled by a council of sixteen men soon after Henri of Guise, chief of the League, had been assassinated, and all those who were suspected of hostility or even indifference to the League were threatened or imprisoned. Palissy, in spite of his great age — he was nearing 80 — was imprisoned about 1588; he died in 1590. Thus ended his life, but not his influence, in spite of efforts to erase it. During his lifetime, he was not widely known, even though his learning was appreciated by the intellectuals of the court. His books were rare and had no immediate influence. Cap (1844, p. xxxvi) notes a single copy of the *Recepte véritable* in the Bibliothèque Royale (now Bibliothèque nationale) and says of the *Discours Admirables* that they have become very rare and therefore fetch a high price. In 1636, Robert Fouët reprinted both works under a changed title and with many unjustified alterations, which were criticized by Cap (1844). With a renewal of Palissy's popularity, Ruault published a new edition in 1777, with notes by Faujas de Saint-Fond and additions by Gobet. The notes are useful, but the reprint has completely muddled the original order of chapters. By the end of the eighteenth century, Voltaire had made his little jokes about Palissy, but Réaumur, Fontenelle, Buffon, Rouelle, Guettard, Venel, and many others had restored him to the place of honor that he deserves. In the next century, Cuvier confirmed their opinion, and ever since he has been admired as an artist and scientist. Perrault (1674, p. 145) mentions Palissy, but only insofar as he discussed the origin of springs; he is not particularly flattering in his opinions of his predecessor.

Sources and Influences

Why does a glazier turned surveyor turned ceramist decide to write a book? Palissy's first book, the *Recepte véritable* (1563) was written in Saintes in response to his desire to repay his benefactors by giving them the benefit of his experience and knowledge. The recipes (*receptes*) are as varied as the curiosity of their author and range from agriculture to geology, the origin of salts, springs, and precious stones. He goes on to plans for an ideal garden, ornamented, of course, with his *rustiques figulines*, and these in turn adorned with biblical quotations. He has some important things to say concerning forests and mines before he goes on to describe the instruments used by the architect, which leads him into a kind of allegorical debate between them. The book also contains a history of the founding of the Protestant church at Saintes and the persecutions it had to suffer. Finally, he describes the construction of a fortress in the form of a spiral, which he thinks is impregnable.

The influences here are fairly easy to trace to the small *coterie* of intellectuals who gathered around Palissy in Saintes and to the growing circle of patrons who prized his artistic works. Antoine de Ponts, Count of Marennes, who was governor of Saintonge among other things, for example, was a cultured gentleman, and Anne de Montmorency was a powerful noble at court. Both of them undoubtedly spoke to Palissy of the new ideas that were being discussed in Paris; the proof is that one of Palissy's first commissions, for the grotto of Ecouen castle, was obviously done in the Italian fashion brought to France by Catherine de Médicis.

Once Palissy had moved to Paris (1567), his sources of information multiplied considerably. There was the small talk of the court and of the royal household and the serious discussions of his many friends to enlighten him. If half of the worthies mentioned by Palissy as subscribers to his lectures were friends with whom he could talk often and at length of science and the arts, then the problem of his sources would be solved quite easily. With his retentive memory and his keen intellect, Palissy would remember the ideas reported to him by his many friends, whether the original was in Latin, Greek, or even Arabic, mull over them in his own mind, and write down his conclusions in the *Discours Admirables*. He may or may not have known Latin and Greek; I am inclined to believe his statement that he knew neither and that he read most of his sources in translation. It is quite possible that he had access to the Royal Library and that printed works published in France with the approval of the King's Council would be quite familiar to him. Palissy's own *Discours Admirables* was published "with privilege of the King," and an extract was printed at the beginning of the book (see LaRocque, 1957, p. 22). There is no mention in this extract of presentation of a copy to the Royal Library, but some member of the council had undoubtedly examined the book before releasing it. In later "privileges," for example, Perrault (1674, pp. xvii–xviii), the printer was required to deposit one copy of the book in the Royal Library, one in the cabinet of the Louvre, and one in the cabinet of the Chancellor of France before offering it for sale. It is most likely that the deposit of at least one copy in the Royal Library was a condition of the granting of royal privilege from the beginning of printing in France.

When these facts are kept in mind, there is little, if any, mystery left about Palissy's sources; he knew some of them in translations printed before 1580; he knew of others through his learned friends; and others were unknown to him (those in German, for example) either because they had not been translated into French or because they were generally unavailable in Paris libraries, or again because no one in Palissy's circle of friends knew of them.

During the translation of the *Discours Admirables* I collected notes on Palissy's sources that it may be instructive to cite here. My main source concerning sixteenth-century works is the catalogue of the Bibliothèque Nationale in Paris. This great library incorporates most of the works salvaged from the various royal libraries, although some of the latter still survive intact outside Paris. How much was lost or discarded as worthless by successive keepers of the King's cabinets we do not know, although we can hazard the guess that it was considerable. In addition, sixteenth-century works were exposed to pilfering for a long time, both by acquisitive courtiers as long as the monarchy lasted and by rioting *sans-culottes* during the Revolution. What is left is impressive enough to give us an idea of the material available to Palissy before 1580. The following includes only the most important authors mentioned by Palissy.

Aristotle is mentioned once on a minor point concerning the collecting of noxious waters in a donkey's hoof, and at second hand, as quoted by Plutarch in the life of Alexander the Great. Had Palissy known Latin he could not have failed to scrutinize Aristotle's works more closely. The fact that he does not refer more extensively to him is an indication that he may indeed have been telling the truth in disclaiming any knowledge of ancient languages.

Arnold of Villanova is mentioned in passing, but the context indicates that either Palissy has not read his works or that he did not think much of them.

Pierre Belon is mentioned as if he had written jointly with Guillaume Rondelet, an almost sure sign that Palissy had read the works of neither of these pioneers of zoology.

The Bible is referred to throughout Palissy's works. He was almost certainly familiar with Philibert Hamelin's French translation.

Cardanus (Girolamo Cardano, 1501–1576) was a slightly older contemporary of Palissy. Palissy had seen Cardanus' *De Subtilitate*, as he says in so many words, but probably in translation.

Fra Francesco Colonna's novel, the *Dream of Polyphilus*, originally entitled *Hypnerotomachia Poliphili*, was published in Venice in 1499 and 1545. There was a French translation by Jean Martin, published in 1546, 1553, and 1600.

Dioscorides is alluded to briefly. The little that Palissy says of him may have been gleaned from quotations in other authors or from hearsay. It does not prove that Palissy had read Dioscorides in the original or in translations.

Du Cerceau. A passage in the *Recepte véritable* shows that Palissy was familiar with Du Cerceau's work. He had seen and studied his plans of fortresses.

Gébert (Abu-Musa-Jābir-ibn-Haiyan, fl. 776) is mentioned with Arnold of Villanova (q.v.), and the same observations apply to both of them. In Palissy's time, several pseudo-Géberts were available in French.

Philibert Hamelin, printer of Bibles, is mentioned by Palissy, who probably quoted from Hamelin's French translation.

Raymond Lull (1232?–1315) is cited, and the context indicates that Palissy had read some of his works, probably a pirated and abbreviated version in French. Lull was in Paris from 1309 to 1311. Palissy certainly had not read the enormous pile of Lull's works; had he done so, he would not have failed to comment on their tediousness.

Paracelsus (1493–1541) is mentioned incidentally, but there is no indication that Palissy had read his works.

Ambroise Paré (1510–1590) attended Palissy's lectures and was a fellow Protestant. Palissy undoubtedly knew his works, some of them published in French in 1545, 1561, 1564, 1568, 1572, 1573, and 1575, but he does not mention them by name or say that he had read them.

Pliny the Elder is quoted by Palissy with such care ("Thirtieth book, chapter 16") that he must have had access to a translation, at least; many were available in French during his time.

Plutarch was also known to him, again quite possibly in a French translation.

The *Roman de la Rose* is one of the most ancient of French poems, begun by Guillaume de Loriis toward the end of the thirteenth century and finished by Jehan de Meung; it contains, according to Cap (1844, p. 192), two alchemical writings in verse. Palissy undoubtedly knew this work in the original French.

Sebastiane (Sebastiano Serlio, 1475–1552) wrote *Rules of Architecture* in Italian. There was a French translation as early as 1547.

Jean Sleidan. The context seems to indicate that Palissy had read Sleidan's works. There was a French translation, published in 1561.

Leonardo da Vinci. Palissy may have known something of Leonardo's ideas through Cardan, who is said to have borrowed freely from Leonardo, or through pirated versions of Leonardo's notebooks, which existed in Italy and France during Palissy's lifetime (Thompson, 1954, quoting Duhem).

Vitruvius is mentioned several times, and the context indicates that Palissy had read his works. There was a French translation by 1547.

Ysidore, that is, Isidore of Seville (died 636), is mentioned briefly, almost in passing (p. 74). Palissy may have known of him only at second hand; if there was a French translation of the Lapidary, I have been unable to trace any mention of it.

We may now consider the sources that Palissy did not mention. Thorndike (1941) felt that Palissy must have known several and should have cited them. These are Conrad von Megenberg's *Das Buch der Natur* (1475), Douglas' *De naturae mirabilibus opusculum* (1524), Savonarola's *Compendium totius philosophiae* (1542), Copernicus' *De revolutionibus orbium coelestium* (1543) and Agricola's works, especially *De natura fossilium* (1546) and *De re metallica* (1556). He also charges that Palissy did not mention but plagiarized Jacques Besson's (1569) work on springs.

These omissions are not the only ones that can be noted in the works of Palissy. For example, according to Thorndike's reasoning, he should have known the work of Hieronymus Fracastoro (1483–1553) who wrote on the true nature of fossils as early as 1517; of that of Alessandro degli Alessandri (1461–1523) whose works describe petrified shells in the Calabrian mountains and account for them by inundation of the continent by the ocean, which he ascribes in turn to some exceptional catastrophe or a change in the axis of rotation of the earth. Likewise, he does not mention Gesner (1565–1566) *De omni rerum fossilium genere*, etc., Kentmann (1565) *Nomenclaturae rerum fossilium*, etc., and Rueius (1547) *De Gemmis aliquot . . . . libri duo*. Palissy never claimed to give a complete review of all works on springs and other topics which he discusses and he should not be criticized for being a poor bibliographer. It was much easier to omit references in his day than it is at present, yet we find living writers missing books and papers which reviewers gleefully call to their attention. Later writers by the score have paid no attention to Palissy even though the titles of their books should have required it.

Posterity

The chorus of praise for Palissy has not been unanimous. In his own country, Brongniart (1844) has been less than kind to him. The following passage is quoted (in translation) from the second edition of Brongniart's *Traité* (1854, p. 63–64):

> Bernard Palissy's book teaches us nothing in ceramics and perhaps very little in physical, chemical and geologic sciences; his stories, his collections, his lectures given to an audience of physicians and men of the world, many of them very distinguished, reveal a man who was curious, active, hardworking, who had extremely varied knowledge, had seen much more than he had read, had been struck by many phenomena which had gone unobserved by others, but which he himself had accounted for inexactly and falsely, and on which he had built theories, as did at that time all the others whom he blamed; therefore he has not distinguished himself from them either by a useful publication of by a wise reserve.

Orcel (1961, p. 13) finds this evaluation unjust. Later, Duhem (1906) in his studies on Leonardo da Vinci, accused Palissy of plagiarism of Leonardo through Cardanus, even though Palissy denies borrowing from the latter. Thompson (1954, pp. 162–163) has analyzed Duhem's indictment and found it very shaky.

Thorndike (1941) is also rather harsh with Palissy and accuses him of plagiarizing Jacques Besson's (1569) work. On this point, also, Thompson (1954, pp. 163–164) has expressed serious doubt.

The rediscovery of Palissy began more than a century after his death, with Réaumur's article in the memoirs of the French Academy of Sciences; Fontenelle, in his history of that organization, really its annual reports (1720), amplified Réaumur's brief account, and Cuvier (1841–1845, Vol. 2, p. 31) also wrote flatteringly of Palissy.

The first republication of Palissy's works seems to have been that of Fouet (1636), according to Cap (1844, p. xxxvii), but it was defective in several respects, as Cap has indicated. The Faujas de Saint-Fonds-Gobet edition (1777), which Cap criticized quite fairly, was the first to consecrate the legend of Palissy as a great scientist and martyr, but, having regard to religious attitudes in France at the time, it made him out as a martyr of science, not of Protestant religion.

The next republication is that of Cap (1844), which is painstakingly faithful to the originals, copiously annotated, and preceded by a fine introduction (pp. iii–xxxix), which makes it one of the best and most reliable editions; it was reprinted in 1961 with a preface by Jean Orcel, who lists some of the literature on Palissy subsequent to Cap's edition. Cap's concluding remarks (p. xxxv) summarize his opinion of Palissy: "It is one of these imposing personalities which permeate their times with an austere and solemn character, a member of the illustrious phalanx that separated the Middle Ages from modern times; beacons of intellect, rising in the midst of a century of darkness, as if to call back the human mind to its noble destiny, and to guide it henceforth in the quest for perfection."

Orcel (1961, p. 2) after more than a century, echoes Cap in his preface to the reprint of the Cap's edition: "Direct heir of the pioneers of this great epoch whose dawn was illuminated by the universal genius of Leonardo da Vinci, Bernard Palissy belongs to their glorious phalanx. Certainly the creative activity of Bernard Palissy does not embrace such vast domains of knowledge as those nurtured by Leonardo da Vinci, 'the complete man, well-balanced, all-powerful in all things, who summarized all the past, anticipated the future' (J. Michelet).

"Nevertheless, Palissy was one of the most remarkable creators of the Renaissance, not only through his quick intellect, his technical competence, and his artistic taste, but also by his high moral qualities."

The 1863 edition by Fick of Geneva is a very small part of the *Discours Admirables* (44 pages) generally overlooked, which I noted in my 1957 translation (La Rocque, 1957, p. 262). A much more nearly complete edition is that of Charavay, with a preface by Anatole France; it was published in 1880, the three hundreth anniversary year of the publication of the *Discours Admirables.* France maintains the attitude of praise and admiration that is general among scientists since 1777.

Audiat (1868) is said to have the most extensive and most satisfactory notes of any edition and it should also be reprinted; so far as I know, it has not been. For other bibliographical materials concerning

Palissy, the reader is referred to my translation (La Rocque, 1957) of the *Discours Admirables* and the notes by Orcel (1961, pp. 18–19).

In the histories of science, Palissy ranks as a definitely minor figure. For example, Zittel (1908, pp. 18–132) notes that "Palissy rightly pointed out that many of the fossil conchylia were identical with living species, and said they must have developed in localities which had previously been under fresh or sea-water." When he says "Palissy's ideas were violently attacked by his compatriots, and he was denounced as a heretic in his philosophical and scientific writings, just as he was a Huguenot and a heretic in his religion," he exaggerates a little, for apparently, true appreciation of Palissy's scientific ideas did not begin until more than a century after his death, and they attracted little attention during his lifetime. The only other mention of Palissy (p. 132) is also regarding his correct interpretation of fossils.

Adams (1938) refers to Palissy in three places. He refers to his concept of "congelative water," which he quite correctly places with other ideas on "lapidifying juice" and other such notions (p. 90). Palissy is credited with recognizing the true nature of fossils, along with Fracastoro, Leonardo da Vinci, Cesalpino, and others (p. 26). Adams gives a summary of Palissy's life and work in a chapter on the origin of springs and rivers (pp. 446–448), which he concludes with the following paragraph: "It may here be mentioned that Conrad von Megenberg in his popular work *Das Buch der Natur*, a printed edition of which appeared as early as 1475, gives nearly the same explanation as Palissy, although not treating of the subject in detail."

Petit and Théodoridès (1962, pp. 292–294) are interested mainly in his zoological ideas, but are less lyrical in their praise than those (e.g., Fontenelle) who revived interest in Palissy. They say: "Nature interests him, certainly, and he has prophetic pages on the destruction of forests; but he treats of the earth and of earths, of waters, minerals and fossils. . . ."

Of Palissy the artist, it is presumptuous of a scientist to attempt to speak with authority. In the Musée de Cluny, in Paris, it is possible to examine great platters attributed to Palissy, ornamented with leaves, animals, and shells, all very faithful to the originals from which they were directly cast. To this admirer of Palissy the scientist, they appear heavy and artificial. Some of the later pieces, of Greek and Roman inspiration, appear definitely uninspired and clumsy. Sauzay (1862) has lovingly reproduced Palissy's works and those of his school in 100 magnificent hand-colored plates, for the appreciative eyes of mid-Victorian Europe. It is just the kind of thing that would go with the era of Napoleon III and Victoria and that would have been discarded in the gay nineties, when people were more interested in "art nouveau" and contemporary schools of painting and sculpture.

Palissy the Protestant martyr had quite a vogue with the pious English public of the late nineteenth century. Morley's *Palissy the Potter* (1852), which went through several editions in England and in America, is an example of this popularity. There were also anonymous works for the mass market and the missionary societies. I have before me *The Story of Palissy the Potter*, published in London by T. Nelson and Sons in 1882 (Anonymous, 1882), a small book of 119 pages with a colored frontispiece, which seems to form part of a series, "Records of Noble Lives," as indicated on the cover. There is frequent reference in the text to "Mr. Morley," but without indication of source. This small book contains nothing new but has an evident propaganda flavor in such passages as the following (p. 60): "Palissy, as we have seen, was a Huguenot — pure, zealous, conscientious; but he took no direct part in the warfare of the Church. When occasion arose, however, he spoke out honestly what he believed to be the truth, and his sense of duty prevented him from sparing the evil-doer. Against the loose-living Papists, the priests, who fed well and drank long, but neglected their pastoral duties, he cherished a very deep aversion, and he did not fear to smite them with heavy censure."

Morley (1852) was more temperate in his account and cites Palissy's two main works at length; it is Palissy who is violent and undiplomatic in his criticism of his fellow citizens, and it is Morley (1869?, p. 207) who comments: "Plain speaking, of the character here indicated, must be added to the list of Bernard's occupations during the period of his prosperity at Saintes. The same originality and force of intellect which produced patrons for the Potter, was serving at the same time to multiply enemies about the austere Reformer, who indulged so freely in the luxury of truth." He reproduces (1869?, p. 307) the story of the King's visit to Palissy shortly before his death in prison, but refrains from comment of any kind when he reports Palissy's death (1869?, p. 308).

References

Adams, Frank Dawson (1938). *The Birth and Development of the Geological Sciences.* Baltimore: Williams and Wilkins Co.; reprinted 1954, Dover Publications, Inc., 506 pp., illus.

Audiat, Louis (1868). *Bernard Palissy: Étude sur sa vie et ses travaux.* Paris: Didier, 480 pp.

Besson, Jacques (1569). *L'art et la science de trouver les eaux et les fontaines cachées sous terre.* Orléans: E. Gibier, in–4°, 84 pp.

Brongniart, Alexandre (1844, 1854). *Traité des arts céramiques ou des poteries, considérées dans leur histoire, leur pratique, et leur théorie.* (1st ed., 1844; 2nd ed., 1854). Paris: Béchat, 2 vols., in–8°, atlas, in–4°.

Cap, Paul-Antoine (1844). *Œuvres complètes de Bernard Palissy, édition conforme aux textes originaux imprimés du vivant de l'auteur; avec des notes et une notice historique par Paul-Antoine Cap.* Paris; J. J. Dubochet, xxxix + 437 pp. Reprinted Paris: Albert Blanchard, 1961.

Cuvier, Georges (1841–1845). *Histoire des sciences naturelles depuis leur origine jusqu'à nos jours chez tous les peuples connus.* Paris, Béchet aîné, 5 vols., completed by Magdeleine de Saint Agy after Cuvier's death.

Duhem, Pierre (1906–1913). *Etudes sur Léonard de Vinci. Ceux qu'il a lus et ceux qui l'ont lu.* Paris, A. Hermann, 3 vols.

Fick, J.-G. (1863). *See Palissy* (1663).

Fouet, Robert (1636). *See* Palissy (1636).

La Rocque, Aurèle (1957). *The Admirable Discourses of Bernard Palissy.* Urbana, Ill.: University of Illinois Press, vi + 264 pp.

Leroux, Désiré (1927). *La vie de Bernard Palissy.* Paris: Librairie Ancienne Honoré Champion. 129 pp., 5 pls.

Morley, Henry (1852). *Palissy the Potter. The Life of Bernard Palissy, of Saintes, his labours and discoveries in art and science, with an outline of his philosophical doctrines, and a translation of illustrative selections from his works.* London: Chapman & Hall, 2 vols. Other editions, Boston and London.

——— (1869?). *Palissy the Potter*, another edition. London: Cassell, Petter & Galpin (The Belle Sauvage Library), viii + 320 pp., frontisp.

Orcel, Jean (1961). "Préface." *Œuvres Complètes de Bernard Palissy*," by P. -A. Cap, pp. 1–17.

Palissy, Bernard (1563). *Recepte véritable par laquelle tous les hommes de la France pourront apprendre a multiplier et augmenter leurs thresors, Item, ceux qui n'ont jamais eu cognoissance des lettres pourront apprendre une philosophie necessaire a tous les habitants de la terre. Item, en ce livre est contenu le dessein d'un jardin autant delectable et d'utile invention qu'il en fut onques veu. Item le dessein et ordonnance d'une ville de forteresse, la plus imprenable qu'homme ouyt jamais parler, compose par Bernard Palissy, ouvrier de terre et inventeur des Rustiques Figulines du Roy et de Monseigneur le duc de Montmorency, Pair et Connestable de France, demeurant en la ville de Xainctes.* Barthelemy Berton, 130 pp.

——— (1580). *Discours Admirables, de la nature des eaux et fonteines, tant naturelles qu'artificielles, des metaux, des sels et salines, des pierres, des terres, du feu et des esmaux, avec plusieurs autres excellens secrets des choses naturelles, plus un traite de la Marne fort utile et necessaire pour ceux qui se meslent de l'agriculture, le tout dresse par dialogues esquels sont introduits la Theorique et la Practique. Par Bernard Palissy, inventeur des rustiques figulines du Roy et de la Royne sa mere.* Paris; Chez Martin le Jeune, a l'Enseigne du Serpent, devant le college de Cambrai, 16 + 361 + 22 pp.

——— (1636). *Le Moyen de devenir riche et la maniere veritable par laquelle tous les hommes de la France pourront apprendre a multiplier leurs thresors et possessions. Avec plvsieurs avtres excellens secrets des choses naturelles, desquels iusques a present l'on n'a ouy parler, par Maistre Bernard Palissy de Xaintes, Ouurier de terre et Inuenteur des Rustiques Figulines du Roy.* Paris: Robert Fouët, 16 + 255 pp.

——— (1636). *Seconde partie dv moyen de devenir riche, contenant les Discours Admirables de la nature des eaux et Fontaines, tant natvrelles qu'artificielles des Fleuves, Puits, Cisternes, Estâgs, Marez et autres Eaux douces, de leur origine, bonte et autres qualitez. De l'alchimie des métaux, de l'Or potable, du Mithridat, des glaces, des sels végetatifs ou generatifs, du sel commun. Description des Marez salans. Des pierres tant communes, que précieuses. Des causes de leur generation formes, couleur, pesanteur et qualites d'icelles, des terres d'argille, de l'art, de la terre, de son vtilité, et du feu, de la marne, et le moyen de les cognoistre. Par M. Bernard Palissy, Inuenteur des Rustiques Figulines du Roy.* Paris: Robert Fouët, 16 + 526 pp.

——— (1777). *Les Œuvres de Bernard Palissy revues sur les exemplaires de la Bibliothèque du Roi, avec des notes par M. Faujas de Saint-Fonds, et des additions par M. Gobet.* Paris: Ruault, 734 pp.

——— (1863). *Discours admirables de l'Art de la Terre, de son Utilité, des Esmaux et du Feu.* Geneva: J. -G. Fick, viii + 44 pp.

——— (1880). *Œuvres de Bernard Palissy, publiées d'après les textes originaux, avec une notice historique et bibliographique et une table analytique par Anatole France.* Paris: Charavay Frères, xxvii + 499 pp.

Perrault, Pierre (1674). *De l'Origine des Fontaines.* Paris: Pierre le Petit, 18 + 353 + 6 pp. (Published anonymously. Later editions of 1678, and others, recognize Perrault's authorship).

Petit, Georges, and Théodoridès, Jean (1962). *Histoire de la Zoologie des Origines à Linné.* Paris: Hermann, 360 pp., illus.

Saint-Fonds et Gobet (1777). *See* Palissy (1777).

Sauzay, Alexandre (1862). *Monographie de l'œuvre de Bernard Palissy, suivie d'un choix de ses continuateurs ou imitateurs dessinée par MM. Carl Delange et C. Borneman et accompagnée d'un texte par M. Sauzay et M. Henri Delange.* Paris: Martinet, 38 pp., 100 col. pls., portr.

Thompson, H. R. (1954). "The Geographical and Geological Observations of Bernard Palissy the Potter." *Annals of Science, 10* (1954), 149–165.

Thorndike, Lynn (1941). *History of Magic and Experimental Science: Vols. 5, 6, The Sixteenth Century.* New York: Columbia University Press, 6 vols., 1923–1941.

Zittel, Karl Alfred von (1901). *History of Geology and Palaeontology* Trans. Maria M. Ogilvie-Gordon from the German edition, Munich, 1899. London and New York: Scribner, 562 pp., illus.

Anonymous (1882). *The Story of Palissy the Potter.* London: T. Nelson & Sons, 119 pp., colored frontispiece.

Reflections on A. G. Werner's "Kurze Klassifikation"

Alexander M. Ospovat

Two events occurred in 1785 to make that year significant in the history of geology: James Hutton read his paper "Theory of the Earth" before the Royal Society of Edinburgh, and Abraham Gottlob Werner finished his "Kurze Klassifikation und Beschreibung der verschiedenen Gebirgsarten."[1] Hutton presented a history of the earth's crust in which he used an old idea, uniformitarianism, as the basis of a detailed new explanation of the changes in the earth's crust since its original formation. Werner offered a theory of the earth that, although it did not exclude uniformitarian ideas, was based primarily on the neptunian origin of the earth's crust.

As is well known, Werner's theory of the earth did not withstand the test of time nearly as well as Hutton's did. At the time it was put forth, however, it had a far greater impact on the geological world than Hutton's work or, in fact, any of the systems proposed by Werner's contemporaries. The "Kurze Klassifikation" guided geological observations and formed the basis for rock classifications throughout one significant era in the history of geology, from 1786 until about 1825. It established petrography as an independent branch of the geological sciences and made the doctrine of geological succession a cardinal principle of earth history. It was widely adopted, adapted, and sometimes just copied. A survey of the geological literature of Western Europe and the Americas shows that the Wernerian classification of rocks,

Research upon which this article is based was done with the support of the Oklahoma State University Research Foundation.

[1] The genesis of Werner's "Kurze Klassifikation" is discussed in my article "The Place of the 'Kureze [sic] Klassifikation' in the Work of A. G. Werner," *Isis, LVIII* (1967), 90–95.

sometimes slightly modified, sometimes not, was used by a majority of the geologists of the day: by Werner's students and followers, by practical geologists who had no interest in theory, and even by geologists who disagreed thoroughly with Werner's theory of the formation of the earth's crust. Thus, for a time, the "Kurze Klassifikation" served, as Werner had intended, as a "foundation for a complete canvass of the universal formation"[2] of rock masses.

It is hard to say just how much the "Kurze Klassifikation" contributed to Werner's personal fame and how much Werner's personal fame contributed to the influence of the "Kurze Klassifikation," but two things are well known: the "Kurze Klassifikation" has long been regarded as a landmark in the history of geology, and the man who wrote it was generally acknowledged by his contemporaries to be the greatest geologist of his time. He was, indeed, as Cuvier called him, the "great oracle of geology." His opinions dominated geological thought and geological publications for more than a quarter of a century. His methods of observation were scrupulously followed by scores of young geologists. His teachings inspired geological research, unprecedented both in quantity and quality; and he numbered among his students many of the greatest geologists of his day: Leopold von Buch, Alexander von Humboldt, Jean D'Aubuisson de Voisins, Robert Jameson, Ernst Friedrich von Schlotheim, and many others.

It is pointless, however, to debate which came first, Werner's fame or the fame of his pamphlet, especially since the evidence indicates that they arose almost simultaneously, but not necessarily as cause and effect. The important point is that the "Kurze Klassifikation" was both the product and the propagator of the neptunian system, which Werner had worked out in great detail over a considerable period of time, beginning perhaps with random thoughts and observations when he was just a boy.

It has been commonplace to assume that neptunistic theories of the earth are necessarily attempts to reconcile the facts of nature with the story of creation as told in Genesis, and Werner's theory, too, has been subject to this assumption. However, there is nothing in his religious training or his education to indicate that Werner would have felt obliged to support Mosaic theory, for he received a very progressive education, one that was in keeping with the Age of Enlightenment, and he was particularly strongly influenced by German Pietism, which stressed the social gospel, the sciences, the cultivation of modern languages, and

2 Werner manuscript—unpublished reply by Werner to a review of his "Kurze Klassifikation" in the *Allgemeine Deutsche Bibliothek*.

the continual exchange between the theoretical and the practical. The Pietistic influence eventually led him to Deism, a belief that he adhered to so strongly that during his student days in Leipzig he was accused of being an atheist.[3] Thus, although it is true that the "Mosaic" geologists of the late eighteenth and early nineteenth centuries took comfort in the fact that the greatest geologist of their day supported the geological theory that was most pleasing to them, it seems that Werner's neptunism was based not on religious bias but on his observations of the region in which he was born and reared.

In Werner's homeland there was no visible evidence of the action of heat or volcanoes. Only the action of water was clearly and unmistakably evident. In the floetz formations, in the floods created by the river Queiss, and in the quarries where so many fossils lay exposed, Werner saw concrete evidence that what was before him could have been formed only "auf dem nassen Weg," in the wet way. And it seems that he had considerable observational evidence for his neptunistic theories long before he started to write the "Kurze Klassifikation" or to teach at the Bergakademie Freiberg, where he began his career in 1775, the year after the publication of his well-known work on the external characteristics of minerals.

Werner's duties at the Bergakademie were to take care of and enlarge the various collections of the school and to teach the course in mining, the course from which he worked out the details of his system, developed the course in geognosy that was to make the Bergakademie one of the most famous schools in the world, and ultimately derived the "Kurze Klassifikation."

If Werner had written and published his geognosy lectures in full, there would have been no need for the "Kurze Klassifikation." But he was reluctant to write for publication, and such a project would have taken a great deal of time. On the other hand, there seemed to be a certain urgency at the time in publishing at least a summary of his neptunistic system. The "Kurze Klassifikation" was the result.

The decades immediately preceding the publication of the "Kurze Klassifikation" had been unusually active geologically, as a perusal of the literature clearly indicates. There had been a great outpouring of articles and monographs that were not concerned so much with geological theory, especially a consistent and universally applicable theory, as with descriptions of investigations of particular and isolated areas.

[3] Shortly after Werner left Leipzig, his friend M. Gehler wrote him about the activities of the small group of friends among whom Werner had been the leader. In this letter, Gehler reports that the group had decided to publish some of the poems written by its members. The one by Werner was entitled "Wie könnt ihr mich noch Atheiste nennen."

As Keferstein points out, there was a need to "combine the particulars and abstract from them a universal system."[4] That there was an awareness of such a need is evidenced by the fact that in 1785 the St. Petersburg Academy of Science offered a prize for the best "*classification exacte & naturelle des divers fossiles dont les couches qui forment la surface de notre globe sont composées.*"[5] Werner did not enter the contest, but he had already been working on the problem, and his work had been completed before the contest was announced. A note to one of the manuscript copies indicates that the work had been completed more than a year before it was published in 1786, under the title "Kurze Klassifikation und Beschreibung der verschiedenen Gebirgsarten," as an article in the *Abhandlungen der Böhmischen Gesellschaft der Wissenschaften.*

Just what was the "Kurze Klassifikation," then? What was its importance, and what did it contribute to the devolopment of geology? Is it really entitled to the reputation that it has enjoyed as a "landmark of geology"?

To answer the first question first, the "Kurze Klassifikation" was a very short work, only 28 pages long, that attempted to classify systematically all of the major rocks that compose the earth's crust. It was derived from Werner's course in geognosy, specifically from the second part of the course. It was important for two main reasons: it was the first classification of rocks that attempted to present a complete, universally applicable system; and it is the only exposition of the Wernerian system actually to come from Werner's own hand. As for its contributions to geology, there were several.

First, it did fulfill the need mentioned earlier to combine the particulars of the various geological investigations and to abstract from them a universal system. And in doing so, the "Kurze Klassifikation" also cleared away many of the confusions and contradictions with which the mineralogy and geology of the time were cursed. In the relatively short period from 1647 to 1775, no fewer than 27 systems of mineralogy had been proposed by various writers,[6] and most of these had also included classifications of rocks. However, since there was no general agreement on nomenclature or descriptive techniques, the difficulties in communication among geologists were almost insurmountable. What

[4] Christian Keferstein, *Geschichte und Litteratur der Geognosie* (Halle: Johann Friedrich Lippert, 1840), p. 65.

[5] *Nova Acta Academiae Scientiarvm Imperialis Petropolitanae*, Vol. III, *Praecedit Historia Eivsdem Academiae ad Annvm MDCCLXXXV* (St. Petersburg: Academiae Scientiarvm, 1788), p. 8.

[6] Frank Dawson Adams, *The Birth and Development of the Geological Sciences* (New York: Dover Publications, 1954), p. 200.

some geologists called granite, for example, was called moorstone, roche feldspathique, whinstone, or sandstone by others. And what some called sienite others called granite, roche amphibolique, or granitelle.[7] Yet geologists were frequently deluded into thinking that because they used the same name they were necessarily speaking of the same rock, when in actuality they might be speaking of two entirely different rocks. Commenting upon this state of affairs, Werner wrote in the "Kurze Klassifikation": "There reigns in mineralogical writings an astounding confusion in the definition . . . of rocks and in the use of their names." He went on to say that clear definitions and a "correct" classification were very much needed, especially since "so very much is being written in and about geognosy and mineralogical geography, and the existing definitions and classifications of these substances by older mineralogists have come to be entirely inadequate." Thus, the "Kurze Klassifikation" made geologists aware of the need for a universally understood geological language and helped to allay the existing confusion and to establish the tendency toward uniformity in nomenclature and classification. In fact, the Wernerian nomenclature itself dominated geological literature for a long time.

As a petrographical work, too, the "Kurze Klassifikation" is of fundamental importance, if for no other reason than that it was the first important work in which the study of rocks was separated from the study of minerals. It is, in fact, the work that established petrography as an independent science. Also, in the "Kurze Klassifikation" Werner formulated the basic principle of petrography, that despite the apparent infinite variety among the rocks of which the earth's crust is composed, the makeup and composition of these rocks are not random and can be determined and defined. Werner was convinced that nature is basically simple, and he went so far as to say that probably all known rocks were included in the "Kurze Klassifikation," which lists only 31 rocks.

Perhaps the greatest contribution of the "Kurze Klassifikation" was that it guided the studies and observations of so many of the young geologists of the day. A study of the geological work of the day reveals a very wide use of Werner's nomenclature, his techniques of description, and his general scheme of classification. Even in the United States, where virtually all geology was purely practical, one finds evidences of Werner's classification in textbooks of geology and in the observations of western explorers.

Yet, in spite of its importance and its fame, the "Kurze Klassifikation" has not been widely read in recent times, and it has in fact been

[7] Robert Jameson, *System of Mineralogy,* (Vol. III; Edinburgh: W. Blackwood, 1808), pp. 101, 139.

considerably misunderstood by geologists and historians of geology. This misunderstanding has been the result largely of the brevity of the work, combined with imperfect understanding of Werner's theory of the formation of the earth's crust, upon which the work is based, and of the difficulties presented by the key word in the title, *Gebirgsarten.*

As is well known, Werner's theory[8] was thoroughly neptunistic. That is, Werner believed that the earth was once enveloped by a universal ocean and that all the important rocks that make up the earth's crust were either precipitates or sediments of that ocean. It is important to note, too, that according to Werner's theory, the ocean stood over an originally uneven surface. During the first period of rock formation, the period that Werner called Uranfänglich, or Primitive, the ocean was very deep and calm at first, and the first rocks were crystalline precipitates, which took their configuration from the uneven surface to which they adhered. Granite, which Werner believed to be the foundation for all other rocks, was formed first. And Werner noted that it can be found at the lowest as well as the highest points of the earth's crust.[9] As the waters of the universal ocean began to recede, gneiss, mica-slate, and clay-slate were formed, showing a decreasing relative altitude of their outcrops. Then, Werner theorized, after the waters had receded somewhat, they rose again, but not to their previous level; and as they began to recede again, they lost some of their original calmness, so that the rocks deposited during this part of the primitive period show some signs of the mechanical action of water, they crop out at progressively lower levels, and the strata, though they are conformable among each other, are unconformable or unconformable and overlapping relative to the strata of older rocks (Fig. 1).[10]

Originally, the second period in Werner's system was the Floetz period. Later, however, perhaps to account for observations that did

[8] The exposition of Werner's theories is based on Werner's own unpublished manuscripts and on manuscripts of lecture notes by his students, which Werner had looked over.

[9] Werner remarked that, even though it is difficult to prove that granite is truly the foundation rock, one has to assume the truth of this, since no geologists have ever reported anything to the contrary. And he explained his apparently contradictory statement that the primitive rocks are found in lower and higher places than the rocks of later periods by pointing out that such a statement can be made only with regard to a particular location and that this would not hold true if, for instance, the Floetz rocks of the Swiss Alps were to be compared with the Primitive rocks of the Erzgebirge (Werner manuscripts, Geogonsie, Vol. VI, Part I).

[10] By unconformable and overlapping, Werner understood a rock mass that, in addition to being unconformable, also covers the outcrops of the rocks on which it rests, as is the case with *c* in Figure 1. He cited the second porphyry formation, which seemed to him to be superimposed upon the primitive clay slate and primitive trap, as an example of a formation that is unconformable relative to the older primitive rocks and also overlaps them.

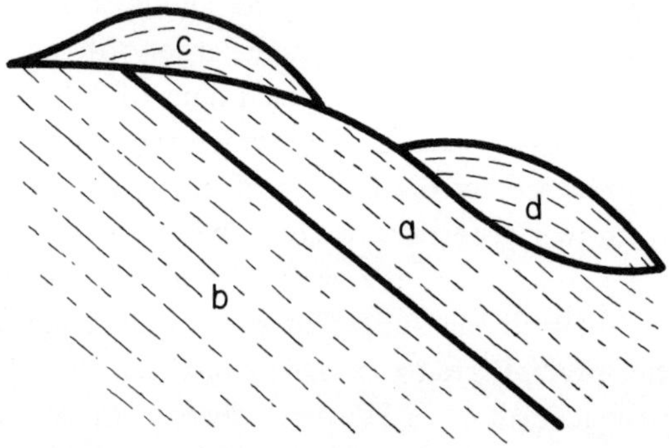

FIG. I. *Tracing from Adolph L. Schippan manuscript, "Lecture notes taken in Werner's course of geognosy, 1814–1815," p. 423.*

not fit into the original theory and to answer criticisms of his system, Werner added a Transition period between the Primitive period and the Floetz period. But as far as the "Kurze Klassifikation" is concerned, this Transition period is irrelevant, since it was added to the system after the "Kurze Klassifikation" was published.

Werner's Floetz period was characterized by storms and violence. While life was developing in great abundance in those regions that were no longer under water, the storms were destroying much of the new life as well as many of the previously formed rocks. For a while the waters receded rather rapidly, depositing the materials that had been destroyed on mountain tops and mountain slopes, but mostly at the bottom of the primitive mountains, thus giving the country a hilly, undulating appearance. Then the waters rose again, reaching a higher level than they had ever reached before. As the waters receded this time, they gradually became calmer and the depositions from them chemically purer, some of them even showing some crystallization. Then again the waters became stormy, destroying some of the rocks that had been most recently deposited from them.

According to Werner, the rocks of the Primitive, the Transition, and the Floetz periods were the only ones deposited from the universal ocean, the rocks that he classified as alluvial and volcanic being the result of local conditions. The gradual receding of the ocean, which began in the Primitive period and continued through the Floetz period, was interrupted by two general inundations, or rising of the waters, one in the Primitive period and one in the Floetz period. Besides these general floods, Werner also postulated various local floods and storms. Since he also believed that the contents of the universal ocean varied from time to time and from place to place, his theory could account for variations from the general principle that the rocks deposited from the universal ocean had been laid down in layers, one above the other. Werner noted, for example, that the first, second, and third gypsum formations do not occur everywhere and that, therefore, the same rocks are not found everywhere deposited one above the other. It is important to note, too, that, although Werner based his classification of rocks upon the period of formation, there is nothing in his theory to preclude the formation of similar rocks at different times. In the Floetz period, for instance, Werner places the formation of three different sandstones at different times. Thus, between the Old Red Sandstone and variegated sandstone, he places the Floetz limestone formation, the Floetz gypsum formation, and the rock salt formation. Furthermore, in the introduction to the section on the Alluvial rocks in the "Kurze Klassifikation," he says that the formation of the last three classes of rocks (the Floetz, the Volcanic, and the Alluvial) falls almost into the same period but that the very last (the Alluvial) can almost (*ziemlich*) be considered the newest. Later, in a never-published reply to a review of the "Kurze Klassifikation" in the *Allgemeine Deutsche Bibliothek*, he emphasized the fact that he had said that the Alluvial rocks could *almost* be considered the newest and went on to explain that since Floetz, Volcanic, and Alluvial rocks are still being formed at the same time today, they must have been formed at the same time in the past, but that he had placed the Floetz period before the other two because all the evidence indicated that the Floetz period had *begun* long before any Volcanic or Alluvial rocks were formed.

Thus, it is clear that Werner's theory was very flexible and that the idea most often presented, that Werner envisioned the rocks of the earth's crust as enveloping the earth in layers much like the layers of an onion, is far oversimplified. The truth is that Werner's theory could account for virtually all the phenomena observable in his day, and it included the possibility of both the simultaneous formation of different kinds of rocks and the formation of similar rocks in different periods of time.

The "Kurze Klassifikation" was based upon the theories that have been outlined, but even a perfect understanding of the theories will not enable the reader to understand the "Kurze Klassifikation" unless he also understands the word *Gebirgsart*, which has been translated by some historians of geology as *rock* and by others as *rock formation*, and by still others as *types of mountains*.[11] Actually, none of these English

[11] Archibald Geikie translated the term as *rocks* in his *Founders of Geology* (New York: Dover Publications, 1962), p. 230. In the English translation of Zittel's *Geschichte der Geologie und Paleontologie*, the term is translated as *rock formation;* see Karl Alfred von Zittel, *History of Geology and Palaeontology to the*

expressions adequately conveys the meaning of the German term as Werner used it.

In the German language, the word *Gebirge* is usually used in a geographical sense, designating a mountain range. But among miners, even today, the term has another, quite different meaning. A German miner always speaks of being in a *Gebirge* as soon as he goes below the surface of the earth, whether the surface is mountainous or plain. Once underground, the miner, in Werner's day, distinguished between different kinds of *Gebirge*, depending upon the kind of equipment and methods he had to use to work the various rocks. This, however, did not mean that every time he encountered a single layer consisting of a different kind of rock he would consider it as a *Gebirgsart*, or a different kind of *Gebirge*, for this term was reserved for extensive rock masses, and the interruption in his working methods by a single layer of different kinds of rocks did not raise the rocks to the dignity of the term *Gebirgsart*. Another consideration in distinguishing between one and another kind of *Gebirge* was the location of the rocks. For example, if on his way down into the earth the miner encountered two sandstones widely separated from each other by layers of other kinds of rocks, he would give each of the sandstones a different and distinct name, as, for example, first sandstone and Rother Todter sandstone. Altogether, then, the German miner used three criteria in designating a rock as a *Gebirgsart:* method of work; extent of the deposit; and location. Werner borrowed the German miner's concept of *Gebirge* and *Gebirgsart*, and in his geognosy developed it further and defined it more exactly. He considered the whole solid earth's crust to be a *Gebirge*, consisting of different parts, or *Gebirgsarten*. Not every rock was a *Gebirgsart*, however, only those that formed sufficiently large or independent units. Thus, Werner wrote in the "Kurze Klassifikation": "In a way [Primitive limestone] may be considered to be subordinated to gneiss, mica schist, and clay slate because it usually occurs in these *Gebirgsarten* and alternately with them. But this seems to be contradicted by reports that in some mountain ranges it makes up entire far-reaching districts or mountain regions and thus qualifies as a distinct *Gebirgsart*." Another criterion that Werner used was also closely related to the miners. Where the miners used work methods, Werner used structure and texture, from which, he believed, he could determine the mode of formation of the rock and to a certain extent the time of its formation. Thus, crystalline

End of the Nineteenth Century, Trans. Maria M. Ogilvie-Gordon (London: Walter Scott, 1901), p. 58. And John Greene, in his *Death of Adam* (p. 78) translated the term as *types of mountains*.

and fine-grained texture indicated to him that the rocks had been precipitated from very deep, calm waters and, therefore, that they had been formed during the Primitive period. Werner also relied upon the relative position of rocks, considering this the most important clue to the time of the rock's formation. The constituent parts of the rocks were only secondary considerations. As Werner's student Leopold von Buch put it when he explained the concept *Gebirgsart* to members of the Berlin Academy of Science: in numbering the houses on a street it does not matter what kinds of materials the houses are built of or what their colors are; the only things that matter are whether the various units in that street fulfill the concept *house* and where the units are located.[12] Thus, the "Kurze Klassifikation" is not a classification of mountains, nor simply a classification of rocks in the narrow petrographical sense. Rather, it is an ordering of *Gebirgsarten* according to their mode and times of formation, and thus it is both a work on historical geology and a work on petrography.[13]

The "Kurze Klassifikation" itself offers no explanation of the overall theory. Like the second part of Werner's course in geognosy, it is based upon the assumption that the reader is already familiar with the theories behind the classification. For instance, when Werner writes that the Primitive rocks ("rock" here is used as a substitute for *Gebirgsart*) show all the characteristics of aqueous formation, he does not say what these characteristics are.

In the description of the various rocks in the "Kurze Klassifikation," Werner very carefully defines their essential mineral constituents. He also describes, as in the case of granite and gneiss, the structure and texture, because the essential mineral constituents are the same, and the difference between the two rocks lies only in the preponderance of one mineral over the other, especially in the way in which they are put together. The way in which the minerals combined was of especial significance to Werner, for he thought that the structure and texture of a rock gave a significant clue to the conditions under which it had been formed and thus would help to reinforce his time-based classification. He did not believe that one rock was transformed into another, but rather that, because of the gradual change in the conditions under which the rocks were formed, gradual changes would result in the rocks

12 Leopold von Buch, *Leopold von Buch's Gesammelte Schriften*. Edited by J. Ewald, J. Roth, and H. Eck (Vol. II; Berlin: George Reimer, 1870), pp. 89–90.

13 Petrography is here understood as the systematic description and classification of rocks. It does not include the microscopic investigation of rocks, since the use of the microscope for geological purposes was introduced several decades after Werner's death.

themselves. These changes in the proportion of the constituent parts and structure and texture are imperceptible when rocks that were formed under similar conditions and at almost the same time are compared. But the changes were cumulative, eventually resulting in what must be considered a different kind of rock. For example, Werner thought that mica schist constitutes a complete compositional transition to gneiss on the one hand and to clay slate on the other.

In accordance with the order of formation according to the Wernerian system, the first section of the "Kurze Klassifikation" deals with the Primitive rocks, of which Werner recognized twelve: granite, gneiss, mica schist, clay slate, porphyry slate (phonolite), porphyry, basalt, amygdaloid, serpentine, Primitive limestone (marble), quartzite, and topaz rock. "Of these," he wrote, "clay slate, serpentine, Primitive limestone, and quartzite are simple [monomineralic]; all the rest are composite rocks. And of the latter, granite, gneiss, mica schist and topaz rock are composite with the parts directly and intimately joined with one another, and porphyry slate, prophyry, basalt, and amygdaloid are composite having a groundmass." The section dealing with the Primitive rocks is the longest of all the sections, comprising 10 of the 26 pages of text.

The next section deals with the Floetz rocks. In the introduction to this section, Werner points out that the Floetz rocks are newer in origin than the Primitive rocks and that "it is highly probable that the Primitive rocks gradually merged into the Floetz." Floetz rocks usually contain fossils, and they are characteristically monomineralic. But unlike the Primitive rock masses, the Floetz rock masses consist of various kinds of interbedded rocks, and of these, "one is always the principal and characteristic one, after this one the particular kind of Floetz formation is always named."

Werner remarked that the Floetz rocks are not nearly so "distinct and determinable" as the Primitive rocks but that limestone, sandstone, coal, chalk, rock salt, gypsum, and iron clay are sufficiently distinguished to be recognized as separate and distinct rocks. In the description of each of these, he cites, next to the principal rock, those that are interbedded with it.

Here it might be pointed out that because of the different natures of the Primitive and Floetz formations, the one consisting of one kind of rock, the other of several, Werner's descriptions of the Primitive and Floetz rocks are different. In describing the Primitive rocks, he placed emphasis upon mineral constitution and structure and texture, thus giving that section a decidedly petrographical character, whereas in the description of the Floetz rocks the petrographical descriptions are overshadowed by the description of the formation in which a particular Floetz rock occurs.

As far as the Volcanic rocks are concerned, Werner distinguished between true Volcanic and pseudo-volcanic. The true Volcanic rocks "owe their entire existence to fire"; the pseudo-volcanic rocks are those that were "only altered by fire and resulted from earth fires. . . ." "Volcanic rocks," Werner wrote, "are usually distinguished by a partly dry, withered, and burst, partly blistered and scoria-like appearance." Among the true Volcanic rocks he lists lavas, pumice, and volcanic ash, which when solidified becomes tuff and trass. Among the pseudo-volcanic rocks he cites lava-like earth slag and porcelain jasper.

The number of Volcanic rocks given is very small. In the absence of microscopic investigations and of any understanding of deformational processes, Werner relied primarily on stratigraphy and morphology and on that basis explained why he included basalt, porphyry slate, porphyry, and amygdaloid among the Primitive and Floetz rocks and not among the Volcanic. "Through several very careful investigations of these rocks and their formations," he wrote, "which I myself made . . . and through reports about these from expert observers in other countries, I have become convinced that these rocks occur in other countries as in Saxony, under the same conditions as other Primitive and Floetz *Gebirgsarten* and show nothing which would hint at an origin through fire." In other words, Werner included among the Volcanic rocks only those that he could trace directly to Volcanoes and earth fires.

The last class in the "Kurze Klassifikation" contains the Alluvial rocks, which "consist almost entirely of parts of destroyed Primitive and Floetz, sometimes even Volcanic formations." Werner subdivided the Alluvial rocks into those that are found in the "deep and wide valleys of high mountain ranges," which he called *Seiffengebirgsarten*, and those that are found in the low country and plains, which he called simply "*Gebirgsarten des niedrigen Landes*." The *Seiffengebirge*, or upland formations, consist of gravel, rubble, and sand, and sometimes also clay or loam, which often contain tin, goldsand, ironstone pebbles, and precious stones. The *Gebirgsarten des niedrigen Landes*, or lowland formations, were subdivided, according to "the mineral masses out of which [the lowland] is formed and those which predominate in it . . ." into sand land — beds of sand, rubble, and gravel, in which occasionally conchylia are embedded; loam land — beds of loams and clays, more or less sandy and sometimes alternating with thin beds of sand;

and moor land — peats and bituminous earth, which sometimes alternate with sand or, more rarely, with clay beds.

At the very end of the "Kurze Klassifikation," there is a note by the author explaining that he had kept the work brief and excluded much detail because he wanted to reserve for himself in the future "a more complete and extended account of this subject." The more extended account never came forth, however, and so the "Kurze Klassifikation" itself remained as Werner's only published exposition of his own system.

Some well-known historians of geology have insisted that Werner's unreasonable attitude, combined with his great reputation, hindered the progress of geology. And Werner's critics, from the last years of his life until the present, have made much of the allegation that his work was not original, that it was "imaginative," not based upon observation, and that it could not account for the obvious facts.[14] The first accusation, it seems to me, is scarcely worthy of consideration. There have been few, if any, instances in the history of science when a scientist has actually abandoned his own theory. And thus the same sort of charge might be leveled at any number of the other great figures in the history of science — at Aristotle, at Ptolemy, at Galen, at Vesalius, at Priestley, at Dalton, at Lamarck, and certainly at Hutton. And who knows how future generations will evaluate Einstein? If every scientist who enjoyed a great reputation and whose theory was ultimately rejected had actually hindered the advance of his science, most sciences would have stood still. In Werner's case, it seems likely that the great surge of geological activity stimulated by his great reputation and his "unreasonable" attitude, particularly in the basalt controversy, did more for the progress of geological observation than any other single factor in the history of geology.

As for the other criticisms, they, too, seem unjustified. Of course Werner did not work in a vacuum. He did not create his theory of the history of the earth's crust without knowledge of the labors of his predecessors and contemporaries. He knew of Steno's work and was acquainted with the writings of Torbern Bergman, Arduino, Pallas, Lehmann, and Füchsel, to name only a few, and he incorporated some of their ideas and concepts into his own work. The law of superposition was a commonplace in his day. In Pallas' work he found confirmation of what he already knew to be true of northern and central Germany: that in all mountain systems granite occurs at the lowest as well as the highest points. And Pallas, too, inferred that it must therefore be the

14 Geikie, *Founders of Geology*, p. 230.

foundation rock. Torbern Bergman put forth the idea that the Primitive rocks are purely chemical and therefore purely crystalline precipitates and that the Floetz rocks are partly chemical and partly mechanical, an idea that Werner worked out in detail. Lehmann spoke of three classes of rock formations: those formed at the time of creation, those that resulted from a general change in the earth's surface, and those that are formed from time to time by accidental causes such as local floods, earthquakes, and volcanic eruptions. To the first two classes he gave the names Primitive and Floetz. Arduino classified the rocks of northern Italy as Primitive, Secondary, Tertiary, and Volcanic. In his *Historia terrae et maris,* Füchsel wrote that every rock stratum corresponds to a certain period of time and that a series of strata consisting of the same kinds of rocks and having the same internal structure can be called a rock formation. Thus, it is clear that many of Werner's concepts were held in common with, if not actually borrowed from, other geologists. But neither Lehmann nor Füchsel, nor any of the others mentioned, wrote a geological work that by any stretch of the imagination could be called a classification of the "various extensive masses of which our solid earth is composed." The "Kurze Klassifikation" was the first classification of rocks to be based firmly upon the principle of geological succession, and it is, in fact, the work that established that principle as the basis for geological research.

Of course Werner's theory was imaginative; all theories are. And it is true that Werner did not travel extensively (neither did Hutton). But the Wernerian system was based upon the physical characteristics of central Germany, which Werner knew thoroughly, and was confirmed, so he thought, by the reports of observations of other geologists in other parts of the world. The "Kurze Klassifikation" itself fitted the observable phenomena in central Germany very well. In a recent article on Werner's geological system, Professor Otfried Wagenbreth, of Weimar, writes:

> The stratigraphic significance of the rock groups as defined by Werner can be derived from a comparison (and it is historically correct) of schematicized tectonic-stratigraphic composite profiles of central and northern Germany, insofar as these show the rock groups and stratigraphic units as they were known in Werner's time according to the relative age . . .
>
> Here is found today as was found then the simplest application of the law of superposition, so that the geological succession of these rocks [the Floetz rocks] as specified by Werner in 1787 [sic] was quite correct and in the nineteenth century needed only to be refined.[15]

[15] Otfried Wagenbreth, "Abraham Gottlob Werners System der Geologie, Petrographie und Lagerstättenlehre," in *Abraham Gottlob Werner Gedenkschrift aus Anlass der Wiederkehr seines Todestages nach 150 Jahren am 30. Juni 1967* (Leipzig: VEB Deutscher Verlag für Grundstoffindustrie, 1967), pp. 116–117.

As for the accusation that Werner's theory simply did not fit the obvious facts, it must be remembered that the obviousness of the facts is always to a large extent dependent upon the theory held by the observer. And, of course, there are severe limits to what any one man can observe. One of the most interesting aspects of this era in the history of geology is that the actual observations of both Werner and Hutton were few compared to those of many of their contemporaries and that the two great opponents to a large extent observed similar formations. Hutton's theory erred on the side of plutonism almost as much as Werner's did on the side of neptunism.

Werner's theory was flexible, so flexible, in fact, that it could account for most of the observable phenomena; as far as the sedimentary rocks are concerned, it is not at all incompatible with modern theory. And in spite of his apparent obstinacy in the basalt controversy, there is considerable evidence that Werner was mild and reasonable, even where his own theory was concerned. The Englishman John Pinkerton reports in his *Petralogy: A Treatise on Rocks* (1811) that Werner had said to him that "a theory is useful to concatenate facts, and render them more clear and pleasing to an audience." "Nor," Pinkerton went on, "with the modesty of a man of real genius, did he conclude his own theory to be unobjectionable."[16]

Thus, it seems to me that Werner has been unjustly disparaged. Members of our own generation, evaluating his work in the light of their own cherished theories, have been all too prone to accept uncritically the biases of Hutton's followers, who in the flush of their triumph wrote at least as emotionally and unscientifically as any of the disciples of Werner. They have apparently been unable to see the paradox in calling a man in one breath one of the founders of modern geology and in another a hindrance to the progress of geology. It is to be hoped that future studies will show more accurately the contribution of the man who, more than anyone else, established historical geology as a science and of the one work that he published on the subject, the "Kurze Klassifikation und Beschreibung der verschiedenen Gebirgsarten."

[16] John Pinkerton, *Petralogy. A Treatise on Rocks* (Vol. II; London: Printed for White, Cochrane & Co., 1811), pp. 4–5.

Diluvialism and Its Critics in Great Britain in the Early Nineteenth Century

Leroy E. Page

In my opinion, the influence of religious preconceptions on the thinking of geologists in Great Britain in the early nineteenth century has been overemphasized. There has been perpetuated the old, but erroneous, idea that the Wernerian theory was strongly influenced by the Bible and that it regarded the flood as a great geological event.[1] This error probably arose because (1) some of Werner's British disciples did introduce the flood into their geological writings,[2] and (2) Charles Lyell misrepresented the subject in his *Principles of Geology*.[3]

Werner's geological views seem to have been no more affected by a desire to confirm the Bible than were Hutton's. Werner, like Hutton, was apparently a Deist.[4] The appeal of the Wernerian geognosy to many was that, unlike previous "theories of the earth," it invoked neither catastrophes nor unfamiliar agents and refused to speculate about the origins of the earth. It seemed to be based, not on imagination nor on the Bible, but on an induction from the facts.[5]

1 See Stephen Toulmin and June Goodfield, *The Discovery of Time* (New York: Harper & Row, 1966), pp. 152–153.
2 Notably Richard Kirwan, *Geological Essays* (London: D. Bremner, 1799), pp. 54–86, and Robert Jameson (see notes 9 and 10).
3 Charles Lyell, *Principles of Geology, Being an Attempt to Explain the Former Changes of the Earth's Surface, by Reference to Causes Now in Operation* (3 vols.; London: John Murray, 1830–1833), Vol. I, p. 69. Lyell's purpose was to free geology from theological influence by emphasizing the harm it had done geology in the past.
4 Personal communication from Professor Alexander M. Ospovat, Department of History, Oklahoma State University, Stillwater, Oklahoma, March 20, 1967.
5 Robert Jameson, *System of Mineralogy, Comprehending Oryctognosy, Geognosy, Mineralogical Chemistry, Mineralogical Geography, and Economical Mineralogy*, (Edinburgh: William Blackwood, 1808), Vol. III, p. 41–42. See, also, a review of William Brande's *Outline of Geology, British Critic, VII* (1817), 553.

Plate III

WILLIAM BUCKLAND (1784–1856)

Thomas I. Sopwith (1803–1879), student and friend of Buckland, "who was Dr. Buckland's companion in some of his tours in search of glacial scratches, made [about 1840] a semi-caricature of Buckland, who, encountered with numerous heavy cloaks, thick travelling boots, bags of fossils, and rolls of maps, presents a figure fancifully like a glacier." [Mrs. Gordon, *Life and Correspondence of William Buckland* (London, 1894), p. 145.]

Specimen No 1 Scratched by a Glacier Thirty three Thousand Three hundred & Thirty Three Years before.
Scratched by a Cart Wheel on Waterloo Bridge the day before yesterday
Prodigious Glacial Scratches
Scratched by T. Sopwith
The Rectilinear Course of these Grooves corresponds with the motions of an IMMENSE BODY the momentum of which does not allow it to change its Course upon Slight Resistances
COSTUME of the GLACIERS

The Wernerians distinguished between the hard strata laid down by the universal ocean and the alluvial or loose beds above the hard strata formed by the breaking up of the latter that occurred either during the retreat of the universal ocean or since then under the action of local causes. They, as well as some Huttonians, tended to ignore the evidence that was to convince many geologists that the alluvial beds were in large part the product of a geologically recent, catastrophic deluge that had seemingly covered much, if not all, of the earth.[6] This evidence was the existence in the alluvial beds of many poorly assorted deposits of gravel and clay that had evidently been transported great distances before being dumped. Also found in association with these were very large erratic blocks that had clearly been carried from distant sources as well as the bones of large animals, such as the mammoth, whose extinction was held not to be explicable by ordinary causes. It was felt, also, that only such a deluge could have formed the larger valleys.[7]

The most popular and influential book on this subject in Great Britain was a work by Georges Cuvier, which in 1813 was published in English under the title *Theory of the Earth*, with a preface and notes by Robert Jameson, professor of natural history at the University of Edinburgh and the leading Wernerian geognost in Great Britain.[8] In his work on geognosy in 1808, Jameson had stated that the rising of the universal ocean that had produced the Newest Floetz Trap strata could be called "a Deluge." He had otherwise ignored the flood and had referred the overlying alluvial beds to local causes.[9] In his preface to Cuvier's work, however, Jameson made much of the close coincidence of geology and the Mosaic history. He supported the interpretation of the six days as periods of indefinite length, and he approved Cuvier's view that the alluvial beds were in large part the product of a deluge.[10] By the time of the third edition, in 1817, Jameson had had second thoughts, and he omitted all reference to Moses. This change

[6] Jameson, *System of Mineralogy*, pp. 26–36, 97, 206–211. See also, Sir George Mackenzie, "Deluge, or Debacle, in Geology," in David Brewster (Ed.), *The Edinburgh Encyclopaedia* (18 vols.; Edinburgh: William Blackwood, 1830), Vol. VII, p. 604, for a statement that the Wernerians ignored the geological deluge (written about 1815).

[7] This evidence was summarized by William Buckland, *Vindiciae Geologicae; or the Connexion of Geology with Religion Explained, in an Inaugural Lecture Delivered Before the University of Oxford, May 15, 1819, on the Endowment of a Readership in Geology by His Royal Highness the Prince Regent* (Oxford: the Author, 1820), pp. 35–38.

[8] Georges Cuvier, *Essay on the Theory of the Earth, . . . with Mineralogical Notes, and an Account of Cuvier's Geological Discoveries, by Professor Jameson*, trans. Robert Kerr (Edinburgh: William Blackwood, 1813).

[9] Jameson, *System of Mineralogy*, pp. 203, 206, 349.

[10] Cuvier, *Essay*, pp. v–vi, viii–ix.

was probably caused by the criticism that his interpretation had received.[11]

Cuvier's diluvial theory maintained that the sea had shifted its location during the flood, overwhelming the previous land and evacuating its former bed, which became the present land. Cuvier also assumed that there had been at least one previous revolution, before the creation of man, in which the sea had deserted its bed in order to occupy the continents. The effect of the flood, therefore, was the return of the sea to the bed it had occupied prior to the first revolution.[12] Another theory, held by William Buckland, maintained that the present continents had existed immediately before the flood. This theory regarded the flood as a temporary incursion upon the land and rejected Cuvier's first revolution.

At this time, both theories were about equally consistent (or inconsistent) with the known phenomena; however, each had its own theological merits. Cuvier's was in agreement with his idea of successive creations: the first revolution destroyed the old forms of life, man and the present species being created afterward (in six days) and inhabiting the present sea bed until the second revolution (the flood). The theory was thus consistent with the biblical statement that every species was rescued from the flood. It also explained why man's remains had not been found in the diluvial deposits, which in this theory were produced, not by the flood, but by the first deluge or by the sea between the two revolutions. The flood itself would have produced few visible deposits, as it was essentially a retreat from the continents. Its main effects would have been the carving of some valleys and the removal of material into the sea. The mammoths and other extinct species would have been overwhelmed by the first revolution. This theory was complicated by an additional deluge; and it assumed that the antediluvian land was not the same as the present land, which seemed to contradict certain statements in Genesis.[13]

[11] Cuvier, *Essay* (3rd ed.; 1817), pp. v–ix. For the controversy started by Jameson's preface, see the *Philosophical Magazine*, Sept. 1815 to Oct. 1816, especially vol. *XLVI* (1815), pp. 225–229.

[12] Cuvier, *Essay* (1st ed.), pp. 171–172.

[13] Buckland, *Vindiciae Geologicae*, p. 31. Cuvier's theory was similar to that of Jean André De Luc of Geneva, tutor to Queen Charlotte of England. De Luc's first revolution was caused by the collapse of a portion of the bottom of the universal ocean, into which all of the ocean rushed, uncovering the first continents. Then, at the time of the flood, caverns under these first lands collapsed, and the sea occupied the resulting basin, leaving dry its former bed, which became the present continents. See De Luc's *Lettres sur l'Histoire Physique de la Terre, Adressées a M. le Professor Blumenbach, Renfermant de nouvelles Preuves géologiques et historiques de la Mission divine de Moyse* (Paris: Nyon, ainé, rue du Jardinet, 1798), pp. 227–229 (Lettre V). These letters were originally published in the *British Critic*, 1793–1795.

The publication of Cuvier's *Theory of the Earth* had been a hopeful sign to those, especially of the clergy, who had been disturbed by the lack of correspondence between geology and Moses, and it suggested to Thomas Chalmers a way of reconciling the two. As Chalmers expressed it:

> Moses may be supposed to give us not a history of the first formation of things, but of the formation of the present system; and as we have already proved the necessity of direct exercises of creative power to keep up the generations of living creatures; so Moses may, for anything we know, be giving us the full history of the last great interposition, and be describing the successive steps by which the mischiefs of the last catastrophe were repaired.[14]

This interpretation was also adopted by John Sumner, later Archbishop of Canterbury:

> We are bound to admit, that only one general destruction or revolution of the globe has taken place since the period of that creation which Moses records, and of which Adam and Eve were the first inhabitants. . . . *But we are not called upon to deny the possible existence of previous worlds, from the wreck of which our globe was organized, and the ruins of which are now furnishing matter to our curiosity.*[15]

William Buckland was appointed reader in mineralogy at Oxford in 1813, succeeding John Kidd; in 1814 he began to lecture annually on geology as Kidd had done before him. Buckland's correspondence with his friend, William D. Conybeare, indicates that in the spring of 1814 both of them were very interested in the relationship between Moses and geology, presumably because Buckland would need to lecture on this subject. In 1815, in order to illustrate his lectures, Buckland published a table comparing the English strata with those on the continent, as arranged by Werner. In this table, which was widely circulated, he introduced the term *diluvian detritus* to refer to that part of the alluvium produced by the flood and distinguished it from the postdiluvian detritus attributable to wind and rivers.[16] Cuvier had previously dis-

[14] Thomas Chalmers, "Remarks on Cuvier's *Theory of the Earth;* in Extracts from a Review of That Theory Which was Contributed to the *The Christian Instructor* in 1814," in *Miscellanies; Embracing Reviews, Essays, and Addresses* (4 vols.; New York: Robert Carter, 1848), Vol. I, p. 191.

[15] John Bird Sumner, *A Treatise on the Records of the Creation, and on the Moral Attributes of the Creator; with Particular Reference to the Jewish History, and to the Consistency of the Principle of Population with the Wisdom and Goodness of the Deity* (2 vols.; London: J. Hatchard, 1816), Vol. I, pp. 284–285.

[16] Mrs. Elizabeth Oke (Buckland) Gordon, *The Life and Correspondence of William Buckland, D.D., F.R.S., Sometime Dean of Westminster, Twice President of the Geological Society, and First President of the British Association* (New York: D. Appleton and Company, 1894), pp. 14–18. Buckland's table, entitled "Order of Superposition of Strata in the British Island," is appended at the back of William Phillips, *A Selection of Facts From the Best Authorities, Arranged so*

tinguished between two kinds of alluvium, belonging to two different geological epochs. The older deposit, which contained only extinct species, had been produced by a great marine inundation, while the more recent deposit contained only existing species and was attributable to the ordinary processes of erosion and deposition.[17] Buckland's diluvian detritus contained both extinct and existing species.

In his inaugural lecture, after his appointment as reader in geology at Oxford in 1819, Buckland asserted that geology supported the Bible and religion, citing Cuvier and Sumner among his authorities, but insisting, in opposition to the former, that "the antediluvian continents were the same with the present."[18] The chair in geology seems to have been created as the result of a compromise, like that proposed by Sumner, in which the Church would encourage geology if geologists, in turn, would support the Bible with respect to the flood. In the summer of 1819, Buckland began field investigations in order to substantiate his theory of the flood; out of these came a paper, which he presented to the Geological Society in December.[19] His best new evidence was that from the Kirkdale cavern, discovered in 1821, on which he published his *Reliquiae Diluvianae* in 1823.[20]

It is a misconception that there was virtually no opposition by geologists in Great Britain to Buckland's diluvial theory in the years immediately following its promulgation.[21] Although most geologists

---

as to Form an Outline of the Geology of England and Wales (London: William Phillips, 1818). Joseph Townsend had earlier (1812) divided the alluvial deposits into "ancient alluvial, which may be called Diluvian," and recent alluvial; see his *Geological and Mineralogical Researches, During a Period of More than Fifty Years, in England, Scotland, Ireland, Switzerland, Holland, France, Flanders, and Spain: Wherein the Effects of the Deluge are Traced, and the Veracity of the Mosaic Account is Established* (Bath: Samuel Bagster, 1824), p. 252 (first published in 1812).

17 Georges Cuvier, "Sur les os fossiles de ruminans trouvés dans les terrains meubles," *Journal de Physique, LXVIII* (1809), 377.

18 Buckland, *Vindiciae Geologicae*, pp. 5, 24, 29.

19 William Buckland, "Description of the Quartz Rock of the Lickey Hill in Worcestershire, and of the Strata Immediately Surrounding It; with Considerations on the Evidences of a Recent Deluge Afforded by the Gravel Beds of Warwickshire and Oxfordshire, and the Valley of the Thames from Oxford downwards to London; and an Appendix, Containing Analogous Proofs of Diluvian Action. Collected from Various Authorities," *Transactions of the Geological Society, V* (1821), 506–544. This was read on December 3, 1819.

20 William Buckland, *Reliquiae Diluvianae; or, Observations on the Organic Remains Contained in Caves, Fissures, and Diluvial Gravel, and on Other Geological Phenomena, Attesting the Action of an Universal Deluge* (London: John Murray, 1823).

21 See Charles C. Gillispie, *Genesis and Geology: A Study in the Relations of Scientific Thought, Natural Theology, and Social Opinion in Great Britain, 1790–1850* (New York: Harper & Brothers, 1959), pp. 108, 111. This view was also criticized by Walter Cannon in *Victorian Studies*, Dec. 1961, p. 126 and note 34.

at this time accepted the evidence for a geologically recent and widespread deluge, only a minority, mostly clergymen, such as Buckland, Conybeare, Sedgwick, and Kidd, identified this deluge with the flood.[22] Geologists had a far more liberal attitude toward the Bible than most laymen, some of whom attacked the current geology (and Buckland in particular) because they believed that the flood was responsible for much more than geologists would allow.[23]

The interpretation of the flood as a quiet rising of the waters that left no distinctive traces was stronger among geologists at this time than has been recognized. Jameson, in 1827, while naming Cuvier and Buckland as authorities for the violent theory of the flood, stated that "numerous writers" had advocated the nonviolent theory.[24] Buffon had argued this view, and many of his arguments were repeated at this time by opponents of the violent theory.[25] It is well known that the Huttonian theory made no use of deluges.[26] Although it is questionable whether Hutton himself believed in the flood,[27] his follower, Playfair, certainly did, and he held to the nonviolent interpretation.[28] The catastrophist Huttonians, Sir James Hall and Sir George Mackenzie, regarded their deluge as having occurred long before the flood.[29]

Many non-Huttonians also refused to identify the geological deluge with the flood. Chalmers objected to Cuvier's violent flood.[30]

22 John Kidd, *Outlines of Mineralogy* (Oxford: J. Parker, 1809), Vol. I, pp. vi, xxi. Kidd was not so positive in his second work: *A Geological Essay on the Imperfect Evidence in Support of a Theory of the Earth, Deducible Either from its General Structure or from the Changes Produced on its Surface by the Operation of Existing Causes* (Oxford: the Author, 1815), p. 163. For Adam Sedgwick, see his "On Diluvial Formations," *Annals of Philosophy*, X (1825), 34–35. See, also, Rev. Joseph Townsend (see note 16).

23 See, for example, Granville Penn, *A Comparative Estimate of the Mineral and Mosaical Geologies: Revised, and Enlarged with Relation to the Latest Publications on Geology* (2 vols.; 2d ed.; London: James Duncan, 1825).

24 See Georges Cuvier, *Essay on the Theory of the Earth, . . . with Geological Illustrations, by Professor Jameson* (5th ed.; Edinburgh: William Blackwood, 1827), p. 437.

25 Count de Buffon, *Natural History, General and Particular . . . The History of Man and Quadrupeds*, trans. by William Smellie, edited by William Wood (20 vols.; London: T. Cadell and W. Davies, 1812), Vol. I, pp. 128–132. This is in the "Proofs of the Theory of the Earth, Article V."

26 John Playfair, *Illustrations of the Huttonian Theory of the Earth* (Edinburgh: William Creech, 1802), p. 400.

27 Hutton seems to deny that God would act so as to destroy life; see his *Theory of the Earth, with Proofs and Illustrations. In Four Parts* (2 vols.; Edinburgh: William Creech, 1795), Vol. I, p. 273.

28 [John Playfair], Review of Cuvier's *Essay on the Theory of the Earth*, *Edinburgh Review*, XXII (1814), 469.

29 See Sir James Hall, "Experiments on Whinstone and Lava," *Transactions of the Royal Society of Edinburgh*, V (1805), 68, and Mackenzie, *Edinburgh Encyclopaedia*, Vol. VII, p. 605.

30 Chalmers, "Remarks," p. 187.

The geologist Robert Bakewell accepted Cuvier's evidence for "the last grand revolution of the globe before the formation of man," but said nothing about the flood, the Bible being ignored in his work.[31] William Smith believed in the geological deluge, but did not relate it to the Bible.[32] Thomas Whitaker, an English clergyman, stated in 1819 that current geological opinion supported the nonviolent view of the flood and cited Linnaeus and Cuvier in his support:

> How little reason there is, from the account of Moses himself, for believing that the general surface of the globe underwent any material change in consequence of that catastrophe. The annihilation of the human race, with a few exceptions, was the object of God, and for that purpose an inundation, without these supposed convulsions, otherwise than as required for producing that inundation, was quite sufficient.[33]

The nonviolent interpretation of the flood was particularly commendable to scientists because of the tradition of not mixing science and religion, for the good of both. William Knight, in 1818, after quoting Francis Bacon on the folly of mixing science and the Scriptures, added:

> This is a passage which cannot be too often remembered, and which, like many others in the same immortal work, is pregnant with the spirit of true wisdom. It seems, as Professor Playfair justly remarks, to have been prophetically addressed to De Luc, Kirwan, and the other cosmogonists of the present day, who have done all in their power to degrade the Sacred Writings by the arguments which they have brought forward in their defence.[34]

John Farey, the pupil and spokesman of William Smith, wrote a letter to the *Philosophical Magazine* criticizing Buckland's inaugural lecture in the following way:

> I remember having seen Mr. Bakewell commended in your Work, for having in the year 1813 abstained, from introducing the Deluge of Moses into his "Introduction to Geology," as the previous Writers had almost invariably done, to the manifest injury of Geology on the one hand, and of Religion on the other: since which, the practice has almost entirely grown into disuse, while the number of writers on Geological subjects, have been greatly on the increase; and I regret therefore to see, the new Geological Professor at Oxford, attempting now to revive the exploded notion, that

31 Robert Bakewell, *An Introduction to Geology, Illustrative of the Structure of the Earth; Comprising the Elements of the Science, and an Outline of the Geology and Mineral Geography of England* (2d ed.; London: J. Harding, 1815), pp. 430–431. Bakewell does state that natural appearances confirm the Mosaic history with respect to the recent creation of man.

32 William Smith, *Strata Identified by Organized Fossils, Containing Prints on Colored Paper of the Most Characteristic Specimens in Each Stratum* (London: the Author, 1816), Introduction (second unnumbered page).

33 [Thomas Dunham Whitaker], Review of Thomas Gisborne's *The Testimony of Natural Theology to Christianity, Quarterly Review, XXI* (1819), 53.

34 William Knight, *Facts and Observations towards Forming a New Theory of the Earth* (Edinburgh: Archibald Constable & Co., 1818), p. 325.

any of the phaenomena at this time *visible*, on or within the Earth, are, with any proper regard to probability, referable to the Deluge of which Moses writes.[35]

Farey believed that universal and violent floods had occurred on the earth, but long before the deluge of Noah, which had been nonviolent. James Smithson, in what is probably a veiled attack on Buckland, wrote in 1824 that, although it had been natural for believers to refer many geological phenomena to the flood, "the success . . . was not such as to obtain the general assent of the learned; and the attempt fell into neglect and oblivion." He then proceeded to attack the latest "revival of this system."[36]

The Baconian tradition of eschewing speculation also reinforced the tendency to ignore the Bible in geological writing. Thomas Thomson, who had been introduced to the Wernerian geognosy by Jameson, criticized in 1819 the reasoning of George B. Greenough on the deluge:

It possesses considerable plausibility as he has placed the arguments. But I have no doubt that Deluc, were he now alive, and in the vigour of his understanding, could write an equally plausible refutation of the whole essay; that Sir James Hall could give an air of plausibility to his doctrine of elevations; and that Mr. Jameson could state very plausible reasons for supposing that the mountains and valleys either existed originally, or have been produced by the gradual action of the weather. The subject does not admit of precise reasoning. I consider it as but of second-rate importance, and am of opinion that those persons who confine the science of geology to such speculations mistake its true nature, and rather injure than promote its progress by calling off the attention of its cultivators from the investigation of facts to loose discussions which are not susceptible of accurate demonstration.[37]

There is evidence that even the *geological* deluge was not considered as well accepted before Buckland. Benjamin Silliman, a Wernerian, although favorably inclined, considered the theory in 1820 as unproved; and Edward Hitchcock stated in 1824 that

. . . the general belief is, that existing streams, avalanches and lakes, bursting their barriers, are sufficient to account for all their phenomena, and not a few geologists, especially those of the Huttonian school, at whose head is Professor Playfair, have till recently been of this opinion.[38]

[35] [John Farey], "Reflections on the Noachian Deluge, and on the Attempt Lately Made at Oxford, for Connecting the Same with Present Geological Appearances," *Philosophical Magazine*, *LVI* (1820), 10–14.

[36] James Smithson, "Some Observations on Mr. Penn's Theory Concerning the Formation of the Kirkdale Cave," *Annals of Philosophy*, *VIII* (1824), 50–51 (dated June 10).

[37] [Thomas Thomson], Review of George B. Greenough's *Critical Examination of the First Principles of Geology*, *Annals of Philosophy*, *XIV* (1819), 373. The growth and extent of the reaction in geology against speculation was noted by W. H. Fitton in his review of the *Transactions of the Geological Society*, vol. II, *Edinburgh Review*, *XXVIII* (1817), 175–177.

[38] [Benjamin Silliman], Review of Horace H. Hayden's *Geological Essays*, *American Journal of Science*, *III* (1821), 48. [Edward Hitchcock], "Review of Buckland's *Reliquiae Diluvianae*," *American Journal of Science*, *VIII* (1824), 332.

William Fitton's review of Buckland's *Reliquiae Diluvianae* was critical, despite Fitton's desire not to offend his friend. He accepted much of Buckland's theory, but he concluded that there was not "sufficient data from which to reason with safety, upon any general question touching the comparison of the antediluvian population, with the actual inhabitants of the globe." He also objected to arbitrary assumptions whose purpose was to make the geological deluge more closely resemble the flood:

> That there has been a deluge, affecting universally all parts of the Earth's surface, and producing every where the same or similar effects, no person who has duly examined the evidence can deny. . . . The only question is, whether that great event . . . is proved by *physical evidence*, to have been *recent*, transient, and *simultaneous;* and upon these points . . . the facts appear to us to afford but imperfect evidence as to *the date*, and still less as to *the duration*, of the submersion. . . . For to speak of the support to the Sacred narrative afforded by extrinsic inquiry, *if the narrative itself be made to form a part of the evidence*, is a mode of reasoning that appears to us to be altogether inadmissible.[39]

Buckland's nemesis was the Reverend John Fleming, of Fifeshire, Scotland. In 1824, at Jameson's request, Fleming began the controversy with Buckland that resulted in the downfall of his theory.[40] Although this controversy is pertinent and interesting, this paper will be concerned only with Fleming's earlier and lesser known writings, which reveal more clearly the connection between his uniformitarianism and Wernerianism.[41]

Fleming studied for the ministry at the University of Edinburgh, where he was exposed to both Wernerian and Huttonian influences. He found the former doctrine more congenial, and through his friendship with the chemist Thomas Thomson, he acquired a love for mineral-

39 [William H. Fitton], Review of Buckland's *Reliquiae Diluvianae*, *Edinburgh Review*, *XXXIX* (1823), 229–232.

40 John Fleming, "Remarks Illustrative of the Influence of Society on the Distribution of British Animals," *Edinburgh Philosophical Journal*, *XI* (1824), 287–305 (dated July 19). Jameson was the editor of this journal; for hints of his devious role in the controversy, see John Duns' memoir in John Fleming, *The Lithology of Edinburgh. . . . Edited, with a Memoir, by the Rev. John Duns* (Edinburgh: William P. Kennedy, 1859), pp. xxxvii–xl. Fleming wrote in 1829 that his friends had assured him that his articles had given "the death-blow to the diluvian hypothesis. Certain at least it is, that, since that time, with the exception of a very few individuals who may still be found on stilts, amidst the 'retiring waters,' the opponents of the hypothesis have become as numerous as were formerly its supporters, and the period is probably not far distant, when the 'Reliquiae diluvianae' of the Oxonian geologist will be quoted as an example of the *idola specus*." See John Fleming, "Additional Remarks on the Climate of the Arctic Regions, in Answer to Mr. Conybeare," *Edinburgh New Philosophical Journal*, *VIII* (1829), 68.

41 For a discussion of this controversy, see my unpublished dissertation, *The Rise of the Diluvial Theory in British Geological Thought*, University of Oklahoma, 1963, pp. 133–150.

ogy. He also, from an early age, had a strong interest in natural history. After leaving the university in 1805, he joined the Wernerian Natural History Society of Edinburgh, whose founder and president was his friend Jameson. He became a fellow of the Royal Society of Edinburgh in 1814, having been proposed by Jameson, Playfair, and David Brewster.[42]

Fleming opposed the Huttonian theory because of its plutonism and because he regarded it as a speculation constructed without examination or knowledge of the rocks. His earliest writings reveal a uniformitarian approach to geology. He believed that "many of the inequalities of the earth's surface owe their existence to the long-continued action of air and water," which is good Wernerian as well as Huttonian doctrine;[43] and he recommended the study of the laws regulating the distribution of recent shells as the best preparation for studying their distribution in the past.[44] He consistently opposed catastrophist explanations, such as "deluging the earth by a comet."[45] Fleming criticized Knight's theory of the earth because it was mostly plutonist hypothesis, but he agreed with Knight (and Playfair) that the scientist must be free to inquire without regard for the Scriptures.[46] While critical of the Wernerian theory in some respects, he defended most of its principles before the 1820's.[47]

During the years 1819 to 1823, Fleming's geological opinions were in a state of flux. By nature he was conservative, always reluctant to alter views once they were firmly established. Yet his writings reveal two opposing tendencies. He tended to favor the Wernerian explanations, and yet he indicated how little faith he put in these and other speculations. His old convictions were shaken by the defection of Wernerians to the vulcanist camp.[48]

42 The best published sources for Fleming's life are the memoirs by John Duns (see note 40) and Alexander Bryson, *Transactions of the Royal Society of Edinburgh, XXII* (1861), 655–680.

43 John Fleming, "On the Mineralogy of the Redhead in Angusshire," *Memoirs of the Wernerian Natural History Society,* Vol. II (1811–1816), pp. 343–345, 349, 352. This was read on Feb. 4, 1815. For criticisms of Hutton, see also in the same volume, "Observations on the Mineralogy of the Neighbourhood of St. Andrew's in Fife," pp. 153–154 (read Feb. 5, 1813).

44 [John Fleming], "Conchology," *Supplement to the Fourth, Fifth, and Sixth Editions of the Encyclopaedia Britannica* (6 vols.; Edinburgh: Archibald Constable and Company, 1824), Vol. III, p. 316 (written before 1818).

45 John Fleming, "Observations on the Orthoceratites of Scotland," *Annals of Philosophy, V* (1815), 200.

46 [John Fleming], Review of Knight's *Theory of the Earth, Edinburgh Monthly Review, I* (1819), 356.

47 Fleming, *Mem. Wern. Nat. Hist. Soc.,* Vol. II, p. 365.

48 [John Fleming], Review of A. Boué's *Essai Geologique sur l'Ecosse, New Edinburgh Review, III* (1822), 372.

Despite his distrust of theory, Fleming was appalled and disgusted by the skepticism of Greenough's *Critical Examination*, which he felt would only injure geology. Since his attitude to Greenough's work was shared by other geologists, his comments are worth quoting:

> Before taking our leave of Mr. Greenough, we feel disposed to state candidly, that his performance is one by no means calculated to advance his own reputation, or promote the interests of geology. . . . He appears to be totally unacquainted with the laws of evidence. With him the testimony of every mineralogist is of equal value; all are supposed to have been equally well informed, and to have studied the subject with equal care. . . .
>
> There has resulted from all this a kind of geological scepticism, which we regard in this instance as the index of a mind unaccustomed to philosophical induction, but which others may consider as the mark of free and independent thinking. It is calculated to disgust the adept, and to perplex the tyro. It has been publicly intimated, that the author is a gentleman of independent fortune, and that he has expended large sums in furnishing an extensive collection of minerals. These circumstances, while they aggravate his errors, and render him more dangerous as an authority, recall an expression of the late illustrious Playfair in reference to De Luc, "to *reason* and to *arrange*, are very different occupations of the mind; and a man may deserve praise as a mineralogist, who is but ill qualified for the researches of geology."[49]

Fleming was also dissatisfied with Cuvier as a geologist, and with Jameson for approving his views:

> Now here we are left to choose between Dr. Hutton and M. De Luc; as the former elevates and the latter depresses the strata. Not a hint of Werner all this time; although we did believe that Cuvier's works were to confirm those of the philosopher of Freyburg. Must we understand that the editor of this preface has read it with approbation, and sanctioned it with his name and his preface too; or are we mistaken in supposing that he is teaching and publishing doctrines the reverse of these? We are ourselves indifferent good Wernerians, but are always ready for conviction; and as we are also lovers of peace, shall be most delighted to find that there is some chance of agreement among the several belligerent powers on these subjects. Werner says decidedly, if we have not misapprehended what we learnt when at college, that there have not been any subversions of strata, either by depression or elevation, but that all rocks were deposited as we now see them, having been crystallized, compound and simple, veins, conglomerates and all, from solutions in the waters of the sea. For our parts, we are ready to believe any thing, provided we are told what it is to be; Werner, or Hutton, or Whiston, it is all one; but we cannot believe one thing one day, and the reverse the next.

With reference to Cuvier's claim to priority in the recognition of the value of fossils in geology, Fleming commented:

49 [John Fleming], Review of Greenough's *Critical Examination, Edinburgh Monthly Review, IV* (1820), 571.

We have a notion that M. Cuvier is better acquainted with his own writings than with those of other geologists. He could not else have imagined that he was among the first to point out the importance of organic remains. . . .

After pointing out that many geologists, including Smith, had recognized this, he took up Jameson's claim that Werner deserved credit:

We must pardon M. Cuvier in this particular case at least. But for Professor Jameson's friendship, we might have made the same mistake ourselves, and imagined that Werner was entirely ignorant of the organic department of geology. If a man will "die and make no sign," how can we know his meaning?

However, he did not agree with Cuvier's estimate of the importance of organic remains in geology in which, Fleming believed, mineralogy was the foundation "and building too." He was also unconvinced by Cuvier's revolutions:

Surely M. Cuvier does not mean to say that, if animals are not found in any particular alluvial strata, it is because their races were destroyed by irruptions of the sea, and that such an absence is a proof in itself of the revolutions and deluges in question. He must know better than we do, that a race of animals may disappear from a particular tract in consequence of many causes, dependent, not on geology, but on their own habits or natures. . . . [He mentions the destruction of forests, disease, famine, enemies, and man.] . . . We do not happen to know what the opinion of Werner was on this subject, or whether, indeed, he had ever considered it. We may possibly have misapprehended his general views, as it is not easy to get at those doctrines of his which have never been properly before the public; but our general impression is, that he does not admit of such interminable revolutions of the sea and land after the deposition of the latest rocks, whatever these may be. As all his opinions must be considered valuable, we rather regret that Professor Jameson has not informed us what the fact is, and we shall be thankful for that information at some future day.

Fleming questioned Cuvier's belief that the genera change with the strata, claiming that the latter was refuted by the evidence in his own book, which showed that the greater part of the genera in the transition limestone "are found all the way up, . . . through the greywacke, and through all the secondary strata, up to the chalk—nay, to the very living ocean itself."

Cuvier's stand against the evolution of man and in favor of his recent origin met Fleming's approval, although he could see none but a sinister purpose in bringing up subjects on which there was such universal agreement:

He argues like a sound naturalist, and well, against the philosophers, who, like Darwin, De Maillet, and others, chose to imagine that man had volunteered from the condition of an oyster into his present service. These personages do not deserve so much attention.

Among all the wildest theories of geologists, there is not one who has ever thought of giving to the human race a higher antiquity than that which is assigned by Scripture, and which is amply confirmed by every thing that we know of the progress of human society, arts, and languages. These were matters fully admitted before Cuvier undertook to prove them; and we can see no purpose in bringing them forward again, unless it be to excite doubts in the minds of those who never examined the question.[50]

This attitude was undoubtedly a common one among scientists, which perhaps explains the conspiracy of silence that existed until Lyell with respect to Lamarck's views in Great Britain.

As late as 1822, Fleming was willing to allow that the flood had produced great changes in life on the earth, although he attributed changes in the earth's surface to nonviolent causes. At the same time, however, he argued that many changes in life were explainable by natural causes, and he soon would question whether any detectable change could be referred to the flood.[51]

It is not necessary here to go into Fleming's subsequent history. Enough has been shown, I think, to suggest that there were strong uniformitarian tendencies in Wernerianism that may have facilitated the later acceptance of Lyell's more comprehensive uniformitarian views. Lyell's work would also express the feelings of the majority of geologists that their science should be kept free from involvement with religion. On this subject, Lyell was to get much more support than on his uniformitarian thesis.

50 [John Fleming], Review of Cuvier's *Essay on the Theory of the Earth* (4th ed.), *New Edinburgh Review, IV* (1823), 383–384, 387–395. Jameson later admitted that Werner had not advocated "the geological diluvian hypothesis," *Edinburgh Philosophical Journal, XI* (1824), 299.

51 John Fleming, *The Philosophy of Zoology; or a General View of the Structure, Functions, and Classification of Animals* (2 vols.; Edinburgh: Archibald Constable & Co., 1822), Vol. II, p. 104.

The Geological Atlas of Guettard, Lavoisier, and Monnet: Conflicting Views of the Nature of Geology

Rhoda Rappaport

The *Atlas et Description minéralogiques de la France* (1780) was once described by Sir Archibald Geikie as the first of the national geological surveys. While it is true that there was, at least in scale, no comparable work until the nineteenth century, Geikie's characterization is nonetheless misleading, suggesting as it does that this early French survey was a direct ancestor of the massive productions of the next generations. Such a comparison is, in fact, no compliment to Guettard, Lavoisier, and Monnet—although Geikie clearly intended it to be—because their *Atlas* falls far short of the revolutionary ideas and methods developed and used so effectively by William Smith, Georges Cuvier and Alexandre Brongniart, and their successors.

To understand and appreciate the Guettard *Atlas*, it is more appropriate to study these maps in their own eighteenth-century context; that context might be summarized briefly as, first, a lack of established cartographic methods to guide Guettard and his collaborators, and second, a certain degree of confusion and disagreement about the aims and even the subject matter of geology. In other words, when the various eighteenth-century maps and the explanatory memoirs published with them are examined in detail, it is clear that geological cartography in this period was a personal affair, largely dependent upon the views of the individual cartographers. There are, to be sure, some problems, ideas, and methods shared by many of these early map-makers, but to compare the differences rather than to chart the similarities is an exercise that can shed light on the development of geological thought in a period of rapid change and controversy. As the collaborative work of three geologists, the Guettard *Atlas* offers a particularly good example of conflicting ideas about the nature of geological science and the purpose of geological maps.

The construction of the *Atlas* began as early as 1746, when Guettard produced his first "mineralogical map" of France. Intended as a preliminary survey, this map was later modified in detail over a period of almost 40 years as Guettard continued to gather information about the location of mineral deposits and rock formations. In 1766, Guettard's work came to the attention of Bertin, Minister and Secretary of State, whose own concern with the efficient use of French natural resources made him look with favor on the project which, until that time, Guettard had carried out alone. The financial support arranged by Bertin and the collaboration of Lavoisier began almost simultaneously, and while the former soon ended, Lavoisier continued to work with Guettard for about 12 years, until 1777. At that date, with 16 quadrangles completed and perhaps an equal number close to completion — out of a projected total of 214 — Antoine Monnet took over the direction of the survey and in 1780 published the set of 31 maps known as the *Atlas*. Additional maps were to appear some years later, bringing the final total to 45 completed quadrangles.[1]

Although more might be said about these events, three points in this brief chronicle are of particular significance here. In the first place, Guettard is obviously the key figure, since it was he who originated, designed, and brought to completion the preliminary map of France; for the next 20 years, he gathered data that were to be useful in preparing the more detailed quadrangles; and it was his pioneer work that attracted Bertin's attention. A second point is the early date at which Guettard began his work, before the publication of such classics as Buffon's "Théorie de la terre" (1749) and Desmarest's geological map of Auvergne (1774); some features of the *Atlas* are therefore more easily understood as the product of an earlier scientific generation to which Guettard belonged. Finally, there are the very different men involved; if one were to apply a label to each, Guettard might best be called a naturalist, Lavoisier a stratigrapher, and Monnet a mineralogist.

1 Denis I. Duveen and Herbert S. Klickstein, *A Bibliography of the Works of Antoine Laurent Lavoisier 1743–1794* (London: Dawson and Weil, 1954), pp. 236–244, with additions and corrections in Duveen, *Supplement to a Bibliography of the Works of Antoine Laurent Lavoisier 1743–1794* (London: Dawsons of Pall Mall, 1965), pp. 129–132. In addition, Guettard prepared five quadrangles describing the province of Dauphiné; these were not published and cannot now be located. See Archives nationales, F^{14}1312, and E. J. A. d'Archiac, *Introduction à l'étude de la paléontologie stratigraphique* (2 vols.; Paris, 1864), Vol. I, p. 294.

They differed widely in professional training, interests, and ability, and the fact that they were engaged in an enterprise without precedent gave each one the hope that his own ideas might guide the survey. At the same time, whatever their hopes, Lavoisier and Monnet were able only to modify to a greater or lesser degree the project as conceived by Guettard and sanctioned by Bertin.

To analyze the maps of the *Atlas*, it is essential to begin by examining the state of geology in the 1740's and the ideas of Guettard. Geological cartography at that date can immediately be dismissed as virtually nonexistent. There is no evidence that Guettard then knew either the early hydrographic map by the Abbé Coulon or the "chorographic chart" by Christopher Packe, and certainly neither of these looks like a possible model for Guettard's distinctive mineralogical map.[2] If we turn to the geological literature before 1746, it is clear, too, that there existed no generally accepted principles, theories, or defined subject matter that Guettard might have been expected to adopt as the basis for his cartographic work. The pioneer treatise of Steno seems, in retrospect, to mark the start of modern geology, but in Guettard's youth, Steno, insofar as his work was known, was only one of several authors whose writings seemed to pull in different directions. It would have taken a mind of some brilliance and daring to attempt to transform into a coherent discipline the achievements of Steno, the speculations of Leibnitz, the paleontological memoirs of Réaumur and Jussieu, and the natural theology of the Abbé Pluche and John Woodward. In 1746, however, Buffon's great synthesis was still three years away.

After the publication of Buffon's "Théorie de la terre," there was discernible in France a new geological climate of opinion that, in some ways at least, is directly traceable to the influence of Buffon. Whatever the merits of Buffon's treatise, it was undeniably a best-seller and it succeeded, too, in bringing together, dramatically and eloquently, ideas that had never quite been combined in a single synthesis and that might, in certain cases, have remained neglected. Although Buffon's synthesis did not go unchallenged, it did provide geologists with a common fund of information and ideas and a common point of reference in the ensuing debates. Indeed, as Jacques Roger has shown, Buffon's synthesis brought to a close one era in the history of geology and opened another during which geologists paid more attention to the solution of specific

[2] I have been unable to locate a copy of Coulon's map, but it is described briefly by Archiac, *Géologie et paléontologie* (Paris, 1866), p. 106. Packe's map, published in 1743, is reproduced by E. M. T. Campbell, "An English Philosophico-Chorographical Chart," *Imago Mundi*, 6 (1949), 79–84.

problems than to the broader issues involved in a "theory of the earth." As a result, Buffon's subsequent masterpiece, *Les Epoques de la nature* (1778), was published some 20 years too late and possessed "un certain air désuet, et comme un parfum de passé."[3]

Among the issues treated by Buffon, those most relevant here are the questions of the extent of the geological time scale and the general structure of the earth's crust. Buffon described the earth's crust as consisting of a primitive — to him, igneous — core surrounded by a series of sedimentary formations. Explicit, too, was the idea that the deposition of sedimentary strata had required a long time and that to chart the history of successive formations would serve to reconstruct the history of both the earth's surface and the forms of life that had inhabited it. Whether or not Buffon's contemporaries took these ideas from his "Théorie de la terre" or from some other source is less important than the fact that the ideas gained wide currency after 1749. Geologists might (and did) disagree about the nature of the primitive core of the earth, about the simple dictinction of Primary and Secondary formations, and about the precise extent of the time scale; in general, however, they could no longer ignore the broad proposition that the earth's crust had been formed in stages, so to speak, and over a period of time.[4]

If the foregoing generalizations about "pre-Buffon" and "post-Buffon" geology are at all valid, one would surely expect geological map-makers after 1749 to reveal some interest in the general stratigraphic structure of the earth's crust. In very different ways, this is certainly true of the maps and memoirs of J. F. W. Charpentier, Nicolas Desmarest, and J.-L. Giraud-Soulavie, to name only three of the later cartographers. Significantly, however, this was true neither of Guettard's first map nor of the revised versions he subsequently produced. Ignoring for a moment the shaded regions on the map (Fig. 1), Guettard has simply dotted the map's surface with symbols representing rock formations and mineral deposits.[5] There is no geological purpose apparent here, no indication of stratigraphic levels, no tracing of the extent of any one formation. Given the geological confusion of his youth, Guettard was at liberty to devise his own cartographic methods

[3] Buffon, *Les Epoques de la nature,* ed. Jacques Roger (*Mémoires du Muséum national d'histoire naturelle,* Série C, Tome X, Paris, 1962), Introduction, esp. pp. lix–lx, cxxix–cxliv.

[4] R. Rappaport, "Problems and Sources in the History of Geology, 1749–1810," *History of Science, 3* (1964), esp. pp. 65–66, 68–69.

[5] Copies of this map are to be found at the Bibliothèque nationale (cf. Duveen and Klickstein, *Bibliography,* p. 240), Ecole des mines, and Academy of Natural Sciences of Philadelphia.

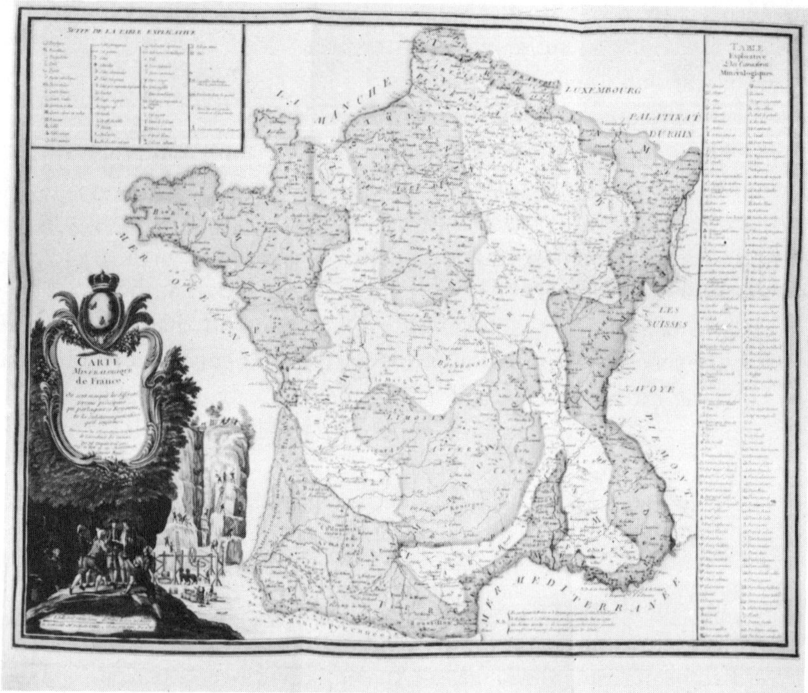

FIG. 1. *Guettard's "Carte Minéralogique de France," 1784.*

and to express on this map his personal views about the nature of geological science.

Guettard's geological ideas are a peculiar blend of his scientific education, religious convictions, broad interests as a naturalist, and a wide streak of utilitarianism. His formal education was in medicine, in which he received his degree in 1742, but his stronger inclination was toward botany. Early field trips with his grandfather, a pharmacist and amateur botanist, had been followed by friendships with the Jussieu brothers at the Jardin du Roi and with Réaumur, then owner of one of the finest natural history collections in France. Guettard became curator of Réaumur's collection, research assistant to Réaumur, and, gradually, a versatile naturalist in his own right. By 1747 he was "médecin-botaniste" to Louis, Duc d'Orléans, and from that date had a chemical laboratory at his disposal, a growing natural history collection to care for, and the beginnings of an international scientific correspondence of considerable scope and volume.

Throughout his career, Guettard showed himself to be a meticulous observer of nature, cautious in his reactions to new ideas, and especially interested in technological progress. In his youth, he was something of a radical in, for example, his early conversion to the Linnaean system of classification and in his desire, so much like Réaumur's, to apply scientific knowledge to the improvement of industrial and agricultural technology. Once committed to these ideas, he was to remain an impassioned defender of Linnaeus and to become the author of numerous memoirs on such subjects as paper-making and the use of kaolin and petuntse in the manufacture of porcelain. At the same time, he showed a distinct aversion to scientific system-building, traceable in part to his fear of materialism and in part to his belief that not enough evidence existed to support any system; he therefore attacked with equal vigor the biological theories of Charles Bonnet and J. T. Needham and the cosmological ideas of Buffon. The building of systems he believed to be premature, doomed to failure, and a waste of valuable time and energy that might better be spent in the gathering of data; but systems were dangerous, too, because wrong ideas tend to pollute both the heart and the soul.[6]

The results of Guettard's beliefs and interests are evident in virtually all his writings. He was a fact-gatherer of inexhaustible energy, extracting information from scientific publications, travel accounts, encyclopedias, almanacs, newspapers, and personal correspondence. His published memoirs are long, erudite, and tedious, often consisting of minute descriptions of a single stone quarry or a particular fossil, and often accompanied by historical introductions summarizing the opinions of experts from Aristotle onward. The talent he most conspicuously lacked was that of generalization, of seeing the implications of his own observations.

There were times, of course, when Guettard did try to bring coherence to his diverse observations, and one such occasion arose in connection with his first mineralogical map. He had noticed during his earliest field trips that there was "a certain regularity" in the distribution of minerals and rocks over the earth's surface, and he decided to plot his observations on a map to see what patterns emerged. By comparing the different quantities of key substances found in each region he traversed or read about, he concluded that France could be divided into three concentric zones or *Bandes*: the Sandy, the Marly,

6 These remarks are based on statements scattered through Guettard's published works and his unpublished papers at the Muséum d'histoire naturelle. The standard biography is still Condorcet's "Eloge de M. Guettard," *Histoire de l'Académie royale des sciences, 1786* (1788), 47–62. See, also, Roger in *Les Epoques de la nature*, p. cxxxix, note 7.

and the Schistose or Metalliferous (see Figure 1). The *Bande sableuse* consisted primarily of sandstones and limestones, the *Bande marneuse* of marl, and the *Bande schisteuse* of metallic ores, schists, slates, granites, marbles, and other less common substances; some clayey and calcareous materials were actually present everywhere, but were more prominent in one zone than another, and the same could be said of fossiliferous deposits. Although these *Bandes* have since been said to follow the general contours of the Paris Basin, it should be emphasized here that stratigraphy was of no concern to Guettard; he nowhere indicates the relationships among the three *Bandes*, and he placed his symbols without reference to the stratigraphic level of each formation or deposit.[7]

What, then, was Guettard's purpose in constructing his map? At one point he remarks that observations such as his will hopefully lead one day to the formulation of "une théorie physique & générale de la Terre," a comment that suggests he was thinking of the structure of the earth's crust.[8] The same interpretation is supported by the fact that Guettard went on in subsequent years to construct mineralogical maps of Switzerland, Poland, the Middle East, and much of North America; each of these maps is marked off in *Bandes*, and the impression is inescapable that Guettard was gradually applying his system to as much of the globe as his observations and reading would permit.[9] In the memoir of 1746, however, Guettard states his more immediate purpose as the practical one of locating important natural resources and learning where to search for valuable mineral deposits, good building stone, and useful agricultural and industrial soils. In 1752, announcing his plans to continue his one-man survey, he described his map as being of especial value to civil engineers, who had sometimes built roads of less durable materials when they were ignorant of the resources of particular French provinces.[10] Admittedly, Guettard was here trying to interest the government in his survey, and utility was eventually to prove a strong selling point. But a strikingly similar point of view is apparent in a later memoir that contains Guettard's only known comment on the vertical type sections constructed by Lavoisier: showing the stratigraphic "composition" of a given region, says Guettard, will enable one "to judge the cost of obtaining one or another of these substances." For the scientist, he continues, these sections are an important

7 Guettard, "Mémoire et carte minéralogique sur la nature & la situation des terreins qui traversent la France & l'Angleterre," *Mém. Acad. roy. sci., 1746* (1751), 363–392.

8 *Ibid.*, p. 363.

9 *Mém. Acad. roy. sci., 1763* (1766), 228.

10 *Journal œconomique*, June 1752, pp. 113–138.

illustration of "the varieties that nature can employ in the arrangement of the [strata] of the earth's mountains."[11]

A well-read naturalist, Guettard was not unaware of the newer currents in geology after 1750. He clearly knew about contemporary studies of sedimentation rates, the evidence for extinction of species, and many other ideas and observations that suggested a long history for the earth's crust, a history that could be charted stratigraphically. Indeed, he was himself the author of three lengthy memoirs on sedimentation and erosion, he came to believe that the earth's crust consisted of superimposed primitive and sedimentary formations, and he frequently emphasized the need for comparative anatomical work to shed light on the problem of the possible extinction of species. In attacking all hypotheses about the formation of the earth, he also stressed his own interest in the *trans*formations of the earth's crust. Furthermore, Guettard understood how to assemble a type section and occasionally gave verbal descriptions of "coupes générales." In connection with one such description, he even remarked that variations in particular outcrops are perhaps only "an accidental disruption in the general structure common" to a given region.[12]

Despite all the evidence that Guettard knew, understood, and contributed to the development of new geological ideas, most of his work reveals, nonetheless, that he tried hard to avoid thinking of the earth as having much of a history. Thus, when his observations presented him with a sizable sedimentary formation, he could only suggest that the whole mass had been deposited at once and that the effect of bedding had then been produced by a drying and cracking "en quelque sorte suivant une direction horizontale."[13] Similarly, he refrained from discussing the age of the earth, equivocated on the subject of the geological effects of the biblical flood, and constantly denied the likelihood of species becoming extinct. In one of his rare perceptive moments, Antoine Monnet summarized Guettard's problem thus: "he resisted [the implications of] his own observations and Moses was always for him a divine and infallible oracle, and his six thousand years of antiquity assigned to the earth was the barrier which halted his [Guettard's] observations in many respects."[14]

[11] *Observations sur la physique, sur l'histoire naturelle et sur les arts, 5* (1775), 361. For an explanation of the meaning of "type section" as used throughout this paper, see later text and note 21.

[12] In manuscript, "Lavoisier et Guettard: Géographie minéralogique de la France," Vol. II, "Valois" (Lavoisier Collection, Cornell University).

[13] *Mém. Acad. roy. sci., 1756* (1762), 237.

[14] Monnet Papers, Ecole des mines, MS 8286, pp. 120f.

The tortuous paths of Guettard's reasoning do much to explain the peculiar nature of his contributions to the *Atlas*. Although still committed to his own system of *Bandes*, he decided to eliminate all theoretical controversy in the construction of the quadrangles of the *Atlas*.

> We have not followed any *systèmes physiques* in the construction of these maps, that is to say, we have not classified terrains as primitive or secondary, as some scientists do, nor as schistose or metalliferous, marly, or sandy.[15]

As a result, the maps of the *Atlas* resemble in their general design Guettard's map of 1746: conventional symbols have again been used to locate the major deposits, and stratigraphy has been ignored (Fig. 2).

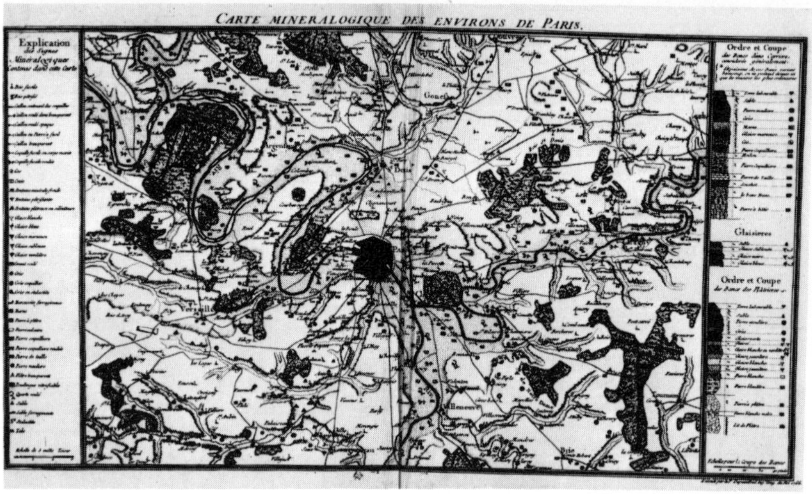

FIG. 2. *Guettard and Lavoisier, "Carte minéralogique des environs de Paris," 1766 (Map no. 40bis).*

The radical differences between Guettard and Lavoisier as geologists are symbolized in Figure 2 by the neat line separating the right-hand margin from the rest of the map. Lavoisier's type sections would undoubtedly have posed a technical problem had he attempted to show this third dimension on the map itself. The problem, however, was not insurmountable, but Lavoisier never had the opportunity to experiment with the technique he had devised.

Lavoisier's training in geology was of a fairly common sort for a young Parisian educated in the 1760's. As he later remarked, his three teachers were Rouelle, Buffon, and Guettard,[16] and the lectures of Rouelle and the writings of Buffon were easily accessible to many of

[15] *Observations sur la physique*, 5 (1775), 360.

[16] *Œuvres de Lavoisier* (6 vols.; Paris: Imprimerie impériale et Imprimerie nationale, 1862–1893), Vol. V, p. 226. See Henry Guerlac, "A Note on Lavoisier's Scientific Education," *Isis*, 47 (1956), 211.

his contemporaries. Rouelle's lectures included some discussion of the then-current theory that the earth's crust consisted of a primitive core, formed by a process of aqueous crystallization, and a series of more recent sedimentary formations. Here Lavoisier learned, too, about sedimentation, erosion, and the nature of marine fossils. In the same period, beginning in 1763, he gained considerable experience in the field during excursions often made in Guettard's company. Guettard also taught the younger man his own system of *Bandes*, and the two geologists are known to have discussed the merits of Buffon's *Histoire naturelle.* Lavoisier was impressed by Buffon's work, despite Guettard's criticisms of it and despite eventual doubts of his own about the accuracy of some of Buffon's geological observations.[17]

During his earliest field trips, in the years 1763 through 1766, Lavoisier devoted much of his time to applying the lessons of his teachers and developing certain practical skills. His field notes show his concern with surveying techniques and with the chemical analysis of minerals and rocks, the latter activity resulting in two memoirs on gypsum (1765 and 1766). In the same period, possibly following the example of Rouelle, he began to use the barometer to measure the altitude and thickness of geological formations.

Lavoisier's insistence on precise measurement is one of the striking differences between his work and that of his teachers. How he intended to use these data is not always as clear as one could wish, although an early note offers some hint: "I am thinking of surveying the bed of the Seine in order to produce a map."[18] His next statement of this kind, probably written in the summer of 1766, offers startling clarification of his intentions; here he announces that he has begun and expects to continue a series of barometric measurements that will yield

> precise information about the former limits of the sea, the bed which it occupied, the former arrangement of land masses; in short, a system . . . [describing] the changes which the earth has undergone.[19]

This tantalizingly brief statement certainly suggests that Lavoisier was working toward a sophisticated view of the evolution of the earth's crust, and later in the same year, 1766, he was to explain his ideas in some detail.

[17] Lavoisier Papers, Académie des sciences, dossiers 419, 420 (notes dated 1763 and 1765).
[18] *Ibid.*, dossier 417 (May 1765). Rouelle or one of his friends had earlier made a series of similar measurements. "Cours de chymie de M. Rouelle," Bibliothèque municipale de Bordeaux, MSS 564–565, pp. 571–572, and "Cours de Chimie de m^{r}. Rouëlle Pour l'année 1761—30 octobre," p. 404, manuscript in collection of the late Dr. Pierre Lemay.
[19] *Œuvres de Lavoisier,* Vol. III, p. 109.

Whether the ideas implicit in Lavoisier's statement were based to any great extent upon those of his teachers or other contemporaries is at present only a matter for conjecture. Lavoisier himself implies that his thinking was stimulated by his teachers and then modified considerably by his observations in the field. Thus, for example, he remarks that Rouelle had taught him the differences between pelagic and littoral deposits — he could easily have learned this from Guettard as well — but, until 1766, he was under the impression that sedimentary formations in any one locality should be *either* pelagic *or* littoral. During his field trips he was therefore surprised to notice how frequently limestone was to be found superimposed on coarse sandstone; his reaction, as he describes it, was to begin to rethink his geological principles and to play with the notion that alternating pelagic and littoral beds were evidence of the "flux and reflux" of the sea.[20]

Once Lavoisier had arrived at this key idea, his barometric measurements took on added significance: they could be used to trace the lateral extent of formations, to estimate the time needed for deposition, and, eventually, to determine the changing "limits of the sea . . . [and] the former arrangement of land masses." But Lavoisier also recognized that comparing two outcrops of the same formation revealed some variations in the number and nature of strata and that to reconstruct the geological past required that he assemble vertical type sections. His own explanation of what such a section represents is worth quoting, since it shows not only his grasp of the problem but also his difficulty in expressing what was then a new concept.

> These are the only two mountains on this part [i.e., quadrangle] of the map which wholly correspond to the [general] section just described; but the remarkable thing is that most of them or, more accurately, all the mountains of this region are almost precise fractions of the same section; so that, to explain myself more clearly, if one were to remove the top quarter, third, or any other fraction of the general section, what remained would describe [in turn] the [structure] of all of these mountains.[21]

Significantly for the *Atlas*, Lavoisier developed these stratigraphic ideas in 1766, the year in which his most active collaboration with

[20] Lavoisier Papers, Académie des sciences, dossier 421 (11–17 October 1766).

[21] *Œuvres de Lavoisier*, Vol. V, p. 75. Although undated, this document concerns a geological map completed and engraved in 1766. Lavoisier always used the phrase "coupe générale," and this is more precise than my own use of "type section." It was pointed out to me by several members of the New Hampshire Conference that, in its modern meaning, a type section is a section observed in the field and used as a standard for comparison with other observed sections. By contrast, Lavoisier's type sections were ideal, composite constructions based on a series of observations; but he used these constructions for a purpose similar to the modern one, i.e., as a standard of comparison.

Guettard began. To see how Lavoisier's ideas were put into practice, it is only necessary to look at the margin of one of the early quadrangles (see Figure 2). This map of the environs of Paris dates from 1766, and similar "coupes générales" are to be found on many of the maps to which Lavoisier contributed; in the more geologically disturbed regions of Alsace, Lorraine, and Franche-comté, he apparently found it impossible to construct type sections, and the margins of those maps show the strata in particular mines or quarries.

Lavoisier's long-range plan soon became to write not a traditional "theory of the earth," but rather a geological history of France in which he would examine the state of the crust during successive epochs. How fully developed this project was in 1766 is not known, since his only detailed explanations of it were written some 20 years later. But it can be said with certainty that the stratigraphic ideas on which he meant to base a geological history of France were worked out in 1766. It is equally clear that Lavoisier later considered his contributions to the *Atlas* as the beginnings of that treatise. Unfortunately, Lavoisier's treatise was never written, although it remained on his list of scientific projects as late as 1792.[22]

It need hardly be said at this juncture that Guettard and Lavoisier inhabited different geological worlds. To Lavoisier, at least, their points of view were potentially reconcilable, and he apparently suggested to Guettard at an early date that they work out a compromise in preparing the quadrangles of the *Atlas*. Exactly what Lavoisier proposed is no longer known, although he later explained that Guettard's system of symbols might simply have been modified slightly in order to reveal the stratigraphic arrangement of the earth's crust.

> In order to indicate the size and extent of each deposit, it is only necessary to link all similar symbols by lines which would show not only the size and extent [of each deposit] but also their various points of intersection....[23]

Whether this or some other method was suggested by Lavoisier in the 1760's, his proposals were rejected and he was permitted only the use of the margins of each quadrangle. It is tempting to attribute Lavoisier's partial failure to his scientific differences with Guettard, and this interpretation is, in fact, a reasonable one when it is recalled that Guettard had only a limited appreciation of the purpose served by type sections

[22] Lavoisier Papers, Académie des sciences, dossier 335C, fols. 17r–17v, and *Œuvres de Lavoisier,* Vol. V, pp. 205–213. See, also, Rappaport, "Lavoisier's Geologic Activities, 1763–1792," *Isis, 58* (1967), 375.

[23] Quoted in Henri Pigeonneau and Alfred de Foville, *L'Administration de l'agriculture au Contrôle général des finances (1785–1787): Procès-verbaux et rapports* (Paris: Guillaumin, 1882), p. 87.

and that he also preferred to eliminate from the *Atlas* any kind of geological "system." But Lavoisier's failure must also be attributed in part to the fact that the survey was a government-sponsored enterprise.

Too little is known about the relationship of Bertin to the *Atlas*. One can only assume that, in view of Guettard's seniority, reputation, and pioneer cartographic work, Bertin would have given greater weight to his opinions than to Lavoisier's. There is indisputable evidence that Bertin did share Guettard's practical views about the purpose of geological maps, and he referred to the survey as a project to be completed quickly for the benefit of "ceux qui s'occupent de la recherche des mines."[24] Furthermore, Bertin was an official in a chronically impoverished regime; much as he might desire economic reform based on scientific knowledge—and his other activities show that this was, indeed, the case—he had neither the funds nor the time to support a complex undertaking that might bear fruit in some 20 or more years. As Lavoisier put it:

> . . . the work had been begun; the minister . . . had only very limited funds at his disposal and he could not even be sure of having them for long; it would perhaps have been unwise to undertake a project so disproportionate to the means available to carry it out.[25]

From Bertin's point of view, then, speed was essential, and Guettard's cartographic method, already in use on Guettard's earlier maps, would serve well enough.

Bertin's financial problems and his desire for a speedy conclusion to the project were also responsible for the appointment, in 1777, of Antoine Monnet as director of the survey. Progress had been slow in the preceding four or five years, and Monnet had suddenly offered to complete the *Atlas* quickly and economically. Since Monnet was already an Inspector General of Mines, making periodic tours at government expense, it is hardly surprising that Bertin should have accepted this proposal.[26]

Although Monnet took up his task so late in the history of the survey, he was still able to impress upon it some of his own geological ideas. Indeed, he had hoped to alter the maps even more radically but was constrained to follow the general pattern established by his predecessors.[27] His principal contribution to the maps was the elimination of the type sections in those quadrangles either begun by Lavoisier and

24 Bertin to Guettard, 14 December 1767, Bibliothèque municipale de Clermont-Ferrand, MS 473, fol. 67.
25 *Œuvres de Lavoisier*, Vol. V, p. 205.
26 Monnet Papers, Ecole des mines, MS 5527, pp. 226–227.
27 *Atlas*, pp. v–vi. The alterations Monnet had in mind are not specified.

completed later or prepared in their entirety by Monnet. This procedure can be fully documented in only one case, in which the Guettard-Lavoisier map and Monnet's revised version both exist (see Figures 2 and 3).[28]

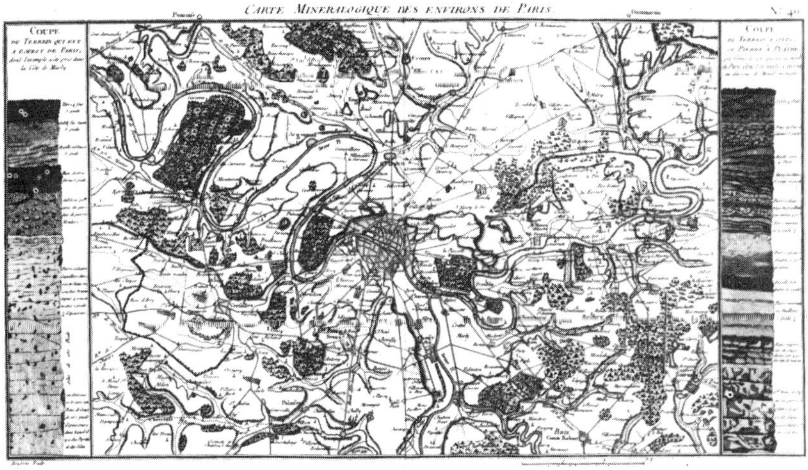

FIG. 3. *Monnet, "Carte Minéralogique des environs de Paris" (Map no. 40).*

Monnet's own geological views were strikingly different from those of Guettard and Lavoisier, but the net result of his beliefs, as far as the *Atlas* was concerned, was a certain sympathy with the aims and methods of Guettard. Monnet's training in chemistry and his early association with influential patrons had, by 1767, obtained for him a post as a traveling student and inspector of mines on behalf of the Bureau du Commerce. A few years later, Bertin began to avail himself of Monnet's talents, finally, in 1776, awarding Monnet the title, "Inspecteur général des mines et minières du royaume." By both training and profession, then, Monnet's interests were largely practical and technological. Thus, although lacking the broader education of Guettard, Monnet did share with Guettard a utilitarian outlook as well as considerable experience in the field.

In his personal and scientific career, Monnet was undoubtedly a maverick. Historians of chemistry know him as one of the last and most violent supporters of the phlogiston theory; historians of mineralogy describe him as a place-seeker and an egotist. He seems to have enjoyed

[28] The only known copy of the *Atlas* to contain both of these maps is now at the Ecole des mines. For a brief description, see Duveen, *Supplement*, p. 131.

attacking his contemporaries both personally and professionally, setting himself up as one of the few progressive, clear-sighted, and unappreciated thinkers and observers of his day. Accordingly, his philosophy of science was sweepingly negative: Nature's law is to have no law.

> Nature never adopts any general rule without also creating many exceptions; it would in fact be contrary to her power and splendor if, once she had traced a path, she could never depart from it.[29]

This remarkable statement was not intended to have theological overtones. Indeed, Monnet prided himself on his emancipation from all religious dogmas and scientific "systems." His aim was pure and accurate observation, untainted by preconceptions.

Monnet's negative philosophy and his practical concerns combined to determine his attitude toward geological type sections. Although he understood Lavoisier's methods, he explained more than once that he preferred to eliminate type sections and to represent the order and nature of the strata *as he observed them* rather than as they might be reconstructed. In his verbal descriptions, but not on the maps themselves, he often pointed out that had he selected another outcrop for his diagram he would have had to include certain observed variations in the strata; and he did recognize that many more such variations would be taken into account in a general section. But, he concluded, a type section cannot represent all the infinite variety in nature, and the whole enterprise is pointless if one's aim is to "indicate precisely the place where one has found a [particular] mineral."[30]

Monnet's other major contribution to the *Atlas* was the text he published to accompany the maps. It was Guettard's habit to issue memoirs along with his early maps, and he probably would have continued the practice had he remained associated with the survey after 1777; in fact, Guettard's unpublished papers include drafts of memoirs on the geology of virtually all of France. Lavoisier, too, may have had some sort of text in mind, perhaps envisioning his treatise on the geological history of France as a commentary on a completed atlas. The text published in 1780, however, was entirely the work of Monnet, despite the appearance of Guettard's name on the title page. As one

[29] Monnet Papers, Ecole des mines, MS 4685, I, 451. Also, Hélène Metzger, *La Genèse de la science des cristaux* (Paris: F. Alcan, 1918), p. 69. For Monnet's career and personality, see Louis Aguillon, "L'Ecole des mines de Paris: notice historique," *Annales des mines*, 8e série, *15* (1889), 433–686.

[30] *Atlas*, pp. 29, 181, 198, and *Collection complette de toutes les parties de l'Atlas minéralogique de la France, qui ont été faites jusqu'aujourd'hui* (Paris, n.d.), p. 2. The latter work is a re-edition of the plates of the *Atlas* with a few additional quadrangles and a prefatory notice by Monnet; the only known copy is at the Ecole des mines (see note 28).

would expect, this work is in part idiosyncratic and in part filled with precise observations that were to draw admiring comments from nineteenth-century geologists.[31]

In conclusion, this analysis of the *Atlas* and its makers has hopefully shed some light — if perhaps a flickering and uncertain one — on the problems of geological cartography in the eighteenth century. It is at least clear that the *Atlas* itself is a hybrid creature that bears the impress of three distinctive approaches to geology as well as three clashing personalities. In a general sense, therefore, the story of this survey supports the original contention of this paper that geology in the eighteenth century was in enough of a state of confusion and change to permit and indeed to invite the individual geologist to treat the science, within broad limits, much as he saw fit. It is equally evident that the *Atlas* is not "typical' of early geological maps since it embodies only three points of view, and one could go on to find additional variations in the ideas and methods of other cartographers of the period.

Within the limits of this discussion, it has also been implied that the authors of the *Atlas* were in many ways representative of extreme points of view. In the decades immediately after 1750, one of the more important links among geologists was a common interest in the general classification of geological formations as Primary and Secondary, or Primary-Secondary-Tertiary, or some variation of these schemes. This simple stratigraphic arrangement, with the addition of more detailed specifications, is to be found in many memoirs and maps of this period. In the case of the *Atlas*, however, Guettard's contributions reveal a quickly outdated nonstratigraphic viewpoint, Lavoisier's type sections go far beyond the conceptions of his contemporaries, and Monnet's observations embody a stubborn refusal to be guided by any theoretical considerations. These peculiarities of the *Atlas* help to explain the fact that this pioneer survey had no imitators and serve also to illuminate the problems faced by all the early geological cartographers, namely, how to construct maps that combine observation and theory, and how to represent three dimensions on a two-dimensional surface. In a period when various solutions to these problems were being suggested, the *Atlas* provides a striking example of unresolved conflict among three geologists.

31 For example, J. J. d'Omalius d'Halloy, *Mémoires pour servir à la description géologique des Pays-Bas, de la France et de quelques contrées voisines* (Namur: D. Gérard, 1828), pp. 236–237, and Brochant de Villiers, in A. Dufrénoy and L. Elie de Beaumont, *Explication de la carte géologique de la France* (2 vols.; Paris: Imprimerie royale, 1841–1848), Vol. I, pp. iv–v.

Lyell on Etna, and the Antiquity of the Earth

Martin J. S. Rudwick

The significance of Charles Lyell in the history of geology has never, perhaps, been summarized more perceptively than it was by Darwin, when he said that "the great merit of the *Principles* was that it altered the whole tone of one's mind, and therefore that, when seeing a thing never seen by Lyell, one yet saw it partially through his eyes."[1] Lyell's purpose in the *Principles of Geology* was to reinterpret the whole range of geological knowledge as persuasively as possible, within a new conceptual framework. The key to this new viewpoint on the natural world was his extension of the time scale of earth history; or rather, his imaginative grasp of the full implications of such a time scale for the interpretation of geology. It was this that enabled others to see the phenomena of geology through new eyes.

But the source of Lyell's own sure grasp of the importance of geological time deserves closer and more critical study. It is fairly clear that his thinking was influenced profoundly by his reading of Playfair's exposition of Hutton, and that, perhaps as a result, he reacted strongly against the outlook of his teacher, Buckland.[2] But to such "literary" influences must be added the direct impact of Lyell's own firsthand experience of geological phenomena. Recently, while following the course of his journey through France and Italy in 1828 and 1829, I became convinced that one of the most decisive events in the formation of his viewpoint was his firsthand study of the volcano Etna. This paper is an attempt to justify that belief.[3]

[1] Darwin to Horner, August 29, 1844: *More Letters of Charles Darwin*, ed. F. Darwin and A. C. Seward (London: John Murray, 1903), Vol. 2, p. 117.
[2] L. G. Wilson, "The Development of the Concept of Uniformitarianism in the Mind of Charles Lyell," *Actes Xme Congr. int. d'Hist. Sci., Ithaca* (1961), Vol. 2, pp. 993–996.
[3] This analysis is based chiefly on Lyell's letters from France and Italy, published in *Life, Letters and Journals of Sir Charles Lyell, Bart.*, ed. K. M. Lyell (2 vols.;

When Lyell crossed the Channel in May 1828 at the start of his first major scientific expedition, an outline for his most influential book was already clear in his mind, and he had even completed the first few chapters.[4] Unlike Darwin embarking on the *Beagle*, Lyell set out already convinced that he possessed a revolutionary conceptual viewpoint; and the object of his journey was to gain firsthand experience of those specific phenomena that would form the most persuasive examples in the service of his argument.

Therefore, he went first to see the extinct volcanoes of Auvergne, which George Poulett Scrope had interpreted persuasively as evidence both for the uniformity of present and past and for the immensity of the geological time scale.[5] Lyell's first major essay on the principles of reasoning in geological science had been built around a review of Scrope's book;[6] in going to see the volcanoes for himself, he was clearly intending to use them for his own extension of Scrope's central theme.

Following Scrope's work closely from Auvergne southward to the Vivarais, Lyell felt increasingly confident that all the phenomena of geology, and not only those of vulcanism, could be interpreted in terms of processes still in operation, and that no differences of kind need be postulated between the geological processes of past and present. The freshwater Tertiary strata of the Limagne and the Cantal were closely comparable to the modern lake-marls he had studied near his home in Scotland,[7] and they clearly represented an extremely long period of deposition. Yet although geologically quite recent, these strata antedated even the earliest volcanic rocks of the Massif Central; and these, in turn, were far older than the fresh but still prehistoric cones and lava flows of Auvergne. This extended sequence of events clearly implied an immensely long time scale.

Even more significantly, however, the freshwater strata seemed to Lyell to have been elevated most strongly where they directly underlay

London: John Murray, 1881), Vol. I, pp. 182–251; on the use he subsequently made of his observations when writing *Principles of Geology* (3 vols.; London: John Murray, 1830–1833) (all citations are to the first edition of each volume); and on my own field work in France and Italy in 1966 and 1967. I would like to thank those present at the New Hampshire Conference for their valuable comments on this paper, which was first read in a preliminary form to the British Society for the History of Science in July 1967.

4 *Principles*, Vol. III, pp. v, viii.

5 G. P. Scrope, *Memoir on the Geology of Central France, Including the Volcanic Formations of Auvergne, the Velay and the Vivarais* (2 vols.; London: W. Phillips, 1827).

6 *Quarterly Review*, *XXXVI* (1827), 437–483.

7 C. Lyell, "On a recent formation of freshwater limestone in Forfarshire and on some recent deposits of freshwater marl." *Trans. Geol. Soc. Lond.*, *II* (1826), 72–96.

the huge volcanic center of the Cantal.[8] This seems to have suggested to him that the connection between vulcanism and tectonic disturbance might be more closely causal than was generally admitted. If the spectacular uplift and folding of strata in mountain regions could be ascribed simply to the long-continued action of ordinary earthquakes, of the kind that commonly accompanies volcanic action, then there would be no grounds for postulating occasional paroxysmal convulsions of the earth's crust, and the keystone of his opponents' model of earth history would be removed.

The full implication of these observations seems to have emerged in Lyell's mind only after he and Murchison had descended from the Massif Central and moved on through Provence to Nice. While Murchison recovered from illness, their enforced rest from field work gave them an opportunity to write up their work for publication and to reflect on what they had seen. "The whole tour has been rich," Lyell wrote, "as I had anticipated (and in a manner which Murchison had not), in those analogies between existing Nature and the effects of causes in remote eras, which it will be the great object of my work to point out."[9]

Within this general program it was now essential to demonstrate the efficacy of ordinary earthquakes as the main agent of elevation. Lyell therefore decided on a radical change of traveling plans. His hypothesis of a causal connection between elevation and vulcanism could only be tested by a firsthand study of active volcanoes, and more particularly by a search for a direct correlation between the degree of recent uplift and the intensity of recent volcanic action. In studying an area of active vulcanism, he wrote confidently, "I scarcely despair of *proving* the positive identity of the causes now operating with those of former times."[10] Therefore, as Murchison told him, "Sicily is for your views the great end: there are the most modern analogies, volcanic, marine, elevatory, subsiding, &c."[11] Regretting that he had not originally planned this extension to the expedition, Lyell wrote to Herschel, Buckland, Daubeny,

[8] C. Lyell et R. S. Murchison [sic], "Sur les Depots lacustres tertiaires du Cantal, et leurs rapports avec les roches primordiales et volcaniques," *Ann. Sci. Nat., XVIII* (1829), 173–214, Pls. XII, XIII; read at Geological Society of London, April and May 1, 1829 (abstract in *Proc. Geol. Soc., I,* (1829), 140–142.

[9] *Life,* Vol. I, p. 199. The style and content of their joint papers on French geology strongly suggest that the inspiration for them came almost entirely from Lyell, although Murchison had of course shared the field work. This conclusion is supported by the comment quoted here; Murchison was evidently less ardent than Lyell, even at this stage, in believing in the adequacy of existing causes, and later his viewpoint became more explicitly catastrophist.

[10] *Life,* Vol. I, p. 199.

[11] *Life,* Vol. I, p. 200.

and Scrope, all of whom had been to Sicily, asking for advice on where to go and what to see.

He and Murchison continued their original plan together as far as the Vicentin, to study the richly fossiliferous Tertiary strata and their interbedded volcanic rocks; for it was already clear to Lyell that any demonstration of the uniformity of present and past would need to take the Tertiary epoch as its chief sample of the past. Even before they began this field work, however, Lyell was already using another key hypothesis, namely, that the percentage of extant species in a Tertiary fauna was some measure of its relative age. This now combined with his tectonic hypothesis; for if Tertiary faunas had changed gradually and in piecemeal fashion into that of the present day, as he believed they had, then the crucial faunal links between present and past were most likely to be uplifted above sea level in the vicinity of active volcanoes.[12]

At Padua, Lyell left Murchison to return to England via the Tyrol, and he himself turned south towards Sicily. From Naples, he climbed Vesuvius and confirmed Scrope's observation that the active cone was similar in structure to the ancient partial cone of Monte Somma.[13] He also visited the "modern" volcano of Monte Nuovo (formed in 1538) and the celebrated half submerged columns of the so-called Temple of Serapis at Pozzuoli, both vivid illustrations of the action of geological processes within historic times. But the observation that excited him most was his discovery of elevated strata, containing the shells of extant molluscan species, on the volcanic island of Ischia.[14] If Ischia had risen some 2,600 feet in the geologically recent past, how much more elevation, he argued, might be expected in the vicinity of the far greater volcano Etna: "I have little fear of bringing a great part of Trinacria [Sicily] into our own times, as it were, in regard to origin."[15]

After receiving letters of advice from Buckland and Daubeny, he wrote, "Between Messina and Syracuse [i.e., on and around Etna] lies all that is of the greatest importance to me."[16] Thus, even in anticipation, before he actually reached Sicily, Etna had come to have a dominant place in his thinking.

On his arrival there, he approached the mountain first from the east, climbing through the orchards and vineyards into the Val del Bove, a huge natural amphitheater five miles across, floored with recent lava

12 *Life*, Vol. I, p. 201.
13 G. P. Scrope, "On the Volcanic District of Naples," *Proc. Geol. Soc., I,* (1827), 17–19.
14 *Life*, Vol. I, pp. 209–213.
15 *Life*, Vol. I, p. 213.
16 *Life*, Vol. I, p. 215.

flows and volcanic cones. This, as Buckland had told him, was a sight more worth his attention than any other in Sicily or perhaps in all Europe.[17] Buckland's interest in it apparently arose from his belief that it supported Leopold von Buch's elevation-crater hypothesis, implying that the whole mountain might have been uplifted in one great paroxysmal event. But in the cliffs surrounding the Val del Bove, Lyell found natural sections that did not at all support this hypothesis. They showed, on the contrary, a series of lava flows and layers of volcanic tuff ("ash"), sloping uniformly outward from the center of the mountain, except where the dip was locally disturbed by buried parasitic cones. This layered structure, cut by basalt dykes that had presumably fed the successive eruptions (Fig. 1), seemed to him to imply a gradual and hence prolonged growth for the whole mountain.

FIG. *1.* *Lyell's sketch of basalt dykes cutting layered volcanic tuff, exposed on the side of the Val del Bove and accentuated by differential weathering* (Principles, *Vol. III, p. 90*). *The tuffs are dipping toward the observer at this point and therefore appear horizontal.*

Soon afterwards, he found the kind of evidence of recent uplift that he had come to expect — marine mollusk shells interbedded with lava flows several hundred feet above sea level; and he hoped that the identi-

[17] *Principles,* Vol. III, p. 83, footnote. In his letters from Sicily, Lyell uses the Sicilian form "Val del Bue."

fication of the species present would "fix the zoological date of the oldest part of Etna."[18]

He was now ready to climb the mountain itself (Fig. 2). Setting

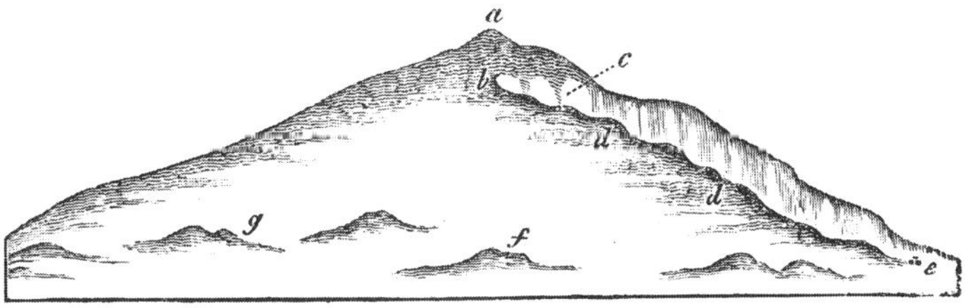

FIG. 2. *Lyell's sketch of Etna (probably from near Catania), showing the Val del Bove on the right (east) flank, the summit cone* (a), *and the parasitic cones* (f, g) *around the base* (Principles, *Vol. III, p. 83*). *The height shown in this sketch is about 10,000 feet, the breadth about 10 miles.*

out from Nicolosi with the local geologist Carlo Gemmellaro as his guide, he at once encountered some of the most recent lava flows, still barren and cindery on the surface but revealing, where they were quarried, a compact stony center indistinguishable from the most ancient volcanic rocks of Scotland. Nearby was the 500-foot double cone of Monti Rossi, thrown up during the great eruption of 1669 (Fig. 3);

FIG. 3. *Lyell's sketch of "Minor cones on the flanks of Etna"* (Principles, *Vol. I, p. 364*). *Number 1 is Monti Rossi.*

this was only one of some 80 major parasitic cones encircling the mountain, in Daubeny's phrase, "like a court of subaltern princes waiting upon their soverign."[19] Each lava could be seen to have flowed down

[18] *Life*, Vol. I, pp. 216–217.
[19] C. Daubeny, *A Description of Active and Extinct Volcanoes* (London: W. Phillips, 1826), p. 203.

the mountain in a broad stream a mile or two across, often flowing around either side of an older volcanic cone and partly burying it. As they climbed higher, through the forest zone and into the zone of grassland, the main string of volcanic cones was reduced by the sheer scale of the mountain to the appearances of small pimples; and on climbing still higher, the view to the south opened out beyond the plain of Catania to the low hills of the Val di Noto toward Syracuse. Finally, in the desert zone of volcanic ash near the summit crater, the whole of eastern Sicily and the tip of Calabria were spread out around them like a map, while almost directly below them the broadly conical form of the mountain was gashed by the Val del Bove (Fig. 4).[20] Later, when

FIG. 4. *Lyell's "View from the summit of Etna into the Val del Bove"* (Principles, *Vol. III, p. 93*). *Note the basalt dykes projecting from the cliffs on the far side.*

he came to describe this scene in the *Principles*, it evoked perhaps the most lyrical passage in the whole book,[21] leaving no doubt of the intensity with which his imagination had been stirred. He was, in fact, only just in time to make the ascent; for the following day Etna was

[20] The ascent is briefly described in Lyell's letter to his sister Marianne (*Life*, Vol. I, pp. 217–218).

[21] *Principles*, Vol. III, pp. 88–90.

covered with the first snow of winter, and the weather held him for three days "a willing prisoner" in Gemmellaro's house, with one of the best geological libraries in Sicily at his disposal.

Lyell's ascent to the summit of the greatest volcano in Europe was the imaginative climax of his whole expedition. No other single experience, perhaps, could have convinced him so persuasively of the essential validity of his whole approach to geology. His ascent of Etna was also the intellectual climax of his journey, but only in retrospect, for its full significance could not be grasped until he could relate it to the geology of the surrounding area. The relation was not easy to make out, for some of the lava flows had extended so far from the foot of Etna that they obscured the foundations on which the mountain had been built up. But beyond the limits of such flows, at La Motta, Lyell found strata with marine fossils, which seemed to underlie the mountain itself (Fig. 5). These strata did not dip uniformly away from Etna, as the elevation-

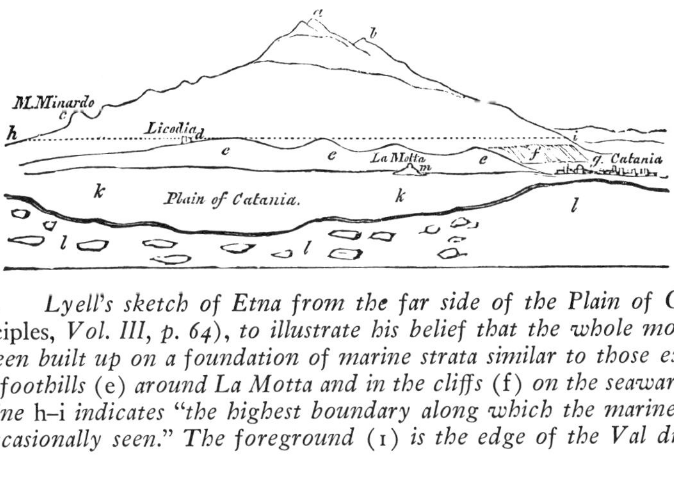

FIG. 5. *Lyell's sketch of Etna from the far side of the Plain of Catania* (Principles, *Vol. III, p. 64*), *to illustrate his belief that the whole mountain had been built up on a foundation of marine strata similar to those exposed in the foothills* (e) *around La Motta and in the cliffs* (f) *on the seaward side. The line* h–i *indicates "the highest boundary along which the marine strata are occasionally seen." The foreground* (1) *is the edge of the Val di Noto hills.*

crater hypothesis would require, and he believed that their interbedded volcanic rocks were also marine in origin and unconnected with the later vulcanism of Etna itself.

Further south, beyond the alluvial plain of Catania, the low hills of the Val di Noto were also formed from marine strata with interbedded volcanic rocks, and Lyell was able to confirm Daubeny's impression that they were Tertiary in age.[22] The most conspicuous rock, through which quite deep valleys had been cut, was a massive limestone with

22 C. Daubeny, "Sketch of the Geology of Sicily," *Edin. Phil. Journ., XIII* (1825), 107–118, 254–269, Pl. IV.

moulds of marine shells; but when Lyell reached Syracuse he found that this was underlain by a more rubbly limestone and then by a blue clay with much better preserved fossils. He was then astonished to discover that most of these shells seemed to belong to species still living in the nearby Mediterranean. According to his hypothesis, therefore, all the thick strata of the Val di Noto were not merely Tertiary, but very recent Tertiary in age; yet in general appearance they resembled more closely the "Secondary" strata of Britain. "All idea of attaching a high antiquity to a regularly stratified limestone, in which casts and impressions of shells alone were discernible, vanished at once from my mind."[23] Yet although these strata were geologically so recent, Lyell now believed that they underlay the huge mass of Etna, which in human terms was of immense antiquity.

From Syracuse, he continued to follow Daubeny's recommended itinerary around the island,[24] though significantly in reverse direction, having begun rather than ended with Etna. Daubeny had noted with regret his inability to discover the relation between the strata of the Val di Noto and those further west; and as Lyell travelled southward along the south coast, it was for evidence on this point that he searched particularly. When he reached Girgenti (now Agrigento), he found the famous Greek temples built on an escarpment of limestone that, as Daubeny had said, was "replete with shells not far, if at all, removed from existing species."[25] Daubeny had conjectured that such strata were equivalent to the Subappenine beds in the north of Italy;[26] but Lyell now recognized that they contained a far smaller proportion of extinct species, and he therefore believed they were "indubitably far, very far more recent" in origin.[27] This limestone was not merely uplifted but gently folded at Girgenti;[28] and Daubeny had reported that at Castrogiovanni (now Enna) in the interior of the island, it reached nearly 3,000 feet above sea level.[29] To Lyell, this alone was remarkable enough, since it meant that central Sicily must have been elevated by that amount

[23] *Principles*, Vol. III, pp. x, xi.
[24] In an Appendix to his paper on Sicily, pp. 268–269.
[25] Daubeny, *Sketch*, p. 117. Daubeny's interests were mainly in volcanic geology, and he did not pretend to have given more than cursory attention to the paleontology of Sicily; nevertheless he knew enough conchology to recognize a recent assemblage when he saw one.
[26] *Ibid.*, p. 259. Daubeny was chiefly concerned to argue for their *Tertiary* age, and he evidently had no thought of being able to subdivide that period.
[27] *Life*, Vol. I, p. 232.
[28] The folding can be seen most clearly from the east of the town, the direction from which Lyell approached it; the temples are built on the southern limb of a syncline and the modern town on the northern.
[29] Daubeny, *Sketch*, pp. 254–255.

since the molluscan species now living in the Mediterranean had come into existence. But this striking evidence of the power of geological agents in the recent past would still be susceptible of a paroxysmal interpretation unless it could also be shown that such a degree of elevation could have occurred by very gradual stages. This required some demonstration that there had been ample time for gradual elevation, although the strata were geologically so recent. It was at this point that the correlation between Girgenti and the Val di Noto was so crucially important (Fig. 6). Lyell thought he could recognize a similarity be-

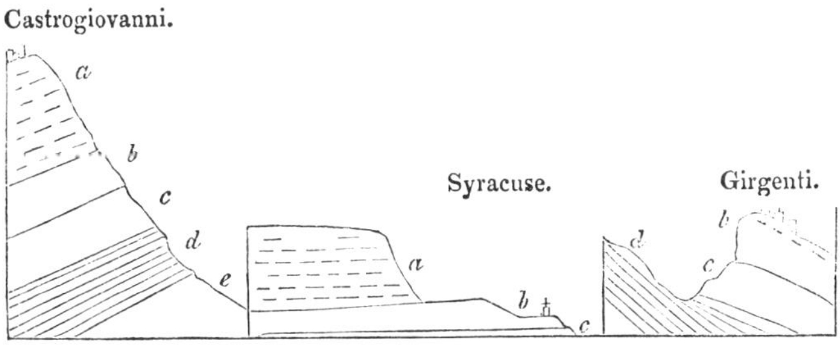

FIG. 6. *Sections showing Lyell's correlation of "Newer Pliocene" strata in Sicily* (Principles, *Vol. III, p. 64*). *The "great limestone" of Val di Noto* (a) *rests on the rubbly limestone of Girgenti* (b), *which, in turn, overlies blue clay with well-preserved shells* (c).

tween the rubbly limestone of Girgenti and the stratum underlying the "great limestone" of the Val di Noto, and both were underlain by a similar blue clay with well-preserved fossil shells. If this correlation was correct, its implications, from Lyell's point of view, were extremely far reaching, for it established that the time required for the elevation of the strata in the center of the island had also been time enough for the accumulation of the huge mass of Etna. He believed that this time could be shown to have been immensely extended by human standards, and hence amply long enough for the elevation to have occurred gradually by a succession of ordinary earthquakes. "I got so astounded by the results I was coming to," he wrote to Murchison later, ". . . that I began to doubt them; and . . . I struck back again to Val di Noto right through the centre of the isle."[30] But this only confirmed the correlation he had made tentatively at Girgenti; and when he crossed the interior once more on his way to Palermo, he was able to study the section at Castrogiovanni and confirm Daubeny's report on the elevation of the strata there.

[30] *Life*, Vol. I, p. 232

Back in Naples, he got his fossil collections identified by Costa; this confirmed his own provisional estimate of the age of the Sicilian strata, and therefore reinforced his confidence in his general conclusions. "The results of my Sicilian expedition exceed my warmest expectations by way of modern analogies," he wrote to Murchison; and significantly, he followed this remark with a brief but revealing statement of the aims and methodological principles on which his book was to be based.[31] It is therefore not surprising to find his Sicilian observations playing a crucial role in the over-all strategy of the *Principles of Geology*.

If it seems questionable to speak of an "over-all strategy," it can only be because the *Principles* have suffered in modern times from a failure to view the work as a whole, and indeed from being more venerated than read. All too many geologists admire it uncritically as the first "modern" textbook of the science, while perhaps remaining puzzled by its curious order of topics. Among historians of science, Lyell is duly applauded for his emphasis on the gradual action of existing causes, which by implication are often assumed to be the only causes that could be "natural" or "scientific" and so escape the "contamination" of theological overtones. But most commonly, Lyell features only as a forerunner of Darwin; attention is concentrated on half of the second volume, and the remainder of the work is dismissed as of merely technical geological interest.

All these views fail to recognize the significance of Lyell's own remark, in his letter from Naples, that the detailed geology in his book would not merely illustrate his dual principles of actualism and uniformitarianism, but would be presented "as evidence strengthening the system necessarily arising out of the admission of such principles." The word *system*, which reappears frequently in the book itself, is crucially important, for the *Principles of Geology*, far from being the first "modern" textbook of geology, can best be regarded as the last major all-explanatory system in earth science, the last important representative of a class of work that Lyell himself strongly condemned.[32] As might be expected of an all-explanatory system, there is indeed an underlying strategy to the book, which makes any part unintelligible without the rest.[33]

31 *Life*, Vol. I, p. 234.

32 Lyell, of course, contrasted his own work with most earlier "systems" on the grounds that his (like Hutton's) excluded "questions as to the origin of things"; but he was able to define his subject in this way precisely because he was championing a uniformitarian and not a developmental model of earth history.

33 Martin J. S. Rudwick, "The Strategy of Lyell's 'Principles of Geology,'" *Isis* (1970), in press.

Thus, the work opens with a "history of geology," which is designed to demonstrate that only those who have adhered prophetically to Lyellian principles have contributed to the progress of the science.[34] This polemical introduction is followed at once by a discussion of the "Progressionist" model of earth history that the *Principles* were designed to refute. After these preliminaries, the remainder of the first volume describes the inorganic or physical "causes now in operation" on the earth's surface; this is designed to show that the effects of such processes, even in historical times, have been far greater than has been generally admitted, and hence that far more of the phenomena of the past can be explained simply by their agency, without invoking any former differences of degree.

The second volume continues the argument on the same lines, dealing with "causes now in operation" in the organic realm. The preparatory nature of the whole of this volume is made quite explicit in the preface, in which Lyell states that it "will be found absolutely essential to the understanding of the theories hereafter to be proposed,"[35] that is, in the third volume. Lyell's opposition to transmutation, and his discussion of the piecemeal extinction of species, cannot be understood except in this context. At this point in the strategy of the work, he simply needs to establish two points: the reality of the species as a discrete unit in natural history, and the probability that these units disappear and — although more mysteriously — appear, in essentially piecemeal fashion, in response to the physical changes he has already described. The remainder of the second volume (after the point at which Darwinian forerunner-hunters give up the chase) continues to set the scene for the third volume, with a discussion of the interaction of organic and inorganic processes, and of the chances that organisms have of being preserved in the fossil record.

The third volume, the climax of the work, begins by summarizing the preceding argument and showing its relevance to the problem of reconstructing the past history of the earth. It is this problem, the reconstruction of earth history, that is the fundamental object of the whole work. The study of present physical geological processes, as set out in the first volume, is necessary for the light it can shed on the interpretation of the past, and for its demonstration that "actual causes" are sufficient to explain all the phenomena of the past if only an adequate time

[34] Lyell's historiographical myth shows remarkable vitality; having been transmitted through Geikie, Zittel, Adams, and others, it is still regularly reproduced in the historical introductions of most modern textbooks of geology.
[35] *Principles*, Vol. II, p. v.

scale is conceded. Likewise, the arguments about species in the second volume are all important simply as the theoretical justification for the *geological clock* that Lyell proposes to apply to the record of the past. The quantitative precision of the physical sciences may be unattainable in a historical science like geology; but Lyell argues, in effect, that at least an approximation to a quantitative time scale can be achieved. Given the continual flux of the physical features of the earth's surface, and the consequent flux in ecological conditions, there should be a statistically uniform rate of change in the organic world. Therefore, it should be possible to calculate the relative age of any fossil-bearing stratum by estimating the ratio between the number of extinct species and the number of still-surviving species that it contains. And since marine organisms stand the greatest chances of preservation as fossils, it is by the abundant marine Mollusca that the geological clock should be calibrated, at least for the Tertiary epoch.[36]

The bulk of the third volume is devoted therefore, to a reconstruction — on the basis of this geological clock — of the whole of Tertiary earth history, divided into four arbitrary periods based on the percentage of fossil molluscan species still extant. The exposition is retrospective, from newer Pliocene to older Pliocene and back through Miocene to Eocene, precisely because, "in the present state of our science, this retrospective order of inquiry is the only one which can conduct us gradually from the known to the unknown."[37] Only by starting with the present state of the earth can the adequacy of present causes be demonstrated for increasingly remote periods of earth history. For the same reason, it is sufficient for his strategy to show the validity of uniformitarianism for the Tertiary epoch; and he is content to end the work with a very brief discussion of the pre-Tertiary history of the earth, simply "to show that the rules of interpretation adopted by us for the tertiary formations, are equally applicable to the phenomena of the secondary series."[38]

36 The consistency with which he proposed to apply this technique is shown by his conjecture (*Principles*, Vol. III, p. 328) that the interval of time represented by the Cretaceous-Eocene faunal discontinuity may be even greater than that comprised by the entire Eocene-Recent series of strata! The significance of this passage seems to have been generally overlooked.

37 *Principles*, Vol. III, p. 62.

38 *Principles*, Vol. III, p. 324. The brevity of this section is not due to lack of knowledge of pre-Tertiary stratigraphy. Although the Transition and Primary rocks were still poorly understood in 1833, Secondary stratigraphy (i.e., Carboniferous to Cretaceous, inclusive) was *better* known than Tertiary. Indeed, Lyell's work was more generally valued for bringing Tertiary stratigraphy up to the standards of Secondary, than for its theoretical argument.

Lyell's experience of Sicilian geology can be seen to fit crucially into this general strategy. Thus, in the first volume, the comparison between the Sicilian Tertiary faunas and those of the present Mediterranean is presented as an *experimentum crucis* establishing the reality of climatic change in the geologically recent past; although the species are the same, the fossil specimens are, on the average, distinctly larger than those now living.[39] This evidence, added to the more debatable evidence from more remote periods of earth history, obliges Lyell to devise a climatic theory that will account for such changes on strictly uniformitarian lines, before beginning the main exposition of his system.

Later in the volume, Etna supplies him with valuable examples of the present efficacy of both aqueous and igneous "causes." Some of the lava flows produced within historical times had already been cut into small gorges by the existing rivers, and cut, moreover, through compact basalt.[40] He had, in fact, seen much more spectacular examples of such gorges in the Vivarais, cut through basalts erupted from fresh-looking volcanic cones; but these basalts, although perhaps geologically recent, were certainly prehistoric.[41] The Etnean examples were more valuable because they unquestionably belonged to what Daubeny (for example) regarded as "the present order of things."[42]

Etna also provided an important argument for the undiminished potency of igneous processes.[43] Its activity within historical times had been more fully recorded than any other volcano except Vesuvius; but it was far larger than Vesuvius, and the scale of some of its eruptions in recent centuries, judged by the extent of the lava flows produced, had been quite as great as any in its entire history. With an example of modern vulcanism on such a scale, it would be difficult to argue that similar processes could not explain such vast ancient volcanoes as that of the Cantal, or to maintain that the intensity of volcanic activity had diminished in the course of earth history. With a prospective eye to his

39 *Principles,* Vol. I, p. 93.

40 *Principles,* Vol. I, pp. 177–179.

41 C. Lyell and R. I. Murchison, "On the Excavation of Valleys, as Illustrated by Volcanic Rocks of Central France," *Edin. New. Phil. Journ., VII* (1829), 15–48, Pls. I–III. Curiously, by the time he wrote Volume III of the *Principles,* Lyell had greatly increased the age he assigned to even the most recent of the volcanoes of Central France, ascribing them to the Miocene period (*Principles,* Vol. III, p. 268).

42 Daubeny, *Sketch, passim.* The undermining of this widespread assumption of a contrast between the present and the past "order of things" was, of course, a major object of Lyell's whole work.

43 *Principles,* Vol. I, pp. 361–371.

use of Etna later in the *Principles*, Lyell also describes its internal strúcture and mode of growth in some detail, briefly refuting von Buch's elevation-crater hypothesis.

In the second volume, Etna figures only briefly, in a vivid thought experiment, in which Lyell argues that even a sudden change in the climate of Sicily would merely shift the ecological zones down the slopes of Etna, and could not produce transmutation of species.[44]

In the third volume, however, the Sicilian Tertiary strata figure prominently as the essential "type" for his newer Pliocene period, dated on his geological clock by their very high proportion of molluscan fossils belonging to species still living in the nearby Mediterranean. These strata demonstrate how "the line of demarcation between the actual period and that immediately antecedent, is quite evanescent."[45] With this role of eliminating the conceptual gap between present and past, it is not surprising to find Lyell devoting more space to the newer Pliocene than to any of the other three periods of Tertiary time, nor to find that the first two-thirds of that space are devoted to Sicily. It is within these four chapters[46] that the crucial role of Etna becomes apparent.

Lyell first describes the thick strata of the Val di Noto (see Figure 6). He stresses the insignificant proportion of extinct species that they contain, their astonishing degree of elevation in the center of the island, and the evidence that they accumulated extremely slowly. These phenomena, he concludes, whetting the appetite of his readers, "suggest many theoretical views of the highest interest in Geology."[47]

Without immediately explaining this remark, he goes on to describe the profile of Etna as seen from the Val di Noto, arguing that these very recent Tertiary strata probably underlie the huge mass of the volcano, which must therefore be later in date (see Figure 5). He has to admit that the reasoning at this point is unavoidably circumstantial; but he is at pains to argue that the volcanic rocks interbedded with these strata represent *submarine* eruptions unconnected with Etna, and therefore do not affect his conclusion.[48]

He then analyzes the internal structure of Etna, showing how the great chasm of the Val del Bove allows that structure to be seen (see Figures 2, 4). He reminds his readers how Etna seems to have been built up during historical times by successive lavas flowing down the

[44] *Principles*, Vol. II, p. 174.
[45] *Principles*, Vol. III, p. 22.
[46] *Principles*, Vol. III, Chs. VI–IX.
[47] *Principles*, Vol. III, p. 75.
[48] *Principles*, Vol. III, pp. 81–82.

slopes of the mountain and gradually burying the older volcanic cones; he then shows that the sections on the sides of the Val del Bove reveal precisely the internal structure that would be expected (see Figure 1). There is no evidence to support von Buch's theory that a chasm like the Val del Bove might be a huge "crater of elevation" formed by the sudden upheaval of the whole mountain. On the contrary, the sections show that Etna has been built up gradually in the manner in which it is still growing, and there is nothing to suggest that its rate of growth has been significantly greater in the past.

Lyell asks, therefore, "whether any estimate can be made of the length of the period required for the accumulation of the great cone" of Etna.[49] He prepares his readers for his conclusion by reminding them of his belief "that confined notions in regard to the quantity of past time, have tended, more than any other prepossessions, to retard the progress of sound theoretical views in Geology."[50] He argues that Adanson's method of estimating the age of a large tree, by taking a sample of its outer growth rings, can be applied by analogy to a volcano, although with less precision. A volcano does not grow with such regularity as a tree, nor are its growth rings (i.e., the successive lava flows) continuous around its circumference. Nevertheless, they can suggest a quantitative estimate of age.

Thus, Etna has a circumference of some 90 miles, so it would need 90 flows, each a mile wide, to raise the sides of the mountain by even the thickness of one flow. Yet the flows recorded in historical times do not even amount to this much change. Moreover, out of some 80 major cones on the flanks of Etna, only Monti Rossi has been erupted in recent centuries. If even a quarter of these 80 cones were formed in the 30 centuries since the first human records of the mountain, that would still imply no less than 12,000 years for all 80. Yet these 80 are only the most recent cones, not yet buried by accumulating lava flows; far more cones must be buried out of sight within the mountain, as the Val del Bove sections demonstrate.

However the facts are viewed, Lyell concludes, "we cannot fail to form the most exalted conception of the antiquity of this mountain." Etna "must have required an immense series of ages anterior to our historical periods for its growth; yet the whole must be regarded as the product of a modern portion of the newer Pliocene epoch."[51] He thought his readers might become "reconciled" to this "staggering" in-

49 *Principles*, Vol. III, p. 95.
50 *Principles*, Vol. III, p. 97.
51 *Principles*, Vol. III, pp. 99, 101.

ference, if they also considered the degree of uplift that the newer Pliocene strata had undergone within the same period of time; for he attributed the uplift to the injection of an even vaster mass of subterranean "lava."[52] This was the logical outcome of the neo-Huttonian tectonic theory that had first brought him to Sicily.

This, then, was the key position of Etna in the strategy of the *Principles.* Lyell had expounded the sufficiency of actual causes to interpret the phenomena of the past if only a vast enough time scale were conceded. He had devised a geological clock, based on the continuous changes in the organic world, by which to show that the Tertiary period alone represented an immense interval of time, even by accepted geological standards. What was needed, to link these aspects together, was a means of calibrating the geological clock, however approximately, by the time scale of human history; this calibration was what Etna could supply.

But such an illustration of the disparity between the human time scale and the geological needed to carry not only intellectual cogency but also imaginative power. Lyell seems to have realized that he had to persuade the imagination of geologists, far more than their intellect, into recognizing the signs of a time scale that was, indeed, almost beyond imagination. It is significant, therefore, that for a frontispiece to his reconstruction of earth history he chose part of his own panorama of the Val del Bove.[53] If anything could effect in his readers the necessary conversion of the imagination, it would be the contemplation of the massive grandeur and immense antiquity of Etna, coupled with the thought of its insignificant place in the whole history of the earth.

[52] *Principles*, Vol. III, p. 107.

[53] Having decided at a late stage to publish the *Principles* in three volumes instead of two, Lyell had to leave this frontispiece incongrously in Volume II, although its relevance is wholly to Volume III (see Vol. II, pp. 303–304). It is significant that he chose the same view of Etna as one of his principal "visual aids" for his lectures at King's College in 1832 (before Volume III was published), commissioning the scene painter Scharf to make a 9-foot enlargement (*Life*, Vol. I, pp. 317, 365). Like the frontispiece to Volume I (the half-submerged "Temple of Serapis" at Pozzuoli), it was carefully chosen for its imaginative impact.

Genesis, Evolution, and Geology in America Before Darwin: The Dana–Lewis Controversy, 1856–1857

Morgan B. Sherwood

In 1855, Tayler Lewis published *The Six Days of Creation.* It claimed to be nothing more than biblical exegesis, a philological interpretation of creation in Genesis, but it contained many sharp criticisms of science and scientists, especially geologists. Lewis referred to the "rashness of science," to the "flippant" geologist with his "millions and billions of years." The author was particularly impatient with reconcilers, the naturalists and lecturers who sought to prove the truth of Genesis with scientific evidence. "Neither . . . does the Bible mean what you, in your little science, and still less Biblical learning, would ascribe to it. Your stale caricatures belong neither to its prose nor its poetry. They are alike alien to its letter and its spirit." Lewis was not afraid to name names. "We must have an honest faith, or none at all. It is wretched self-deception, when we fancy we have a belief grounded on the Scriptures, which after all rests for its main support on Buckland, or Lyell, or Hugh Miller." The language and theories of science were inadequate for the purpose of biblical interpretation; the language was defective, the theories transitory. Science, Lewis felt, dealt only with what he called, "phenomena—of *appearances.*" "We can get along very well without geology; our intellectual and moral dignity would not have been impaired had no such science ever existed."[1]

James Dwight Dana — sensitive to attacks on his friends, on his work as a geologist, and on the piety of scientists — seems to have taken

[1] Lewis, *The Six Days of Creation, or, The Scriptural Cosmology with the Ancient Idea of Time-Worlds in Distinction from Worlds in Space* (Schenectady: G. Y. Debogert, 1855), pp. 1, 2, 16, 107–110, 163, 336, 393.

all this as a personal affront.[2] His case against Lewis was stated in four articles, entitled "Science and the Bible," appearing in *Bibliotheca Sacra*, two in 1856 and two in 1857. Lewis' rejoinder, published before the appearance of Dana's final articles, was called *The Bible and Science*, which made it clear that, to him, Scripture, not science, deserved precedence.[3]

Tayler Lewis was born on March 27, 1802, in New York State. He graduated from Union College in 1820, and in 1825 he was admitted to the bar. Apparently, the legal profession did not suit his temperament, for he later took the chair of Greek at the University of the City of New York. His Phi Beta Kappa address in 1838, at Union College (where he received his LL.D.), is an early indication of his interest in the relation of religion to science, and the address called the young man's talent to the attention of educators. In 1850, he accepted the chair of Greek at his alma mater. He stayed at Union College for the remainder of his life, with the title Professor of Oriental Languages and Lecturer on Biblical and Classical Literature.

Lewis was a classicist of no small ability. He was proficient in Greek, Latin, and Hebrew. Mathematics and astronomy occupied much of his attention, and apparently he enjoyed studying Euclid in Arabic. But Lewis was not a professional theologian, not even a doctor of divinity, although he did tie his classical studies and his scientific avocations to the problem of faith. One of his early books was entitled *Plato*

[2] Lewis' book may have been especially galling to the "dean of American geologists," because Dana had his own scheme of reconciliation. His cosmogony was borrowed from his lifelong friend, Arnold Guyot. About 1850, two years after Guyot immigrated to America, he gave his views on Genesis to Dana at the latter's home. Guyot's cosmogony was not published until the 1880's. By then, it had undergone modification. See Dana, "Creation; or, The Biblical Cosmogony in the Light of Modern Science," *Bibliotheca Sacra, XLII* (1885) 204; Guyot, *Creation* (Edinburgh, 1883 [?]). For a description of Guyot's earlier views, see J. O. Means, "The Narrative of the Creation in Genesis," *Bibliotheca Sacra, XII* (1855), 83–130; second part, *Ibid.*, 323–338.

[3] Dana, "Science and the Bible," *Bibliotheca Sacra, XIII* (1856), 80–129; Article II, *Ibid.*, 631–656; Article III appeared in volume *XIV* (1857), 388–413; Article IV (called Article III, concluded), *Ibid.*, 461–524. The third article deals with Lewis' second book, which had been published in the meantime. Pages 112–128 of Article I outline Dana's cosmogony. The source will hereafter be cited as Dana, Article I, II, III, or IV. (*Bibliotheca Sacra* had, in 1855, a circulation of about 2,000; see E. A. Park to Dana, November 16, 1855, Dana Correspondence, Yale University Library, New Haven.)

Lewis, *The Bible and Science, or The World Problem* (Schenectady, 1856). In the preface of *The Bible and Science*, Lewis claims that the pages of *Bibliotheca Sacra* were closed to him. In article form, perhaps they were, but the Andover periodical did print a long letter from Lewis, answering some of Dana's charges; see "Letter from Professor Lewis," *Bibliotheca Sacra, XIII* (1856), 471–476.

Against the Atheists; it had a warm public reception.[4] Lewis became a controversialist whose main concern was to correct what he believed to be the impertinence of science in the face of religion.

Dana was born in Utica, New York, on February 12, 1813. He attended Utica Academy and later Yale, which he left in 1830 to become a mathematics instructor to United States midshipmen cruising the Mediterranean. After the cruise, he spent two more years in New Haven, as assistant in chemistry to Benjamin Silliman. In 1837, at the age of 24, Dana wrote his *System of Mineralogy*. Numerous editions were issued during his lifetime. Seldom has any scientific textbook survived so long. (A revised edition appeared in 1941.)[5] In 1838, he was induced by Asa Gray to accept the post of mineralogist-geologist on the Wilkes Expedition. The expedition gave him an opportunity to study the natural history of the Pacific area, a task he undertook with great zeal. Dana married Silliman's daughter, Henrietta, in 1844, and two years later he became an editor of the *American Journal of Science*. In 1850, Dana was appointed to Yale's chair of "Natural History," a title that was later changed to "Geology and Mineralogy." He did not assume his duties at Yale until 1855. In the meantime, he struggled to organize and classify the numerous specimens collected on the Wilkes Expedition. Soon after the controversy with Tayler Lewis, Dana suffered a nervous breakdown. He never completely recovered and was forced to limit his activities severely. Nevertheless, in 1862 he published his *Manual of Geology*, which reappeared periodically until 1895. The *Manual* alone would have established his reputation as a geologist.

Dana was probably the most celebrated geologist in the United States. Asa Gray, in a review of the *Manual*, said, "We suppose that no naturalist in this country has achieved a more distinguished position. . . ."[6] Three lines of geological investigation were of particular interest to Dana, and each took form during his early travels. First, the problem of vulcanism and mountain building occupied his attention. Second, his observations of corals led him to conclude that coral islands grew on submarinal volcanic peaks. This differed only slightly from Darwin, who favored a general land subsidence. Third, his study of Crustacea led him to elaborate the principle of cephalization, or the development

4 William Wells, "Tayler Lewis: In Memoriam," *Methodist Quarterly Review*, *XXX* (1878) 604–631. An example of Lewis' earlier attitudes is Lewis, "The Bible, Everything or Nothing," *Biblical Repository*, *IV* (1848), 100–125.

5 D. J. Struik, *Yankee Science in the Making* (Boston: Little, Brown, 1948), p. 292.

6 Gray, "Review of James Dana's Manual of Geology," *North American Review*, *XCVII* (1863), 372.

from the lower to the higher animals of complex central nervous systems, culminating in the brain of man.

Although Dana engaged in countless detailed researches, Louis Agassiz felt that as a young man Dana was given to excessive generalization, and on one occasion Gray brought him up short for arguing by analogy rather than demonstration.[7] Dana himself, in explaining his preference for geology, admitted a fondness for problems broad in scope. Furthermore, he was a theist and a Christian. To him, religion was a prime motivating force. He wrote, as Thomas Huxley would say, "with one eye on fact and the other on Genesis."[8] An address delivered by Dana before the Yale alumni in 1856 is permeated with references to God and His work.

> The word nature, though another expression of God's works, appears to . . . [some] to smack of atheism, and all education that touches on the useful, to be tainted with the mammon of unrighteousness. They overlook the fact that all works on science in our language, endeavor to uphold the sacred Word, those opposing it being exceedingly few; and that infidelity proceeds not from science, but from one fatal and prolific source, man's depravity.[9]

It should not be surprising that Dana felt compelled to take issue with the critics of science.

The first blow was struck by Lewis, with his *Six Days of Creation.* It was a relatively light slap. Were it not for the mild abuse of geology and certain men of science, a reply would not have been forthcoming.

[7] Agassiz to Milne Edwards, May 31, 1847, in E. C. Agassiz, *Louis Agassiz, His Life and Correspondence* (2 vols.; Boston: Houghton Mifflin, 1886), Vol. II, p. 436. Dana, "Asa Gray," *American Journal of Science, XXXV* (1888), 197, 198.

[8] G. P. Merrill, "Dana, the Geologist," *Bulletin of the Geologic Society of America, XXIV* (1913), 67. Dana's geological contributions may be traced in his *Manual* and in the numerous articles that appeared in the *American Journal of Science.* His work is summarized in Merrill, above; in H. S. Williams, "James Dwight Dana and His Work as a Geologist," *Journal of Geology, III* (1895), 601–621; and in Joseph LeConte, "Commemorative Address," printed in D. C. Gilman, *The Life of James Dwight Dana* (New York: Harper & Bros., 1899), pp. 248–260. Gilman's old book is the only full-length biography of Dana available. Biographical material may also be found in C. E. Beecher, "James Dwight Dana," *The American Geologist, XVII* (1896), 1–16; E. S. Dana, "James Dwight Dana," *American Journal of Science, XLIX* (1895), 329–356; O. C. Farrington, "James D. Dana as a Teacher of Geology," *Journal of Geology, III* (1895), 335–340; Bernard Jaffe, *Men of Science in America* (New York: Simon & Schuster, 1946; originally published in 1944), pp. 258–278; G. P. Merrill, *The First One Hundred Years of American Geology* (New York: Hafner, 1964; originally, New Haven: Yale University Press, 1924); and A. T. Hadley, J. M. Hooker, E. O. Hovey, and W. N. Rice, writing (with Merrill) in the *Bulletin of the Geologic Society of America, XXIV* (1913), 55–69. A skillful recent article is W. F. Sanford, Jr., "Dana and Darwinism," *Journal of the History of Ideas, XXVI* (1965), 531–546.

[9] The address was printed as "Science and Scientific Schools," *American Journal of Education, II* (1856), 350, 351.

Dana, however, was jealous of both his piety and his science, and "thought it wise to start him [Lewis] on his way down hill — so gave him a push, believing it right that he and his errors should find a common label in the public estimation."[10] In the controversy, side issues, irrelevancies, and downright name-calling were not uncommon. Lewis strove initially to establish the nature of creation in Genesis by philological means. Dana set up a rival interpretation that differed only in details and was based on geological "evidence." But the controversy was often as constructive as a bilingual debate with two monolingual debaters who had one word in common: *infidel.* Each applied it to the other.

Three issues stand out amid the confusion.[11] First, a difference over the role and activity of God during creation; second, differences about the controversial book, *Vestiges of the Natural History of Creation;* and third, the central proposition of the controversy — the propriety of scientific interpretations of Genesis.

The first issue had two facets: the nature of creation in the beginning, and the nature of God's work after the initial act of creation. Lewis felt that the Hebrew word for *create* did not necessarily indicate creation out of nothing.[12] Dana asserted that to disbelieve that the "beginning" was indeed the beginning, was to do violence to the spirit of the text.[13] After the work of creation had begun, Lewis thought repeated creative or supernatural acts were necessary to raise nature to higher degrees of perfection and thereby save her from "death," which would be inevitable if left to herself too long. Between these supernatural spurts of creative energy, nature was left alone to operate by some natural power imparted to her by God.

From a higher world than the natural, there must have been from time to time a sudden flashing in of the *extraordinary,* of the supernatural . . . a new

[10] Dana to Gray, February 8, 1856, manuscript collection in the Gray Herbarium, Harvard University; note copy from A. H. Dupree, Berkeley, California.

[11] Before examining the main propositions in the controversy, a serious misconception should be dispelled. The controversy was not a revival of the old dispute between Moses Stuart, of Andover, and Benjamin Silliman. (Gilman, Dana, p. 183; B. J. Loewenberg, "The Reaction of American Scientists to Darwinism," *American Historical Review, XXXVIII* (1933), 698 *n.*) Lewis did *not* argue for the 24-hour, literal interpretation of "day" in Genesis. He was certain philologically that the Scriptures did not force an exact acceptance of the modern solar day. Though he thought the "days" were indefinite periods, he did rap the geologists for casual guesses of extreme length. But this does not make him a literalist in the usual sense of the word (Lewis, *Six Days,* pp. 12, 163, 172, 173, 194; Lewis, *Bible and Science,* p. 143). Even Dana acknowledged Lewis' interpretation of the "days" (Dana, Article I, p. 92).

[12] Lewis, *Six Days,* pp. 50, 141.

[13] Dana, Article I, p. 103.

face, a new law, a new idea, call it what you will, accompanied with new methods or laws for its subsequent growth or development, and then leaving it to their undisturbed operation.[14]

To Dana, nature was God's constant work: He continually sustained it and directed it.

We may find, in nature, the cyclical law of all natures, but not the cyclical alternation of the natural and supernatural, appealed to in the "Six Days"; this deterioration of nature at intervals, because she is left to go alone, is against all analogy in a growing individuality, and without any proof in facts.[15]

The geologist also detected a developmental doctrine sandwiched between Lewis' "extraordinary flashings."

By 1856, Dana had become an outspoken opponent of the developmental hypothesis as outlined in the *Vestiges of the Natural History of Creation,* published anonymously by Robert Chambers in 1844.[16] Chambers adopted Laplace's nebular theory regarding the earth's origin, and appropriated many of the developmental ideas of Jean Baptiste Lamarck. The view that ontogeny recapitulates phylogeny was also annexed. According to the *Vestiges*, life originated through a chemico-electric process by which primitive germinal vesicles were produced. All the various forms of life "developed" or evolved from the primitive germ, from the simplest to the next more complex, and so on.[17]

The *Vestiges* was an imaginative, popular synthesis of the sciences,

[14] Lewis, *Six Days*, pp. 98, 99.

[15] Dana, Article III, p. 497; also Dana, Article I, pp. 99, 100. The statement appears partly to contradict Dana's belief (1862) that "creations and extinctions of species were going on through the whole course of history . . . ," which sounds like special or multiple creationism; so does the statement by Lewis. But in the above quotation, Dana seems more concerned with the possibility in Lewis' work of periodic "deterioration" or "death." Nevertheless, Dana (and Lewis, too) was capable of logical inconsistency at very high levels of abstraction. See, for example, W. F. Sanford, "Dana and Darwinism," pp. 531–533.

[16] In London. The book created a sensation in Great Britain and the United States. According to H. F. Osborn, *From the Greeks to Darwin* (New York: Macmillan, 1894), pp. 215–216, *Vestiges* went through 10 editions in nine years. A sequel, designed to clarify points in the original, was also widely read. See *Explanations: A Sequel to "Vestiges of the Natural History of Creation"* (New York, 1846). Controversy over the *Vestiges* is treated in C. C. Gillispie, *Genesis and Geology: A Study in the Relations of Scientific Thought, Natural Theology, and Social Opinion in Great Britain, 1790–1850* (Cambridge, Mass.: Harvard University Press, 1951; reprint, New York: Harper Torchbook, 1959); A. O. Lovejoy, "The Argument for Organic Evolution before the Origin of Species, 1830–1858," in B. Glass, O. Temkin, and W. L. Strans, Jr. (Eds.), *Forerunners of Darwin, 1745–1859* (Baltimore: The Johns Hopkins Press, 1959), pp. 356–437 (reprinted with changes from *The Popular Science Monthly, LXXV* [1909]); and in Loren Eisely, *Darwin's Century* (Garden City, N.Y.: Doubleday, 1961), pp. 132–140.

[17] Chambers, *Vestiges*, pp. 145, 149, 155.

for, as Professor Gillispie has indicated, Chambers — a man with a wide-ranging curiosity — was "more enthusiastically interested in the meaning of science than in science itself."[18] Unfortunately, the "science" in the *Vestiges* was often misapplied or erroneous. For example, the transmutation of species was supported by a story about a gentleman who sowed oats but reaped barley.[19] In denouncing the volume, scientists frequently found themselves concurring with the most conservative clerics in the country.[20] The conservative religious position in 1845 can be illustrated in a review written by Tayler Lewis, who considered the *Vestiges* an example of "blank atheism, cold cheerless, heartless, atheism."[21]

But 10 years later, Dana thought Lewis attached a modicum of respectability to the developmental theory. In *Six Days of Creation,* Lewis stated: "A development theory which has no divine organization, or acknowledges only one divine organization . . . is as near to atheism as it can be. . . . But a development theory in the sense of species from species, as well as of individual from individual, may be as pious as any other." He went on to apply the possibility of development to man. A new man might have been made from an old organism, he claimed; the Bible gives no distinct information on the subject. "There is . . . nothing monstrous or incredible in the idea that what had formerly been the residence of an irrational and groveling tenant might now be selected as the abode of a higher life . . . might be made to assume an erect heavenward position. . . ."[22] "In other words," said the Yale geologist, "the monkey may possibly have been curtailed behind and straightened up into a man."[23] Plainly infidel, snorted Dana! Dana cited Charles Lyell, and maintained that geology, almost alone, deserved credit for exposing the *Vestiges* as phony science.[24] "Geology declares, unequivocally, that new forms were new expressions, under the type-

18 Gillispie, *Genesis and Geology,* p. 159; I have used the Harper Torchbook edition of this book.

19 Anonymous, "Review, Sequel to the Vestiges of Creation," *American Journal of Science, I* (1846), 252. For the opinion of another American scientist, see Asa Gray, "Explanations of the Vestiges," *North American Review, LXII* (1846), 465–506. In Europe, Hugh Miller was particularly effective in his assault on Chambers; see, C. C. Gillispie, *Genesis and Geology,* pp. 170–181.

20 For the tenor of some theological opinion, see "Short Notices," *Princeton Review, XXIV* (1852), 146–151; and J. O. Means, "Narrative of the Creation," p. 83–130.

21 Lewis, "Vestiges of the Natural History of Creation," *American Review: A Whig Journal, I* (1845), 525–543.

22 Lewis, *Six Days,* pp. 216, 248, 249.

23 Dana, Article 1, p. 98.

24 *Ibid.,* p. 93.

idea, by *created* material forms, and not by forms educed or developed from one another."[25] Dana also dredged up Lewis' old review in the *American Review*. A charge of inconsistency was implied, for in the review Lewis had been hostile toward the *Vestiges*. There was not much the philologist could or did say in reply. Had he had access to Dana's correspondence, he could have reproached the geologist with the same accusation. In a letter to Asa Gray, dated 1846, Dana had said, "*should the* [*Vestiges*] theory *prove* true, it would not affect the truths of religion."[26]

Looking backward, the surface positions of the two men were curious indeed. Lewis, the biblical exegete, seems to defend a theory of natural evolution from attack by Dana, the scientist. The analysis, however, is not that simple. Lewis claimed neither to defend the *Vestiges* nor to adopt its theory. The *Vestiges* was not introduced for its own sake, but because of the bearing it had upon the main point in *The Six Days*, the question of the length of the creative days. He calls the *Vestiges* a "terrible book" in one place, and compliments Hugh Miller for his effective attack on the volume.[27] In many ways, Lewis was very near Miller's position. The latter, comparing the appearance of species by individual creative acts with their appearance by evolution, had stated: "Either way it matters little physically or morally . . . it is only your little religious sects and scientific coteries which make a pother about the matter."[28] This was essentially the view of Lewis when he wrote that a development theory may be as good as any.[29]

Lewis would not concede to science the right of biblical interpretation. Dana insisted that science had thrown light on the Bible, "light which could have come from no other source."[30] Here was the central disagreement in the controversy, a point to which Dana only casually addressed himself, although all of his articles were essentially a protest against the other view. Lewis was explicit. The first sentence of his *Six Days* reads: "The Leading Design of the following work is to set forth the Biblical Idea of Creation, philologically ascertained, or

25 Dana, Article II, p. 654. For additional discussion, see Article I, pp. 97–102; Article IV, pp. 468, 519; and Dana, "Science and Scientific Schools," p. 364. Also see Lewis, *The Bible and Science*, pp. 204, 205.

26 Dana to Gray, April 27, 1846, quoted in A. H. Dupree, *Asa Gray, 1810–1888* (Cambridge, Mass.: Harvard University Press, 1959), p. 148.

27 Lewis, *The Bible and Science*, pp. 199, 185. Lewis, "Letter," p. 474.

28 Miller, *The Foot-Prints of the Creator* (Boston: Gould and Lincoln, 1854; from the 3rd London edition), pp. 41, 42. See, also, note 19.

29 Lewis, *Bible and Science*, p. 186.

30 Dana, Article I, p. 106.

'Creation as Revealed,' in distinction from any scientific or inductive theory of the Earth."[31]

To Lewis, biblical interpretation by a scientist was presumptuous and contemptible. "As well might the fly on the dome of St. Peter's speculate on the origin, history, and final destiny of that magnificent structure, from the almost invisible strata which her microscopic eyes have traced in a few inches of the marble on which she sits."[32] The colorful metaphor was employed in 1845. In his *Bible and Science*, the opinion was reinforced in unmistakable language: "Among all the wonders science reveals, there is nothing so truly wonderful as the fact that some of its Professors can stand in the presence of these four great scriptural ideas—the Word, the Spirit, the Ineffable Working, the Divine Repose, and yet babble away about their rock-written revelation."[33] Linked to these opinions was Lewis' conviction that science occupied a place in the public mind out of all proportion to its worth. He bemoaned "the everlasting sing-song of the steam engine, the daguerreotype, and the magnetic telegraph, as though the rapid transmission of thought were of vastly more importance than the quality of the thought transmitted."[34]

Dana's four articles, as a whole, were his answer to the assertion that interpretation must be left to biblical exegetes. He never really confronted the issue directly. Although he admitted that much mischief had resulted from the misapplication of scientific analogies, he seemed to feel that his analogies were not misapplied. "This placing in antagon-

31 Lewis, *Six Days*, pp. 1, 3.

32 Lewis, "Vestiges," p. 540. See, also, Lewis, *Bible and Science*, p. 59.

33 *Ibid.*, p. 226.

34 *Ibid.*, p. 50. In the passage, Lewis equated science with technology, a habit not uncommon during the period, or even now. Perry Miller, in a brilliant but not entirely convincing essay, seems to see in the early national period of American history the development of an intellectual commitment to technology and "practical" science. See Miller, *The Life of the Mind* (New York, 1965, pp. 269–313.) In a similar vein, Hugo Meier links an emerging emphasis on technology during the period with the idea of democracy, and remarks that "There was no escaping the exceedingly pragmatic view of science in the United States by 1830." See Meier, "Technology and Democracy, 1800–1860," *Mississippi Valley Historical Review*, *XLIII* (1957), 618–640. Given the rapid spread of American industrialization during the century, such attitudes deserve close attention; but the interpretations do not account adequately for progress made in the geographical sciences—geology, biology, meteorology, and so forth—or for specialization and professionalization in science. For a broader view of what was going on in the sciences, see Nathan Reingold, *Science in Nineteenth Century America* (New York: Hill & Wang, 1964); and A. H. Dupree, *Science in the Federal Government* (Cambridge, Mass.: Harvard University Press, 1957). The status of geology and the increasing professionalization of science must be appreciated in order to fit the Dana-Lewis controversy into its proper historical niche.

ism God's word and his works, is only fitted to make the young scout the former; for they know the latter has its great truths, having the best of all evidence."[35]

Perhaps Dana was as interested in defending science as in defending the right to interpret the Bible with the evidence of science. Whatever his motivation, the main source of friction was nothing more than a matter of intellectual jurisdiction. Who should interpret the Bible, and how should it be interpreted? The controversy is a striking American example in support of Professor Gillispie's view that conflict between religion and science or between theology and science was not a simple black-and-white affair, that difficulty arose from the presence of religion *in* science,[36] or more precisely in this case, religion *in scientists.*

There were imaginary points of friction in the controversy, imaginary in that the two men were closer to one another than each cared to admit publicly. Again and again, Lewis belittled the failure of science to fathom "first causes," "origins," "primal forces," and "interior laws."[37] Such matters, he insisted, were outside the realm of science. Dana agreed. Just as often, and in just as many places, he proclaimed the limitations of science.[38] The two men were also agreed on the place of man in the cosmos. Lewis took a cut at Agassiz's postulate on the origin of man: "If any fact in creation is clearly revealed . . . it is that the origin of the present human race was from a single pair."[39] In the same spirit, Dana lambasted J. C. Nott and G. R. Gliddon's book, *Types of Mankind,* but defended Agassiz. "Professor Agassiz's short contribution to the volume is altogether out of place, as we believe he himself now regards it."[40] Furthermore, although there was bickering

35 Dana, Article I, p. 91. See, also, Article IV, p. 494.

36 Gillispie, *Genesis and Geology,* pp. vii, ix. The situation may be phrased differently. The question of intellectual jurisdiction arose when some clerics, theologians, and scriptural exegetes felt their authority was undermined by the attempts of scientists or science-minded clerics to apply the evidence of science to biblical interpretation. Lewis provided a clear statement of this matter of "status": "The Church must become less scientific, less philosophic, in its defences, and *more scriptural,* if it would hold the place of true power" (Lewis to Rev. [?] R. Smith, September 3, 1856, Lewis Correspondence, letterbook page 176; from notes and quotations by Professor John C. Greene, 1940, taken from letterbooks in the possession of a descendant of Tayler Lewis, Schenectady, N.Y. Hereafter cited as Lewis Correspondence). Lewis also objected to a geologist reviewing his book in a theological journal (Lewis to Messrs. Park and Taylor, January 31, 1856, Lewis Correspondence).

37 Lewis, *Six Days,* pp. 69, 192, 220, 326, 404, 406.

38 For example, see Dana, Article I, p. 82; Article III, p. 396; and Article IV, pp. 474, 490, 495.

39 Lewis, *Six Days,* p. 248.

40 Dana, Article II, p. 644. Agassiz told Dana that he did not regret his contribution to *Types of Mankind* (7th ed.; Philadelphia, 1855); see William Stanton, *The*

between Dana and Lewis over trivial points, the two rival cosmogenies were approximately the same. They differed mainly in method. Dana's system was an application of scientific conclusions and the scientific language to Genesis; Lewis' system was a philological interpretation. Despite all the agreement, the strong language used by both Lewis and Dana was bound to attract attention.

Many of Dana's scientific friends complimented him on his exposition. Benjamin Peirce, the mathematician, found Dana's articles the "happiest possible reply to the attacks upon the religion of science." The astronomer G. P. Bond was "much gratified" by Dana's statements. Agassiz felt greatly indebted to Dana "for fighting so earnestly the cause of our independence versus clerical arrogance." The Swiss-American naturalist was "sorry to find that this clerical spirit is still alive, as bitter, vehement, and overbearing as in the worst times of religious bigotry."[41]

In 1863, Asa Gray, in a review of Dana's *Manual*, took another tack. Gray regarded Dana's cosmogonical views "with curious interest, but without much sympathy for the anxious feeling which demands such harmonies. We have faith in revelation, and faith in science, in each after its kind; but, as respects cosmogony, we are not called upon to yield an implicit assent to any proposed reconciliation of the two."[42] This would eventually be the view of modern science. Rapid empirical and theoretical advancement in geology during the nineteenth century admitted that discipline to the intellectual establishment, and with the specialization and professionalization, most geologists found it irrelevant and unnecessary to construct elaborate "reconciliations" of their science

---

Leopard's Spots: Scientific Attitudes toward Race in America, 1815–1859 (Chicago: University of Chicago Press, 1960), pp. 233–234 *n.* And in another book by Nott and Gliddon, *Indigenous Races of the Earth* (Philadelphia, 1857), Agassiz repeated many of the same ideas. He identified eight human "types" (or races), which were stamped with distinct differences since the beginning; the unity of mankind, he said, was "contrary to all the results of modern science"; quoted in Edward Lurie, *Louis Agassiz, A Life in Science* (Chicago: University of Chicago Press, 1960), p. 264. For more about *Types of Mankind*, see Stanton, *The Leopard's Spots*, pp. 161–173.

41 Peirce to Dana, July 11, 1856, and Bond to Dana, March 18, 1856, in Gilman, *Dana*, p. 156; Agassiz to Dana, July 1856, *Ibid.*, p. 324. See, also, Agassiz to Dana, January 30, 1856, *Ibid.*, pp. 184, 185.

42 Gray "Review of James Dana's Manual of Geology," *North American Review*, October 1863, p. 375. Compare an opinion expressed by Lewis in 1856: ". . . science, as science, can never find the supernatural. She can never find a God—she can never get out of Nature or natural causality without bringing in something that does not belong to her. . . . Let everything keep its own place is my motto. . . ." (Lewis to Benjamin M. Martin, August 12, 1856, Lewis Correspondence, letterbook page 146.)

with traditional religions. But in 1856 and 1857, Lewis, rather than Dana, seemed more in agreement with Gray's attitude: "each after its kind."

The controversy lasted two years, then interest waned. There were more important concerns: publication of *Origin of Species*, and the sectional conflict that culminated in Civil War. By the early 1870's, Tayler Lewis favored a biblical exegesis that smacked of evolution, but he was an outspoken foe of Darwinism because it lacked divine inspiration and was an "unfinished" doctrine with "physical machinery . . . evolving *for the sake of evolving*." He still refused to concede to scientists the right to interpret the Bible from physical evidence.[43]

Dana gradually became an evolutionist, but a qualified Darwinist.[44] The reasons for his reluctance to accept quickly Darwin's theory are difficult to isolate. On the scientific side, abrupt transition rather than gradation between species was the geological record, a record broken and incomplete in the mid-nineteenth century.[45] On the religious side, evolution challenged some of his theological assumptions. On the personal side, he fell ill in 1859 and remained incapacitated for a number of years. On the social side, his two friends, Gray and Agassiz, were locked in combat over Darwinism.

Perhaps there was a psychological side as well. In the controversy with Lewis, Dana had carefully outlined his views publicly, with conviction and certitude; it may therefore have been difficult for him to retreat or readjust without a long and cautious reappraisal of his position. At any rate, by 1885 he still had not changed his mind on one subject; in that year he defended the right of scientists to engage in biblical interpretation.[46]

[43] Lewis, *The Light by which We See Light; or, Nature and the Scriptures* (New York: R.C.A., 1875), p. 176. Lewis, "The One Human Race," *Scribner's Monthly, III* (1860), p. 739. Nevertheless, Dana and Lewis apparently began to correspond after their own "reconciliation" (they did not know each other personally during the controversy). See Lewis to Dana, March 13, 1871, Lewis Correspondence; also, Lewis to Alfred Owen, September 24, 1860, *Ibid.*, letterbook page 544; and Dana to A. Guyot, January 30, 1875, in Gilman, *Dana*, pp. 330 331.)

[44] Dana's gradual acceptance of Darwinism is traced in W. F. Sanford, "Dana and Darwinism," pp. 531–546.

[45] Dana, "Thoughts on Species," *Bibliotheca Sacra, XIV* (1857), 854–874; the article also appeared in the *American Journal of Science, XXIV* (1857), 307–316. Dana, *Manual of Geology* (4th ed.; New York: American Book Co., 1895), pp. 1030–1033.

[46] Dana, "Creation; or, the Biblical Cosmogony in the Light of Modern Science," *Bibliotheca Sacra, XLII* (1885), 202.

Porcelain and Plutonism

Cyril Stanley Smith

Although Hutton himself was scornful of those who "judge of the great operations of the mineral kingdom from having kindled a fire and looked into the bottom of a little crucible," the experiments of Sir James Hall, beginning with the crystallization of glass and lava into stony masses, played a very important role in convincing geologists of the general correctness of Hutton's theory of the earth.

The purpose of this note, and the excuse for bringing as exotic a material as porcelain into down-to-earth geology, is to examine some of the ideas on the crystallization of matter that had themselves begun to crystallize during the eighteenth century. It was difficult for men to see the identity between the matter in well-formed crystals and that in confused aggregates of crystal grains, and it was hard to abandon the early belief that crystals grew only from aqueous solutions; both stumbling blocks favored the neptunists.

The beauty of crystals in their geometric perfection resulted in an overemphasis on external characteristics and tended to obscure the significance of their internal order. Crystals seemed to be a special form of matter, rare in nature, and produced in the laboratory only from a few substances, under unusual and carefully controlled conditions. The fact that virtually all solid inorganic matter is crystalline was not commonly appreciated even at the beginning of the present century. In the late eighteenth and early nineteenth centuries, it was a question for debate among leading scientists. Other aspects of solids that were slow to come out were the relationships between the simple compounds required by Daltonian chemistry and the prevalence of nonstoichiometric compositions, and, conversely, the facts that crystals could be extremely small and that matter was often structurally heterogeneous when it was visibly uniform. In both the eighteenth and nineteenth centuries, the idea of the molecule provided all the flexibility that chemists needed. Mineralogists could classify by composition and by external shape; mathematicians were bemused by space groups; and it was not until after the introduction of x-ray diffraction, when crystallography could move from external symmetry and effectively grapple with internal structure and imperfections, that the irregular shapes of crystal grains could be properly appreciated by anyone but an empirical observer. If any early scientist saw that a random aggregate of well-shaped crystals could not fill space, he was more likely to abandon the crystallinity than he was to accept as a crystal anything that did not have regular polyhedral shape.[1]

Hall's important experiment, so controversial to geologists at the time, was, I believe, a natural outcome of what had been one of the main activities of eighteenth-century chemists. Back of the spectacular theoretical achievements of Lavoisier and Dalton were centuries of refinement in analytical techniques that established the identity of things and the conservation of some chemical units through numerous changes of state. Much, perhaps most, of this knowledge had been developed by assayers who had established multitudinous separatory reactions for metals and metallic minerals. Some discoveries came under medicinal stimulus. But parallel to all this was the study of the chemical and thermal behavior of earths and rocks that was to a large degree incited by the desire to duplicate Oriental ceramics.

Some Chinese porcelain came to Europe in the Middle Ages, and it began to have a strong influence on European taste from the end of the fifteenth century on. During the sixteenth and seventeenth centuries, some beautiful soft-paste wares in Oriental styles were made in Italy and France, but they were not true porcelain and were fired at relatively low temperatures. Successful factories were established at St. Cloud near Paris in 1693 and at Meissen in Saxony in 1713 to exploit

[1] Even today the dominant role of the interfaces between crystals in determining their shape is slow of acceptance among geologists. Contacts between crystals in metamorphic or metasomatic rocks are more frequently a result of interfacial energy balances than they are of the sequence in which the various crystal species were nucleated. See, for example, R. L. Stanton, "Mineral Interfaces in Stratiform Ores," *Trans. Instn. Mining & Metallurgy*, 74 (1964), 45–79. In sedimentary rocks, which have not yet been studied in detail from this viewpoint, it appears that interfacial energy excludes aqueous solutions from between calcite-calcite and calcite-silicate intercrystalline contacts, but maintains open capillary channels between crystals of quartz and many silicates, as well as pushing aside mineral particles to prevent their entrapment in pyrite and garnet crystals growing in shales and schists.

the discoveries of François Morin and J. F. Böttger (who used kaolin with marble or alabaster, respectively) and at Chantilly, which, after 1725, produced some aesthetically superb work in the Chinese manner but in soft ware with a tin-oxide glaze. A new wave of research was precipitated by the receipt in France of letters written in 1712 and 1725 by François Xavier d'Entrecolles, a Jesuit missionary in China. He gave a fairly complete description of the process as carried out on a mass-production basis in 3,000 kilns at Ching-te-chen (Kiangsi Province), and he gave the Chinese names, *petuntse* and *kaolin*, for the two essential materials, though he could not identify them in European terminology. (*Petuntse* is actually feldspar.)

The great versatile scientist René Antoine Ferchault de Réaumur studied samples that d'Entrecolles had sent from China, and in papers published in 1727 and 1729 reported that *kaolin* was infusible but *petuntse* was relatively easily melted. He incorrectly identified the former as talc or a talc-like stone and the latter as a fusible kind of flint, but he correctly saw that the essential nature of Chinese porcelain lay in its being a mixture of a very fine-grained infusible earth and a glass. Similar, indeed, were the porcelains then being made in France and Germany, but these could be melted in a laboratory furnace, while the Oriental product was quite infusible.

There followed a frantic search for superior potters' materials and for means of characterizing them. Böttger's successor at Meissen, J. H. Pott (one of Stahl's students), is said to have carried out more than 30,000 experiments in trying to discover the secret of true porcelain; he published many of the results in 1746 in his *Lithogeognosia.* He used a high-temperature wind furnace with a tall chimney, and on the basis of the effect of fire on various earths, classified them as calcareous, gypsum-like, argillaceous, or vitrifiable. The last — siliceous materials — he found would act as a flux for most of the others when mixed with them, rendering them more easily fused and giving a glass on cooling. Many natural rocks fused by themselves. Pott reported the color and appearance of the fracture of the resulting material and usually the hardness, but made no chemical analyses, for techniques in this field did not then exist.

Twenty years later, Jean Darcet, or D'Arcet, published a book *Mémoire sur l'action d'un feu égal, violent et continué . . . sur un grand numbre de terres . . .*, in which he reported a large number of experiments similar to those of Pott, to whom he obviously owed much. He did his heating in "porcelain" crucibles in closed saggers, heated in the large porcelain furnace at the factory of the Count de Lauragais. Darcet showed that mixtures were commonly more easily melted than were pure minerals, and that a siliceous mixture would usually result in a glass. Most pertinent in the present connection are his experiments on the melting of a blackish stone coming from Auvergne that had been given to him as a product of ancient volcanoes and that had a prismatic crystallization (i.e., faulting) like those at the Giant's Causeway and elsewhere. This melted to give a coffee-colored enamel like the glaze on one kind of Oriental porcelain, with a perfectly glassy fracture. Several other rocks acted similarly, and he also found that he could easily melt lava, which gave a glass like common bottle glass. He thought that stones of volcanic origin (which, like spars and granites, had been originally at least partially subject to attack by fire) could well have been formed by the coming together and general melting of several materials of a different nature, contrary to lava, which had been more completely liquefied. "However this may be, it is natural enough to conclude that the fire which is used in baking a true porcelain is by far superior to that which works in the terrible overturnings in the terrestrial globe."

Neither Pott nor Darcet suggested that their melted and solidified materials were crystalline, and they ignored the possible effect of rate of cooling. Their studies showed systematically what many potters and metallurgists knew from experience, namely, that many rocks could be melted and that quartz or flint would flux calcareous rocks, clays, earths, and many metallic minerals to give a melted mass with a vitreous look. Since the effect of rate of cooling was at this stage ignored, the experiments led to the belief that rocks that had been molten inevitably became glassy. When the plutonic origin of rocks was suggested, it therefore appeared to be necessary to invoke some other agency in addition to heat.

By 1750 chemists had, of course, become familiar with the formation of crystals in the laboratory, though they generally regarded them as a product of a solution in water or some other liquid: free growth from solution allowed a material to take its natural geometric form. Virtually no attention was paid to the structure of either natural or artificial minerals that were not well shaped. Whether based on aesthetics or on science, this preoccupation with well-formed polyhedra certainly delayed inquiry into internal constitution. Even Haüy's great contribution to crystallography was external, and his structural units were essentially homomorphic little crystals.

After the middle of the eighteenth century, the idea slowly spread that crystallization could follow the removal of heat as well as that of an aqueous solvent. Grignon (1761) complained of chemists' overemphasis on aqueous systems and pleaded with men in industry who used fire on a large scale to counterbalance this by themselves adopting an observant and analytical attitude toward their processes. He extracted crystals from many glassy slags and remarked that art most closely approaches the operations of nature in the furnaces of the ironmaster, which imitate the products of volcanoes. He believed that if all the forms of matter that precipitate in volcanic craters were introduced into ironmaking furnaces, they would give the same results. He then described the crystallization of various glassy slags and showed their similarities to lava, chalk, etc.: "Both fire and water can modify matter, but the important operations are reserved to [the former] which is the most powerful and most active agent in nature, achieving in an instant results for which water would require centuries."

Though Romé de Lisle in 1772 still believed that water was a principal and perhaps the only agent in nature for forming crystals, the popular Macquer (1766), in a remarkably clear article, says that crystallization resulted whenever the integrant parts of a body that had been separated by the interposition of any fluid became disposed to unite again and form solid regular and uniform masses. (The primitive integrant molecules were of unknown shape, but he believed that all those of any one body were of a "constant uniform and peculiar figure.") Thereafter, chemists produced crystals by the partial solidification of many molten materials. The beautiful crystals of metals[2] produced by Grignon (1775) and Guyton de Morveau (1776) made it quite clear that "fire is to metals what water is to salts," and the old objections subsided.

The parallel with the neptunists' views on the origin of rocks is clear. If it was hard to visualize the crystallization of a molten metal, it was even harder to think of crystals forming in solid glass. But as it turned out, it was observations on glass that first suggested to geologists that basalt might be a crystallization of lava.

The devitrification of glass (not, of course, at first understood as crystallization) has a long history. It must have been observed long before it was recorded, for it occurs whenever common glass is heated for

2 For a discussion of the history of understanding crystallization in metals, see Smith (1960). The work of the experimental geologists in the nineteenth century provided the background for microscopic studies of engineering metals and alloys, in which the crystals are generally invisible to the naked eye and never well shaped. The great metallographer Osmond frequently cites Vogelsang, Daubrée, Fouqué, and Lévy, who had been making artificial mineral crystals; and much of the early metallographic terminology was adapted from that of the mineralogist.

an hour or so at a low red heat. Manufacturers of glass could hardly have avoided some crystallization of their product whenever it was allowed to cool in large pots or bulky furnaces. The earliest reference that I have found is in a twelfth-century manuscript of the *Mappae Clavicula*, which gives a garbled recipe for turning "fragile glass into the nature of a stronger metal" for use as a mortar, by cooking it after it has been coated with a lute of ashes and blood. The use of lute suggests that the process had been discovered by the heating of a retort or other laboratory vessels, which were habitually coated with blood-cemented ashes or clay to protect them from thermal shock and to provide support during prolonged heating at high temperatures. The suggestion that the glass will become like a stronger "metal" follows glassmakers' terminology, though it may also be an echo of the legendary malleable glass of the Roman artisan reported by Pliny and other ancient writers.

Caspar Neumann, who died in 1737, is reported by Lewis (1763) to have observed that the bottom of a glass retort used for distilling milk acquired the appearance of porcelain, which he attributed to the fine white earthy matter of the milk being forced by heat into the glass. In his 1756 English translation of Neumann's work, Lewis refers to experiments of his own and promises a full account of them, a promise that he made good seven years later.

Before this, however, Réaumur's work on the devitrification of glass had been published and had aroused considerable interest. In his 1727 and 1729 papers on porcelain, Réaumur had concluded that Oriental ware was a semivitrified material composed of a refractory earth and a glass. He suggested that there might be two general ways of achieving the right structure, one being to catch a vitrifiable material in an imperfectly vitrified state during its transition from terra cotta to glass on heating, and the other being to fire a paste composed of a highly refractory material mixed with one that is easily vitrified. In 1740 he announced a third method (Réaumur, 1740). He had found that common glass, when annealed for a long time, lost its vitreous character and became internally white and fine-grained, like porcelain, and at the same time more refractory as well as more resistant to thermal shock and to chemical action. He says that he had made the first discovery of the porcelain process 20 years before publication, i.e., about 1720, in the midst of his research on malleable cast iron, which constitutes Réaumur's other famous contribution to the practical arts (Réaumur, 1722). He was stimulated to publish when a man named Montamis, in the service of the Duke of Chartres, showed him a porcelain-like object that Montamis had made by annealing glass after he had observed the change in character of the bottom of a flask that had been exposed to fire. Réaumur read his paper before a public meeting of the Académie des Sciences at Easter 1740, where it created so much excitement that it was published in the delayed *Mémoires* of the Academy for 1739, which actually appeared in 1741.

Réaumur's process for making malleable cast iron involved the annealing of a brittle white cast iron, packed in bone ash mixed with a little charcoal. His stated motive for making it was to cheapen works of art so that everyone could possess works as elaborately decorated as those produced by the best smiths. He was equally proud of the cheapness of his porcelain. "The methods of attaining it are so simple that there is no one who cannot transform all the bottles in his cellar into bottles of porcelain." He wanted no fine vases but only utilitarian ware.

He was sensitive to the importance of the structural changes that were deducible from the appearance of the textures of the fractured surfaces of test pieces. He was the last scientist for well over a century who had a serious concern with the structure of materials on this level. The essentially Cartesian corpuscular views that had guided his work on steel and iron served equally for porcelain:

> "Glass . . . has a polish, a lustre, that is never seen in the fracture of true porcelain. Porcelain is granular, and it is partly by its fine grain that the fracture of porcelain differs from that of terra cotta, and finally it is by the size and arrangement of the grains in them that the kinds of porcelain differ from each other and become closer or less close in nature to glass" (1739, p. 375).

Just as steel could be made either by adding something to wrought iron (Réaumur supposed "sulphurs and salts," for carbon was not suspected until 1774) or by removing something from cast iron, so, also, porcelain could be approached in different ways: to his two earlier ways of achieving the mixed structure, he now added the devitrification of glass. If metals can be returned to the metallic state after being converted to their calces or dissolved in glass, why should not the sand and stones that gave rise to glass be restorable?

Réaumur made his porcelain by heating glass packed in a refractory powder in saggers in an ordinary potter's kiln. After experiments with many cements, he came to prefer a mixture of sand and gypsum, and he thought that the physical change was accompanied by a chemical one, as in the cementation of iron. He observed that the change in nature of the glass started at the surface (a fact supporting the supposed chemical change) and grew inward in the form of silklike fibers that were composed of extremely fine grains (Fig. 1). It could be made entirely granular, like ordinary porcelain, under some heat treatments.

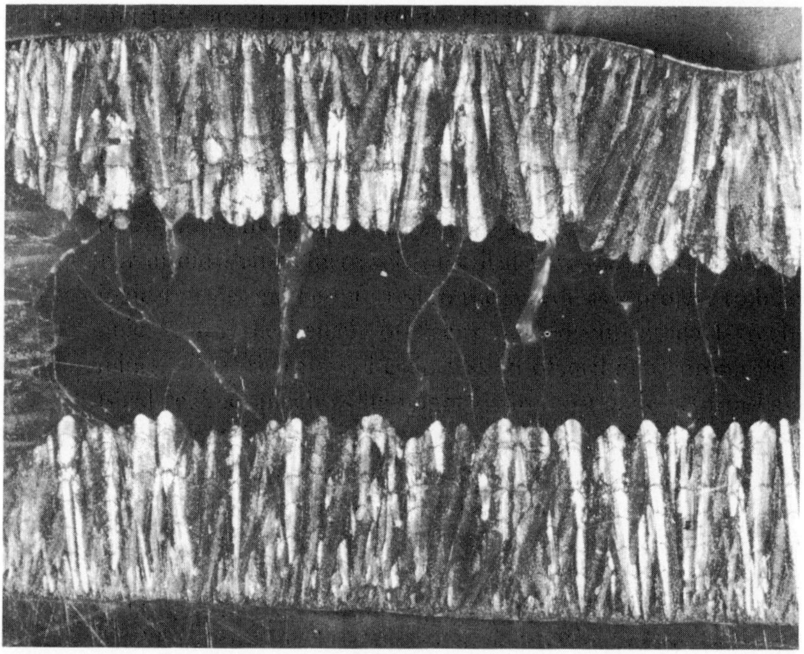

FIG. 1. *Photomicrograph showing the devitrification of glass, proceeding from the surfaces in the manner described by Réaumur. The sample is 4mm. thick. (A lithia-alumina-silica glass, annealed 230 minutes at 1100°C. Photograph courtesy Paul E. Dougherty.)*

Any kind of glass except the best crystal (lead glass) could be devitrified to give porcelain under proper conditions, but the ordinary colored glass used for bottles and chemical matrasses worked best.

Réaumur calls his product "porcelain by transmutation, by revivification, or porcelain from glass." It would have been an anachronism for Réaumur to have used the word "crystallization," but he did see his devitrification as a return to the state of stoniness characterizing the materials that went into the glassmaker's pot.

Réaumur complained that chemists, with their meltings, calcinings, and distillings, had neglected cementation as a means of studying bodies, for he thought this to be

> . . . perhaps the method of operation that comes closest to that of nature, which makes its mixtures only slowly and imperceptibly, and, similarly, decomposes bodies only little by little and only very slowly. Everything is mixed together too suddenly by fusion, and often the materials, before being mixed, have undergone too much change. . . . The heat that a solid body endures during a long period dilates its parts, separates them; it opens thousands of passages through which can seep volatile particles that continuously detach themselves from the materials that touch the body on all

sides; or particles from the body itself escape: its composition is altered, is perceptibly changed" (pp. 380–381).

The most detailed scientific study of Réaumur porcelain ever published is that of William Lewis (1763). He confirmed Réaumur's observations in all respects but gave far more experimental details on the rate at which the reaction occurred in several different kinds of glasses. He found that if the material was overannealed, the elongated fibers began to break up into grains that grew larger until eventually the whole mass became friable, like a mass of loosely adhering white sand. The fibrous state was the hardest, striking sparks with steel and cutting glass, and was resistant to corrosion by acid or alkaline liquors. If the outside of a piece of glass was transformed by a short heat treatment and the temperature then raised, the inside would run out, for the fusibility of the transformed material was much less than that of the original glass. Lewis showed that the action of many different cementing materials was identical except for superficial coloration. Glass simply heated with no surrounding powder devitrified in the same way, although slowly, and it slumped out of shape before the reaction had given it stiffness. He saw that the loss of alkali was not significant, although some alkali did diffuse from the glass into the surrounding sand.

Lewis' work was experimental, and he had little to say on theory beyond stating that the change to the fusible fibrous material formed at the beginning "may depend in part on an alteration produced by the heat in the glass itself considered as a compound or in the nature of its alkaline ingredient"; and at the end, "Thus glass whose production has been commonly supposed to be the utmost limits of the power of fire, has its earth and its salt, which one degree of fire had so firmly united, almost disjoined by another."

For over a century, Réaumur's porcelain was regarded as an interesting material. It is discussed in most books on pottery and porcelain in the nineteenth century. Typical are Lardner (1832) and Knapp (1849), who mention Réaumur porcelain in a laudatory fashion, but they add nothing to Lewis' understanding of it and provide no indication that it was in current production. Knapp mentions that "the idea is well worthy of more attention than has really been expended on it." Even in 1883, Weeks thought that annealed glass should replace porcelain in most of its uses.

For all this, nothing seems to have happened industrially to this supposedly fine invention. The principal collections of art porcelain contain no examples of it, which is perhaps not surprising, since

Réaumur's porcelain was aesthetically unimpressive.[3] Although internally it matched the whiteness of the best Chinese ware, its surface was dark and rough. Its physical properties however — its high melting point (St. Cloud "porcelain" could be melted in a crucible of Réaumur's devitrified glass), its resistance to thermal shock and to the action of both acids and alkalies — suggested its use for chemical apparatus, and both Lewis and Macquer extol its virtues for this purpose; but even here it failed. Liphardt (1785) discusses its use for laboratory utensils at some length. Ehrmann (1786), studying the effect of the newly discovered oxygen torch on materials, "sacrificed" a retort made of Réaumur porcelain that had come to him from the Strasbourg chemist, J. R. Spielman. If historic samples exist, they are more likely to be in museums of science than of art. Despite the praises of its properties in the literature, it must have been an inferior material (probably because of the plane of weakness existing in the region where the surface-nucleated crystallization met in the center), and there is no indication that it was ever produced commercially in appreciable quantities. It was utterly forgotten in the twentieth century until it came back in a vastly improved form, indeed as a totally new material based on extremely fine controlled nucleation, in the Pyroceram of Donald Stookey of the Corning Glass Company (Stookey, 1959).

The words *crystal* or *crystallization* appear nowhere in either Réaumur's or Lewis' paper on devitrification. The change was regarded as a loss of glassiness, not a gain of crystallinity.

With James Keir's paper (1776), glass, crystals, and geology merged for the first time. In 1771, Keir had published his English translation of Macquer's *Dictionary of Chemistry*, which contains the suggestion that the transformation in Réaumur porcelain was due to the diffusion of vitriolic acid from the surrounding gypsum to form a kind of salt within the glass. Shortly afterward, Keir established a glass factory of his own in Stourbridge, Worcestershire. In a mass of glass that had been allowed to cool slowly, he observed a white crust and numerous hexagonal crystals. In reporting on these to the Royal Society, he begins with as good a discussion of the structure of polycrystalline bodies as any that I have found in eighteenth-century literature:

[3] See, however, the recent article describing a box made in Austria around 1750 that seems to be of Réaumur Porcelain: R. von Strasser, "A Box of Schackert Porcelain" *Journal of Glass Studies 9* (1967), 118. Von Strasser cites a book by Robert Schmidt, *Die Brandenburgische Gläser* (Berlin, 1914), describing attempts in Potsdam to make porcelain from glass in the years 1750 to 1783. (Note added after completion of manuscript.)

In many substances, when broken, the parts appear to have some determinate figure. This determination of figure, or grain, as it is called, is obvious in bismuth, regulus of antimony, zinc, and all other metallic bodies which may be broken without extension of parts; and although the ductility of gold, silver, lead, and tin, prevents the appearance of the peculiar grains when pieces of these metals are broken, yet we have reason to believe, that, by exposing them to proper circumstances, they also would shew a disposition to this species of crystallization, as it may be called, by a further extension of that term; for Mr. Homberg has observed that when lead is broken while hot, in which state it is not ductile, a granulated texture appears. Perhaps all homogeneous bodies, in their transition from a fluid to a solid state, would, if this transition were not effected too hastily, concrete into crystals or bodies simularly figured. Instances of such crystallization have occurred to me in glass which had passed very slowly from a fluid to a solid state; and the form, regularity, and size of these vitreous crystals have varied according to the circumstances with which their concretion had been accompanied. I send along with this paper a few specimens of this crystallized glass, together with a drawing of some of the most remarkable crystals.

Some of these regular crystals occurred in window glass, composed of "sand, kelp, calcereous earth and lixiviated vegetable ashes," and also in common bottle glass of similar composition except for the incorporation of slag from an iron blast furnace. These glasses form the best crystals, Keir believed, because they are more fluid and less tenacious, and the minute particles of which crystals consist can more easily "apply themselves to each other with less resistance from the medium: Perhaps also the greater portion of calcareous and other earthy particles may dispose these glasses to crystallize more than others, which contain a larger quantity of saline and metallic fluxes." He saw that, despite the loss of weight (nearly 2 percent in one experiment) the alteration was a change of texture, not basically chemical. "The properties of bodies depend merely on the different arrangements of their integrant parts or on their modes of crystallization." There are many other instances of such nonchemical property changes, for example, the hardening of steel.

Many of Keir's glassy crystallizations were spherulitic aggregates of fibrous crystals. Some, however, were well-formed hexagonal crystals, and it was these that led him to speculate as follows:

Does not this discovery of a property in glass to crystallize reflect a high degree of probability on the opinion that the great native crystals of *basaltes*, such as those which form the Giant's Causeway, or the pillars of Staffa, have been produced by the crystallization of a vitreous *lava*, rendered fluid by the fire of volcanoes?

This opinion is further confirmed by the following considerations. The prismatic and other regularly-shaped *basaltes* have been almost always found to be accompanied with *lava*, pummice-stones, and other vestiges of the fire of the volcanoes whenever they have been carefully examined by intelligent naturalists, as has been shewn by M. Desmarets, in his Memoir on the

Basaltes of the province of Auvergne, in France; *Mém. de l'Acad. des Sciences*, 1771. Basaltic columns have even been discovered, according to the same author, among the productions of volcanoes now existing, as of those of Mount Etna and of the Isle of Bourbon.

He was, of course, quite wrong in attributing the hexagonality of basaltic faulting to crystallization, but its geometry had misled many others.[4]

The devitrification of glass continued to interest chemists—especially French ones—for some time. Dartigues (1804), the proprietor of a glassworks, carried out extensive tests on different kinds of glass and came out very firmly for the theory that devitrification is crystallization. He ended up by referring to lava and its devitrification, and thought that the laboratory facts could provide explanations of the phenomena of geology on which there had not been agreement because "nothing can make people believe that stones had previously been glass." Other important papers were those of Guyton de Morveau (1810) and Pelouze (1855).

F. Leydolt (1852) described and illustrated several examples of true crystallization in glass. He then went on to apply to the structure of glass a technique that he had found useful in studying meteorites. After etching glass with hydrofluoric acid, he found large crystals, which were, of course, superficial growths of reaction product; his paper is of more interest for the fact that it contains a beautiful plate printed from an electrotype made directly over a crystallized sample—an early scientific use of the electrotype process that had been developed a decade earlier. Similar misleading studies of crystals forming on etched glass were done by Wetherill (1866).

Nineteenth-century chemists, so impressed with the simple atomic ratios that compounds had to have according to Daltonian theory, were extraordinarily resistant to the idea that there could be crystalline solid solutions, and they were even reluctant to accept the idea of microscopic heterogeneity or of diffusion in the solid state that was necessary to give rise to it. Pelouze, for example, affirmed that crystals arising in a vitreous mass must have a composition essentially the same as the

[4] The hexagonality of basalt prisms results, of course, from the unidirectional thermal gradient, which produces a network of contraction cracks that adjust to the most economical geometry as they advance through material that is so finely crystallized as to be uniform on the scale of the crack. (The prisms are not uniquely hexagonal, as the pentagons and septagons in the reject pile of Staffa blocks at the end of the Zuider Zee dam show.) The crystalline appearance of the prisms nevertheless provides a nice illustration of structural principles, of the scale-free relationships that occur in aggregates or segregates of anything.

medium in which they grew, although Dumas (1855) argued against this. Studies of the growth of crystals in glasses continued, however, and their formation in borax beads became a useful branch of blowpipe analysis. Nevertheless, the question of the devitrification of glass remained mainly a practical rather than a scientific problem until Stookey's studies of nucleation in the 1950's precipitated a new flurry of interest on a different structural level. (But this has little to do with eighteenth-century geology, to which we return.)

Concern with the nature of basalt and lava remained active after Keir's paper. In 1781, Joseph Priestley wrote to Josiah Wedgwood to thank him for a sample of basalt and to report experiments on the amount of air that could be extracted from it and from lava. After melting in a retort and cooling, basalt had become a perfectly black glass. Priestley also asked Wedgwood's opinion on Keir's view that basalt "though a perfect glass at first . . . might become the substance it now is by length of time," and wondered if Wedgwood's pottery kilns would be convenient for an experiment on the subject.

In 1785, James Hutton presented the first outline of his theory of the earth, which had such great influence after its publication in fuller book form (1795) and elaboration and championing by Playfair (1802). Although Hutton does not mention Thomas Beddoes, his mature ideas were probably stimulated by the latter's paper "On the Affinity Between Basaltes and Granite," presented before the Royal Society in 1791. This was an astute discussion of crystallization. Beddoes expressed the interesting view that the results of chemical analysis (then properly so much in vogue) would often serve to perplex mineralogists rather than to reduce their science to order, and he discussed the whole question of how the transition from a state of fusion or solution to a solid may give rise to various homogeneous or heterogeneous internal structures:

> Nor is it perhaps difficult to assign highly probable reasons why a mixture of different earths with more or less of metallic matter, in returning from a state of fusion to a solid consistence, may assume sometimes the homogeneous basaltic, and sometimes the heterogeneous granitic internal structure. No fact is more familiar than that it depends altogether on the management of the fire, and the time of cooling, whether a mass shall have the uniform vitreous fracture or an earthy broken grain arising from a confused crystallization. The art of making Réaumur's porcelain consists entirely in allowing the black glass time to crystallize by a slow refrigeration; and the very same mass, according as the heat is conducted, may, without any alteration of its chemical constitution, be successively exhibited any number of times as glass or as a stony matter with a broken grain.

Beddoes' disregard of chemistry led him to the erroneous conclusion embodied in his title, but his emphasis on crystallization was badly needed in geological thinking at the time. He described the joint crystallization of minerals in a two-phase eutectic (not, of course, so named).[5] Beddoes was careful to warn that his ideas did not test the igneous hypothesis any more than the opposite one, for regardless of whether the disunion of the parts in forming initial solution "had been effected by the repulsive power of fire or the intervention of water, it is just as easy to conceive of heterogeneous earthy crystals shooting from different points of a uniform liquid" according to either supposition. However, he did believe that both granite and basalt arose from the injection of "a fused mass raising, rending and shivering the incumbent strata, while its heat hardened them into laminated stone." This, and not the infiltration of water as proposed by De Saussure, appeared to be the true origin of granite veins.

We come, at last, to the main point of this paper. As is well known, Hutton's theory of the earth derived much of its support from the series of laboratory experiments duplicating geological phenomena that began with the work of Sir James Hall; and Hall's experiments were a direct outgrowth of the earlier work on glass. One of Hutton's main contentions was that basalt represented intrusions of igneous material essentially identical with lava. It would seem that there would not have been much occasion for argument had not Guettard's original misconception of the nature of basalt been fortified by the dogmatism of Werner and his followers, but there was certainly something in plutonism that cut across the grain of conventional thinking, and we will therefore first consider the scientific basis of the objections to it.

The versatile Irish chemist and geologist Richard Kirwan (1733–1812) was inspired to much of his best work by criticism of others. His objections to Lavoisier's antiphlogistonic chemistry and revised chemical nomenclature are well known to historians of chemistry. In 1793, Kirwan had advanced many arguments of a chemical nature against Hutton's preliminary statement of his theory, and he could not let Hall's later experimental support of it pass unchallenged. William Nicholson, editor of the *Journal of Natural Philosophy, Chemistry and the Arts*, reprinted, or rather preprinted, in his April 1800 issue Hall's paper that had been read before the Edinburgh Royal Society, and followed this in the June issue with a rather bitter attack on it by Richard

[5] That the idea was not a simple one to grasp can be seen from the fact that Hutton advanced the obvious co-crystallization in graphic granite as a conclusive argument against its igneous origin.

Kirwan. Kirwan, conceding that "fanciful and groundless as the Huttonian theory seems to me to be, it may, like the researches for the philosopher's stone, be highly useful for suggesting new experiments," based much of his argument on assumptions as to the nature of crystals and crystallization. He placed great emphasis on the association of perfection with crystallization. In the case of granite, he thought that crystallization would have to result in regular crystals of quartz, feldspar, and mica, formed and stratified in the order of their melting points, a structure never found in nature. He objected to Hall's use of the word "crystallization," and even to the term "crystallites" for imperfect crystals.

> To the vague term of crystalization I must however object, for as those stones [basalts] in their original state present no regular crystals, but are at most internally and imperfectly crystalized, so they must be when reduced from a glassy state to one resembling the original, and thus discover [i.e., reveal] rather a *nisus* toward crystalization than perfect crystals, which latter the term crystalization [as] generally applied would lead us to expect.

The consolidation should not be called crystallizations,

> for according to every sense in which this term [crystallization] has ever been employed, whether that operation was perfect or confused, it denotes at least an union of particles previously dispersed through a liquid medium, they must, therefore, be at liberty to move through this medium in order to coalesce and reunite to each other . . . but in Sir James' experiments we find the consolidation to take place in a fragment of glass which still retained its solid state, and consequently the particles were not at liberty to move toward each other. This consolidation must, therefore, evidently have arisen from some internal change in the constitution of the glasses in which it was observed (pp. 154–155).

The change, he thought, was the volatilization of soda, which several investigators of Réaumur porcelain had discussed. He did not perceive that volatilization could have nothing but a superficial effect if the particles had no internal mobility, and he thought he had entirely destroyed the validity of the analogy between glass and basalts. Kirwan did, however, accept the fact that chemical combination is affected by temperature and that separation can occur on cooling, and he was therefore not surprised by the loss of the close vitreous texture that occurred on slowly cooling melted whinstone, or at the appearance of the looser texture of a mere stony substance. Except for the emotional overtones of the word "crystallization," this is essentially Hall's theory; but in Kirwan's mind, it did not support the Huttonian theory. At best, to an unprejudiced mind it might render the origin of whins ambiguous

> by making them assume the appearance of Neptunian origin when in fact they owe it to fusion; but it is only an appearance, for natural whins are

accompanied with circumstances, and contain substances which contradict that appearance and prove it to be deceitful (p. 156).

After listing various aspects of the external characteristics of whins, Kirwan concluded that there could be no doubt that the production of even the basaltic pillars of Staffa and the Giant's Causeway had occurred in the moist way. As a parting shot, he reiterated his belief that lavas must carry with them a substance capable of maintaining their heat and fluidity, a substance that burns in contact with the atmosphere.

His opponent, Sir James Hall (1761–1832) was a man of very wide interests. Hall was active in the Royal Society of Edinburgh from its beginning. His concern with structure led him not only to geology but also to studies of details of Gothic buildings, which he believed developed from structural antecedents in simpler materials. He introduced Lavoisier's chemistry to the members of the Society and debated phlogiston with Hutton, whom he seems to have admired greatly, despite constant disagreements on scientific questions. In 1788, he discussed the origin of pumice. On January 4 and March 1, 1790, he presented a paper[6] supporting the plutonic origin of granite, written "a few weeks" after he had seen a large pot of green bottle glass at the Leith glassworks that had been allowed to cool slowly and had assumed a stony structure. Hall reported experiments analogous to Réaumur's, reversibly converting bottle glass by slow cooling into a stony substance that would revert to a glass if it were rapidly cooled after being more strongly heated. On March 5 and June 18, 1798, just a year after Hutton's death, Hall presented a detailed paper to the Society. In a preliminary report on this (Hall, 1798), he specifically related his ideas on crystallization to Réaumur's porcelain, which clearly was his inspiration, although reference to Réaumur is omitted from the fuller publication (Hall, 1800, 1805).

The advice of a geologist friend had caused Hall to drop the granite of his early work and concentrate on basalt, choosing for experiment several samples of whinstone, a Scottish term for a variety

[6] Hall's ideas of 1790 were published only in summary form, for he had no wish to offend Hutton, whom he later said "was impressed with the idea that the heat to which the mineral kingdom has been exposed was of such intensity as to lie beyond the reach of our imitation, and that the operations of nature were performed on so great a scale, compared to that of our experiments, that no inference could properly be drawn from one to the other." Although Hutton believed that a congelation or crystallization had taken place with more or less regularity to produce the stony and crystallized structure common to all unstratified substances from large-grained granite to the fine-grained and almost homogeneous basalt, he thought that the texture of basalt arose entirely from the pressure to which the parental lava had been subjected. Hall saw that slow cooling was responsible, not pressure.

of stones elsewhere known as "basaltes, trap, wacken, *grünstein* and porphyry." By heating these in a crucible in "the great reverberating furnace at Mr. Barker's iron foundery," he easily converted whinstone from a quarry in the neighborhood of Edinburgh into a black glass, but if the crucible were very slowly cooled it yielded "a substance different in all respects from glass, and in texture completely resembling whinstone. Its fracture was rough, stony and crystalline; and a number of shining facettes were interspersed through the whole mass." Seven different varieties of whin behaved similarly (Fig. 2).[7]

FIG. 2. *Fragment of crucible containing a piece of Rowley Rag (basalt) melted by Sir James Hall. These samples once belonged to Michael Faraday and are now in the Science Museum, London. (Photograph courtesy Director, The Science Museum, London. British Crown Copyright.)*

Hall made the important observation (related to Lewis' partial melting of partly devitrified glass) that the glass made by melting and cooling whinstone became soft enough to yield readily to the pressure of an iron rod at a temperature far below that at which the original rocks softened. Glass that had been recrystallized in the laboratory was

[7] Some of Hall's experimental samples still survive. In an excellent article on Hall's scientific contributions, V. A. Eyles (1961) gives photographs of melted whinstone samples that are preserved in the Geological Museum in London and photomicrographs of three of them that, though crystalline enough, are not strikingly similar to the original whinstone. Hall's experiments continued to be shown to students even in the present century—Professor Thomas Turner displayed a crucible containing glass made from the local Lickey Hills' basalt when discussing slags before the author's class in metallurgy at the University of Birmingham in 1922. Another cross connection with metallurgy is provided by H. C. Sorby's Hall-like experiments on syenite that were done in the same year, 1863, in which Sorby first revealed the true microscopic structure of iron by careful polishing and etching.

intermediate. The temperatures, measured with a Wedgwood pyrometer, are detailed in Table 1. Crystallization of a typical whinstone

TABLE 1 Table of Softening Temperatures of Crystalline Basalt and Glass (Hall, 1800)*

Substances	Original softened	Glass softened	Crystallite softened
No. 1. Whin of Bell's Mills Quarry	40	15	32
2. Whin of Castle Rock	45	22	35
3. Whin of Basaltic Column, Arthur's Seat	55	18	35
4. Whin near Duddingstone Lock	43	24	38
5. Whin of Salisbury Craig	55	24	38
6. Whin from the Water of Leith	55	16	37
7. Whin of Staffa	38	14½	35
No. 1. Lava of Catania	33	18	38
2. Lava of Sta. Venere, Piedimonte	32	18	36
3. Lava of La Motta	36	18	36
4. Lava of Iceland	35	15	43
5. Lava of Torre del Greco	40	18	28
6. Lava of Vesuvius, 1785	18	18	35

* The numbers represent the temperatures at which the samples were soft enough to yield to the touch of an iron rod. They are expressed in degrees Wedgwood, one degree of which corresponded to a 1/240 change in length of a little test cylinder of a standard clay. The pyrometer had a horribly nonlinear scale, although it was historically important as the first experimental means of measuring high temperatures and was useful in firing pottery, for it combined time and temperature in a manner analogous to the ware itself. Hall's temperatures range from about 800° to 1,200°C. The glass samples were made by melting the whin or lava "by the application of a strong heat and subsequent rapid cooling." The "crystallite" samples are the whins or lavas first turned to glass, subsequently annealed at a lower temperature or slowly cooled to make them crystalline.

The samples were analyzed chemically by Robert Kennedy (1800, 1805). Except for the presence of about 5 percent volatile matter (mainly water and CO_2) in the whinstones, they were essentially all the same in composition, 46 to 51 percent "silex" (i.e., SiO_2), 16 to 19 percent "argil" (i.e., Al_2O_3), 14 to 17 percent "oxyde of iron" (i.e., Fe_2O_3), and lesser amounts of lime, soda, and HCl.

took place in the range 21 to 28 degrees Wedgwood. A minute was time enough to start crystallization in the range of 21 to 23 degrees, regardless of whether temperature was rising or falling. At higher temperatures, up to 28 degrees, the reaction was slower but gave more perfect crystallization; at lower temperatures, down to 21 degrees, the reaction became faster and gave imperfect crystallites. Hall clearly understood the nature of isothermal transformation.

Having established that whinstone would turn to a lava-like glass, and that this could be converted to a stony material either on slow cooling or on prolonged sojourn at intermediate temperatures, Hall then proceeded to examine lava itself. He selected six kinds of lava, five of which were already partially crystalline, and converted them all into homogeneous glasses by heating and rapid cooling, then caused them to become crystalline by treatment at a lower temperature. On a visit to Vesuvius in 1785, Hall had taken samples of lava directly from the molten stream, and these, because of their rapid cooling, were completely glassy; he made them crystalline by annealing. He measured the temperature of softening in each case (see Table 1). With his data, and by invoking the effect of time and temperature on crystallization, he was able to explain all of the glassy and stony characteristics of lava and basalts. Since the glass was more fusible than crystallized material of the same composition, it was no longer necessary to invoke, as Kirwan and Dolomieu had done, some foreign substance or combustible to keep it molten.

Geology had become, at least in part, a laboratory subject.

Kirwan's view on the nature of crystals and his overemphasis on perfect shapes was quite typical of mineralogists and chemists at the time; indeed, something like it continued among crystallographers well into the twentieth century. Concern with imperfect reality was slow to come. Its roots were in the work of Réaumur, who knew of the irregular grains visible in fractured surfaces, and whose work on the structure of iron and steel (Réaumur, 1722) later bore fruit in modern metallography and the ability to control the structure of materials in line with the desired properties. Similarly, Réaumur's observation on the structure of his glass-porcelain precipitated thoughts about the crystallization of lava, although he himself had no inkling that the structural parts he observed were related to the striking geometric crystals of the mineralogist. The similarity was in internal structure, and only practical observation could keep the messy facts alive until theory had arrived at the state at which it could properly disentangle them. Réaumur and his porcelain, Hall and his whins, serve beautifully to illustrate the interaction between scales of structure and scales of thought which gives so much life to science.

References

Beddoes, Thomas (1791). "On the affinity between Basaltes and Granite," *Phil. Trans. Royal Soc., 81* (1791), 48; abridged *Transactions, 17*, 8–18.

Blake, J. F. (1892). *Catalogue of collection of metallurgical specimens formed by the late John Percy . . . now in the South Kensington Museum* (London, 1892).

Darcet, Jean (1766). *Mémoire sur l'Action d'un feu égal violent et continu . . . sur un grand nombre de terres, de pierres & de chaux métalliques . . .* (Paris, 1766).

Dartigues (1804). "Mémoire sur la dévitrification du verre et les phénomènes qui arrivent pendant sa crystallisation," *Annales de Chimie, 50* (1804), 325–342.

Dumas, J. B. A. (1855). Discussion on paper by M. Pelouze, *Comptes Rendus, 40* (1855), 1327–1329.

Ehrmann, Friedrich Ludwig (1786). *Versuch einer Schmelzkunst mit Beihülfe der Feuerluft* (Strasbourg, 1786). French trans. with added papers by Lavoisier and Laplace: *Essai d'un Art de Fusion à l'Aide de l'Air du Feu, ou Air Vital . . .* (Paris, 1787).

Eyles, V. A. (1961). "Sir James Hall, Bart.," *Endeavour, 20* (1961), 210–216.

Gerhard, Carl Abraham (1773–76). *Beiträge zur Chymie und Geschichte des Mineralreichs* (2 vols.; Berlin, 1773–1776).

——— (1788). *Abhandlung über die Umwandlung und den Ubergang einer Erd und Stein-Art in die Andere* (Berlin, 1788).

Grignon, Pierre Clément (1761). "Mémoire sur des crystallisations métalliques, pyriteuses et vitreuses artificielles, formées par le moyen du feu" (written in 1761). In his *Mémoires de physique, sur le Fer . . .* (Paris, 1775), pp. 476–481.

Guyton de Morveau, L. B. (1776). "Observation de la cristallisation du fer," *Journal de Physique, 8* (1776), 348–353.

——— (1810). "Diverses observations relatives à l'art de la verrerie," *Annales de Chimie, 73* (1810), 113–146.

Hall, James (1790). "Observations on the formation of granite" in report of meeting for 4 January and 1 March 1790, summarized in pp. 8–12 in the History of the Society, *Trans. Royal Soc. Edinburgh, 3* (1794).

——— (1798). "Curious circumstances upon which the vitreous or stony characteristics of whinstone and lava respectively depend . . . ," [*Nicholson's*] *Journal of Natural Philosophy . . . , 2* (1798), 285–288. (Abstract of paper read before Royal Soc. Edinburgh on March 5 and June 18, 1798. The full text appeared in 1800 and 1805 — see next two items.

——— (1800). "Experiments on whinstone and lava," [*Nicholson's*] *Journal of Natural Philosophy . . . , 4* (1800), 8–18 and 56–65.

——— (1805). "Experiments on whinstone and lava," *Trans. Royal Soc. Edinburgh, 5* (1805), 43–75.

Hutton, James (1785). "Theory of the earth, or an investigation of the laws observable in the composition, dissolution, and restoration of land upon the Globe" (paper read March 7 and April 4, 1785), *Trans. Royal Soc. Edinburgh, 1* (1788), 209–304.

——— (1795). *Theory of the Earth* (2 vols.; Edinburgh, 1795).

Keir, James (1776). "On the crystallisations observed in glass," *Phil. Trans. Royal Soc., 66* (1776), 530–542.

Kennedy, Robert (1800). "A chemical analysis of three species of whinstone and two of lava," [*Nicholson's*] *Journal of Natural Philosophy . . . , 4* (1800), 407–415 and 438–442.

——— (1805). "A chemical analysis of three species of whinstone and lava," *Trans. Royal Soc. Edinburgh, 5* (1805), 76–98.

Kirwan, Richard (1793). "Examination of the supposed igneous origin of stony substances," *Trans. Royal Irish Acad.*, 5 (1793), 51–81.

——— (1800). "Observations on the proofs of the Huttonian Theory of the Earth, adduced by Sir James Hall, Bart.," [*Nicholson's*] *Journal of Natural Philosophy . . .*, 4 (1800), 97–102 and 153–158.

Knapp, Fredrich Ludwig (1847–1853). *Lehrbuch der Chemischen Technologie . . .* (2 vols.; Braunschweig, 1847–1853). Eng. trans., *Chemical Technology* (Philadelphia, 1849).

Lardner, D. (Ed.) (1832). *A Treatise on . . . the manufacture of Porcelain and Glass* (London, 1832).

Leroux, F.-P. (1867). "Sur la trempe de quelques borates," *Comptes Rendus*, 64 (1867), 126–128.

Lewis, William (1756). "Vegetable Earth." Note appended to chapter on artificial vegetable productions, Sect. IV in William Lewis (Ed.), *The Chemical Works of Caspar Neumann* (London, 1756).

——— (1763). "Experiments on the conversion of glass vessels into porcelain, and for establishing the principles of the art," in his *Commercium Philosophico-Technicum or Philosophical Commerce of Arts* (London, 1763), pp. 230–255.

Leydolt, F. (1852). "Ueber die Krystall-Bildung in gewöhnlichen Gläse . . . ," *Sitzungberichte der K. Akademie der Wissenschaften, Vienna* (Math.-Naturwiss. Klasse), 8 (1852), 261–275.

Liphardt (1785). "Einige Versuche über das Reaumurische Porcellan," *Crell's Chemische Annalen*, 1785(*ii*), 132–138.

[Macquer, Pierre Joseph] (1766). *Dictionnaire de Chymie, contenant la Théorie & la Pratique de cette Science . . .* (2 vols.; Paris, 1766). Eng. trans. by James Keir, *Dictionary of Chemistry . . .* (2 vols.; London, 1771).

Mappae Clavicula. The text of a twelfth-century version of this Latin compilation of recipes is given by Sir Thomas Phillipps in *Archaeologia*, 32 (1846), 183–244. A translation by J. G. Hawthorne and C. S. Smith is to be published.

Neumann, Caspar (1756). *The Chemical Works of Caspar Neumann*, edited by William Lewis (London, 1756; 2nd ed., 2 vols., 1773).

Pelouze, J. (1855). "Mémoire sur la dévitrification du Verre," *Comptes Rendus*, 40 (1855), 1321–1329.

Percy, John (1875). *Metallurgy . . . Introduction, Refractory Materials and Fuel* (Rev. ed.; London, 1875), pp. 51–54.

Playfair, John (1802). *Illustrations of the Huttonian Theory of the Earth* (Edinburgh, 1802).

Pott, J. (1746). *Chymische Untersuchungen welche fürnehmlich von der Lithogeognosia . . .* (Potsdam, 1746). French trans. *Lithogéognosie ou Examen chymique des pierres et des terres en général . . .* (Paris, 1753).

Priestley, Joseph (1781). Letter from Joseph Priestley to Josiah Wedgwood, August 8, 1781. In Robert E. Schofield (Ed.), *A Scientific Autobiography of Joseph Priestley* (Cambridge, Mass.: The M.I.T. Press, 1966), pp. 204–205.

Réaumur, R. A. F. de (1722). "*L'Art de convertir le Fer Forgé en Acier et l'Art d'Adoucir le Fer Fondu . . .*" (Paris, 1722).

——— (1727). "Idée générale des différentes manières dont on peut faire la porcelaine, et quelles sont les véritables matières de celles de Chine," *Mém. Acad. Sci.*, 1727, pp. 185–203.

——— (1729). "Second Mémoire sur la porcelaine, ou suite des principes qui doivent conduire dans la composition des porcelaines de différens genres . . . ," *Mém. Acad. Sci.*, 1729, pp. 325–344.

——— (1739). "L'Art de faire une nouvelle sorte de Porcelaine . . . ou de transformer le verre en porcelaine," *Mém. Acad. Sci.*, 1739 (published 1741), pp. 370–388.

Romé de Lisle, J. B. (1772). *Essai de cristallographie* (Paris, 1772), p. 321.

Saussure, Horace Bénédict de (1779–1796). *Voyages dans les Alpes; précédés d'un essai sur l'histoire naturelle des environs de Genève* (4 vols.; Neuchatel, 1779–1796).

Smith, Cyril Stanley (1960). *A History of Metallography* (Chicago: University of Chicago Press, 1960).

——— (1967). "The Texture of Matter as viewed by Artisan, Philosopher, and Scientist in the Seventeenth and Eighteenth Centuries," *Atoms, Blacksmiths, and Crystals* (Los Angeles: William Andrews Clark Memorial Library, UCLA, 1967).

Sorby, Henry Clifton (1863). "On the microscopical structure of Mount Sorrel Syenite, artificially fused and cooled slowly," *Geol. and Polytechnic Soc. West Riding of Yorkshire, 4* (1863), 301–304.

Stookey, S. D. (1959). "Catalysed crystallisation of glass, in theory and practice," *Industrial and Engineering Chemistry, 51* (1959), 805–808.

Thy, Nicholas Christiern de, Le Comte de Milly (1771). *L'Art de la Porcelaine* (Paris, 1771 [colophon 1772]). Part of the *Descriptions des Arts et Métiers* of the Académie des Sciences.

Watt, Gregory (1804). "Observations on basalt, and the transition from the vitreous to the stony texture which occurs in the gradual refrigeration of melted basalt . . ." *Phil. Trans.*, 1804, pp. 279–314.

Weeks, Joseph D. (1883). *Report on the manufacture of glass* (New York, 1883), 114 pp

Wetherill, C. M. (1866). "On the crystalline nature of glass," *Amer. Journ. Sci. and Arts* [*Silliman*], *91* (1866), 16–27.

NOTE: The punctuation of the quotations from the various sources in English has been modernized. Quotations from works in other languages are from published translations where listed; otherwise they are new translations by P. Boucher and the author.

Nicolas Desmarest and Geology in the Eighteenth Century

Kenneth L. Taylor

Nicolas Desmarest is often remembered as the scientist who first stated the volcanic origin of basalt, and as the worthy successor of Guettard in studying the volcanoes of Auvergne. In connection with his volcanic researches, he has also been credited with an important discussion of the development of landforms through systematic degradation of geological formations, and he has even been named as an early exponent of the uniformitarian principle in geology. Moreover, Desmarest's geological thought has sometimes been regarded as one of the origins of a vulcanist school that in time clashed with Wernerian neptunism in one of the more important controversies in the history of geology. This paper offers a brief survey of Desmarest's geological work, with a view toward determining whether this summary accurately describes his significance in the history of geology.[1]

[1] The most readily available account of Desmarest's life and work is in Sir Archibald Geikie, *The Founders of Geology* (2nd ed.; London and New York: Macmillan, 1905; reprint, New York: Dover, 1962), pp. 140–175. It is based to a great extent on Georges Cuvier's eulogy, delivered before the Academy of Sciences in 1818: "Eloge historique de Nicolas Desmarets," *Recueil des éloges historiques lus dans les séances publiques de l'Institut Royal de France* (3 vols.; Strasbourg and Paris: F. G. Levrault, 1819–1827), Vol. II, pp. 339–374. Cuvier's biographical sketch is, in turn, highly dependent on a manuscript outline prepared by Desmarest's son, Anselme-Gaëtan (1784–1838), "Notes et renseignements sur la vie et les ouvrages de mon père," now in the Bibliothèque de l'Institut de France, Fonds Cuvier, MS 3199. Another *éloge*, given by Augustin-François Silvestre to the National Society of Agriculture (of which Desmarest was a member), deals less with geology than with Desmarest's career in industrial technology: "Notices biographiques sur MM. Journu-Auber, Cotte, Allaire, Desmarets et Tenon, membres de la Société Royale et Centrale d'Agriculture de Paris: Lues à la séance publique de la Société, le 28 avril 1816," *Mémoires d'agriculture, d'économie rurale et domestique, publiés par la Société Royale et Centrale d'Agriculture* (1816), pp. 80–123. The many biographical notes and articles that have appeared in various dictionaries and other compilations add little to Cuvier's and Silvestre's versions of Desmarest's life. The only significant addition to the biographical literature, aside from Geikie's, is by Alfred Lacroix, "Nicolas Desmarest," *Figures de savants* (4 vols.; Paris: Gauthier-Villars, 1932–1938), Vol. I, pp. 7–18.

The study of volcanoes in France began with the dramatic realization by Guettard and Malesherbes that extinct volcanoes existed in Auvergne. According to Desmarest's account, his own discovery of basalt among the volcanic rocks of Auvergne had an impact on him just as decisive as the first recognition of dark Volvic rock had had on his predecessors. Desmarest was first in Auvergne in 1763 on one of his tours to gather industrial information for Daniel-Charles Trudaine, the Director of Commerce; he made his observations during a natural history excursion. This discovery so intrigued him, he relates, that he was determined to investigate the matter thoroughly.[2] It is a fact that his scientific career was significantly altered by this episode. His belief that the Auvergne basalt was a volcanic rock gave him a taste of a kind of scientific achievement he had not known before, and it no doubt played a role in his election to the Royal Academy of Sciences in 1771, after more than a decade of frustration.[3] Behind Desmarest's insistence that the sudden discovery of basalt in an unexpected context was decisive for his career is the suggestion that before this incident he had had no special interest in or knowledge of volcanic geology.

Desmarest had done scientific work that may be called geological since his mid-twenties, yet none of it bears any marks of a genuine interest in volcanic phenomena. His first scientific paper, an essay on the supposed former land connection between England and France

[2] Desmarest, "Mémoire sur l'origine & la nature du basalte à grandes colonnes polygones, déterminées par l'histoire naturelle de cette pierre, observée en Auvergne," *Mémoires de mathématique et de physique, tirés des registres de l'Académie Royale des Sciences*, in *Histoire de l'Académie Royale des Sciences. Avec les mémoires de mathématique & de physique, pour la même année. Tirés des registres de cette académie, 1771* (1774), 706–708. Hereafter, this journal is cited as *Mémoires de l'Académie Royale des Sciences*. Much later than this, Desmarest reflected on the importance of his earliest studies in Auvergne in "Mémoire sur les prismes qui se trouvent dans les couches horizontales de plâtre et de marnes des environs de Paris, et sur leur analogie avec les prismes du basalte," *Mémoires de la classe des sciences mathématiques et physiques*, in *Mémoires de l'Institut National des Sciences et Arts. Sciences mathématiques et physiques, IV* (*an* XI [1803]), 219–220. Hereafter, this journal is cited as *Mémoires de la classe des sciences mathématiques et physiques*.

[3] Desmarest's attempts to gain entrance to the Academy date back to at least as early as 1757, when one of his friends and mentors in his native Troyes, Jean-Baptiste Ludot (1703–1771), an eccentric savant with some influential friends, wrote to Pierre Bouguer on his behalf in hopes of having the young aspirant placed in a vacant slot (copy of letter from Ludot to Bouguer, December 20, 1757, Bibliothèque Municipale de Troyes, MS 2584.3). There followed many years of disappointment, when Desmarest often thought that election was close, but success finally came when he was backed by a winning combination of academicians, including d'Alembert and Condorcet, as well as influential outsiders, such as Turgot, Archbishop Loménie de Brienne, and the Duchesse d'Enville. For a glimpse of some of the politics involved in Desmarest's election, see Charles Henry (Ed.), *Correspondance inédite de Condorcet et de Turgot, 1770–1779* (Paris: Charavay Frères, 1883).

that won first prize in the competition of the Academy of Amiens in 1751,[4] referred to volcanic phenomena only incidentally, in a discussion of the causes that might possibly have ruptured the isthmus. In this essay, Desmarest was concerned principally with the influence of ocean currents and the destructive and constructive power of waves beating on the shore, as means of upholding the thesis that a land bridge joining France and England was destroyed naturally. Considering the possibility that it was volcanic action that had removed the isthmus, Desmarest dismissed this as a serious likelihood on the grounds, first, that volcanoes are irregular phenomena that occur by accident, rather than through ordinary processes; second, that volcanoes, when they do occur, leave chaotic traces of their action, which are not to be observed on the bottom of the English Channel (the *Dissertation* was accompanied by two maps and a cross-section of the English Channel and a part of the North Sea, prepared by the geographer Philippe Buache, which gave the ocean depths at intervals of ten fathoms, showing smooth contours for the bottom); and third, that the source (*foyer*) of a volcano required to create the English Channel would have been, of necessity, quite deep, which was "contrary to experience."[5] As authorities for the alleged shallowness of volcanic sources he cited Johann Anderson's natural history of Iceland and Pierre Bouguer's account of his travels in Peru.[6]

In his second paper on a geological subject, Desmarest addressed himself to the mechanism of earthquake propagation, a timely subject because of the recent occurrence of the destructive Lisbon earthquake of November 1, 1755.[7] In this pamphlet, which was intended as an

4 *Dissertation sur l'ancienne jonction de l'Angleterre à la France, qui a remporté le prix, au jugement de l'Académie des Sciences, Belles-Lettres & Arts d'Amiens, en l'année 1751* (Amiens: Chez la Vve Godart, 1753). This essay was republished in the late nineteenth century, under the auspices of a company that hoped to stir up interest in the construction of a tunnel beneath the English Channel: *L'ancienne jonction de l'Angleterre à la France, ou le détroit de Calais, sa formation par la rupture de l'isthme, sa topographie et sa constitution géologique* (Paris: Isidore Liseux, 1875).

5 *Dissertation*, pp. 128–129.

6 Johann Anderson, *Histoire naturelle de l'Islande, du Groenland, du détroit de Davis*, trans. Gottfried Sellius from the German edition of 1746 (2 vols.; Paris: S. Jorry, 1750). Bouguer's travels are recounted in several papers in the Academy's *Mémoires* and in *La figure de la terre, déterminée par les osbervations de MM. Bouguer et de La Condamine envoyés par ordre du roy au Pérou pour observer aux environs de l'équateur, avec une relation abrégée de ce voyage qui contient la description du pays dans lequel les opérations ont été faites* (Paris: Chez Charles-Antoine Jombert, 1749).

7 *Conjectures physico-mechaniques sur la propagation des secousses dans les tremblemens de terre, et sur la disposition des lieux qui en ont ressenti les effets* (n.p., no pub., 1756). It was published anonymously, but evidence from numerous independent sources shows beyond doubt that Desmarest was the author.

explanation of earthquakes for a certain gentleman of eminence (probably the Attorney-General Guillaume-François Joly de Fleury),[8] Desmarest did not assign to volcanic phenomena any more importance than he had in his earliest publication. Here he was primarily interested in the distribution of mountain ranges throughout the world, and in the probability that an interlocking network of mountain ranges constitutes the medium through which earthquake disturbances travel. For his geographical ideas, some of them highly speculative, Desmarest was indebted to Buache, the geographer who is perhaps best remembered for his insistence on the importance of drainage basins as units is geographical inquiry. Buache emphasized the distinct nature of different drainage areas, but it was the high ground separating one region from another that particularly interested Desmarest. Between these natural basins stand solid mountains, which Buache believed were composed of hard material unlike that of the intervening areas, forming a closed system of interconnecting chains extending beneath the sea and through low-lying areas where their presence may be difficult to detect. Buache extended his concept far enough so that Desmarest had only to adopt it for his use. To Buache, the interest and value of this interlocking mountain system, apparently borrowed from Athanasius Kircher, lay in its function as the structure that holds the earth together and defines the main features of the continents and oceans.[9] However, Desmarest, adopting Buache's idea, turned it to a different use, as a propagator of seismic shock. Most of his essay is devoted to establishing the plausibility of this means of earthquake propagation, which he compared to the delivery of shock through a row of billiard balls.[10]

8 Aside from its deferent tone, the text contains no hint of the gentleman's identity, but a letter from Desmarest to his friend Pierre-Jean Grosley, in Troyes, indicates that Joly de Fleury had solicited the essay (letter dated March 27, 1756, Bibliothèque Nationale, Fonds Français, Nouvelles Acquisitions, MS 803, fol. 91–92).

9 Buache's geographical ideas can be examined in his "Essai de géographie physique où l'on propose des vues générales sur l'espèce de charpente du globe composée de chaînes de montagnes qui traversent les mers comme les terres, avec quelques considérations particulières sur les différents bassins de la mer et sur sa configuration intérieure," *Mémoires de l'Académie Royale des Sciences, 1752* (1756), 399–416; reprinted (without the maps in the original) in *Revue de géographie, XXII* (1888), 293–303. Buache's geography has been treated by Ludovic Drapeyron, "Les deux Buache, ou l'origine de l'enseignement géographique par versants et par bassins," *Ibid., XXI* (1887), 6–16, and by J. Thoulet, "L'étude de la mer au XVIIIe siècle: De Maillet, Buache et Buffon," *Mémoires de l'Académie de Stanislas, VI* (1908–1909), 214–256, as well as in the "Eloge de M. Buache" by Grandjean de Fouchy, *Histoire de l'Académie Royale des Sciences*, in *Histoire de l'Académie Royale des Sciences. Avec les mémoires de mathématique & de physique, pour la même année. Tirés des registres de cette académie, 1772*, pt. 2 (1776), 135–150.

10 *Conjectures*, p. 18ff.

What is of interest to us for the moment, however, is the later section of Desmarest's tract, in which he discussed the possible sources of earth tremors, a problem he regarded as more difficult than that of the disturbance's propagation. He did not commit himself to any hypothesis of earthquake source, but only mentioned briefly the widespread view that earthquakes result from the explosive expansion of the air produced by underground fires. He believed this opinion to have its difficulties, but these did not touch on the existence of burning materials underground, which Desmarest accepted without question. He did not relate the incidence of earthquakes to regions of volcanicity, although, of course, his main thesis related them to mountainous areas. He rejected the often-repeated hypothesis of the underground connection between separate regions of volcanic action — Vesuvius, Etna and the Lipari Islands had frequently been associated in this way in literature since antiquity.[11] But these do not represent extensive remarks about volcanoes, and all in all Desmarest seemed impressed neither with their inherent interest nor with their potential power to cause the earthquake shocks whose dissemination concerned him. He included a passing reference to the volcanoes of Auvergne (which he had not yet seen), but only to remark that they were burned out and incapable of causing damage: ". . . ainsi la nature est tranquille dans le beau pays que nous habitons. Heureux!"[12]

1757 witnessed the publication of the seventh volume of the *Encyclopédie*, including two geological articles by Desmarest: "Fontaine" and "Géographie physique."[13] The former article, dealing with springs and related phenomena, did not touch on volcanoes at all. The second article, however, briefly approached several questions relating to volcanism and heat in the earth, but did not deal with them at any great length, even though Desmarest's writing was here characterized by a degree of generality and scientific temperateness that had been less in evidence in his earlier efforts, which, it may be noted, constituted arguments for particular theories. In the *Encyclopédie*, Desmarest was concerned to deal with his appointed subjects in a comprehensive fashion, and here, for the first time, he formulated for publication his opinions on

[11] This idea was expressed, in fact, in a book edited by Desmarest, a posthumous collection of the wisdom of the cleric Longuerue (1652–1733): *Longueruana, ou recueil de pensées, de discours et de conversations, de feu M. Louis du Four de Longuerue* (2 vols. in 1; Berlin [i.e., Paris]: no pub., 1754), Vol. I, p. 182.
[12] *Conjectures*, p. 60.
[13] "Fontaine," *Encyclopédie, ou dictionnaire raisonné des sciences, des arts et des métiers*, Vol. VII (Paris: Briasson, David, Le Breton, Durand, 1757), pp. 80–101; "Géographie physique," *Ibid.*, pp. 613–626.

the methods and principles of the science of the earth. He distinguished between the primary and secondary segments of the earth's crust, and followed the general opinion of his time in declining to link the earliest strata with any volcanic action.[14] The existence of modern volcanoes was noted, as was the presence of extinct volcanoes, which could be recognized by their shape and the burned nature of their ejecta.[15] Again he declared the source of volcanic heat to be at a shallow depth. In general, though, Desmarest again gave no sign of special concern with volcanoes.

Between the publication of the *Encyclopédie* articles in 1757 and Desmarest's first visit to Auvergne six years later, he produced three more geological writings that have survived. The first of these, some geological observations made in Champagne in 1760, consists mainly of descriptions of local soils and includes considerable discussion of the destructive effects of running water upon land.[16] Here he viewed the cumulative effect of small-scale changes, such as the rounding of pieces of gravel in running water, as amounting in time to alterations of great magnitude — the destruction and removal of entire sets of thick beds, for example.[17] The other two writings, which have never been published, show that Desmarest's attention continued to focus on the alteration of geological features. They are accounts of Desmarest's travels in the southwest part of France in 1761 and 1762, when he was employed to help carry out a statistical survey of agricultural resources for Charles-Robert Boutin, the new Intendant of the *généralité* of Bordeaux.[18] As might be expected from geological excursions in the Gironde and neighboring districts, Desmarest nowhere discussed any aspects of volcanic

14 "Géographie physique," p. 622.

15 *Ibid.*, p. 624.

16 "Observations géologiques," in Pierre Samuel Du Pont de Nemours (Ed.), *Œuvres de Mr. Turgot* (9 vols.; Paris: A. Belin and Delance, 1808–1811), Vol. III, pp. 376–447. These notes were erroneously attributed to Turgot, who was Desmarest's benefactor and superior as Intendant in the *généralité* of Limoges, where Desmarest was Inspector of Manufactures from 1762 to 1771. Part of the notes were reprinted as "Observations géologiques, faites par Desmarets, en 1760. dans la partie de la Champagne qui forme actuellement le département de l'Aube," *Annuaire administratif et statistique du département de l'Aube* (1841), pt. 2, pp. 29–48.

17 "Observations géologiques," pp. 395, 402, 404.

18 "Voyage dans une partie du Bordelois et du Périgord," Bibliothèque Municipale de Bordeaux, MS 721. "Remarques de Mr. Desmarest (de l'Académie des Sciences) sur la géographie physique, les productions & les manufactures de la généralité de Bordeaux, lors de ses tournées depuis 1761 jusqu'en 1764," Archives départementales de la Dordogne, MS 26. The first of these manuscripts is anonymous, but circumstantial evidence, including proof that he purchased a horse and chaise in Bordeaux the day before the journey's beginning (Archives départementales de la Gironde, C 2891, no. 4), indicates that Desmarest is the author. The title of the second is in error, as Desmarest's travels in Gironde ended in 1762.

geology in these documents. What attracted most of his attention, aside from the nature and quality of the soils, was the destruction of rock and soil and the configuration of rivers in beds of their own creation. In both of these manuscripts, Desmarest frequently remarked on the influence of water on terrain, employing the terms *degradation* and *transport*, and relating changes in type of deposit to "revolutions" in the past. He was quite specific about the action of rivers in removing material from one place to another in creating the terrain's main features. Noticing torrential deposits, for example, high above the present level of a river bed, he wrote:

> At first glance it seems rather peculiar that these torrential deposits should be placed on the highest summits on the banks along our river-channel, and at over 200 or 300 feet above the present level of running water: the only reasonable explanation for this circumstance is in supposing that since [these deposits were laid down] the water has excavated its channel and that before the deepening of the valleys the products of torrential floods . . . occupied the heights in question, which served as beds. . . .[19]

He also stated clearly that this type of process was continuing in the present, although perhaps not at the same rate. Thus, Desmarest conceived of a kind of subaerial denudation, and saw valleys as carved by their own streams, ideas that he used to advantage in his studies in Auvergne, as shall be seen momentarily.

The overwhelming impression gained of Desmarest's geological work before 1763 is that up to that time he had no particular interest in volcanic geology, although he had shown a solid interest in the natural history of the earth in general for over a decade. On those occasions when he had referred to volcanoes or to subterranean heat, it was without much detail and often in a way not directly related to the subject at hand. What ideas he had formed about volcanoes he had probably derived from the texts of other naturalists and travelers. This general impression is sustained from what is known of Desmarest's correspondence.[20] It is certain that his opportunity for direct observation of volcanoes was very limited. He traveled little until 1757, when he began to carry out industrial investigations for Trudaine in various parts of France, but he did not go abroad until 1765, except for a brief excursion to San Sebastián. That his travel was restricted in these early decades is

[19] "Remarques de Mr. Desmarest," fol. 4.

[20] What I have been able to find of Desmarest's correspondence, numbering somewhat over 100 letters, dates mainly from the period 1756 to 1772. His principal correspondents were Pierre-Jean Grosley, Turgot, François Pasumot, and a certain Gonthier, of Troyes. The extant correspondence consists rather largely of letters received by Desmarest.

borne out by the fact that it was only in 1761, when he was 36 years of age, that he saw the ocean for the first time in his life.[21]

What captured Desmarest's attention in Auvergne, and converted him almost instantly into a student of volcanic geology, was basalt — not mere ordinary basalt, but "prismatic" basalt, formed into columns and other regular shapes. Desmarest had been familiar with this form of basalt, not from immediate experience, but from second-hand acquaintance with the most famous example of the phenomenon, the Giant's Causeway in County Antrim, Northern Ireland. The Giant's Causeway had been well known since it had been described in scientific fashion in the late seventeenth century, and its fame on the Continent grew enormously after 1740, when two engravings made from Mrs. Susannah Drury's paintings of the location were widely circulated.[22] But these curious columns, regarded as unique in the world (much significance was placed in the fact that Giant's Causeway columns are articulated, or divided into sections at regular intervals, whereas the basalt columns of Saxony are monolithic), were not visibly associated with volcanoes, active or extinct, and it occurred to no one that they might be of anything other than aqueous origin, except for those who assigned to them a human or supernatural cause (a cause implied by their name). The Giant's Causeway was an ornament of some consequence, and the discovery of similar forms of basalt in the heart of France was, in itself, sufficient reason for Desmarest to be excited by his find. But he was not content to claim a similar formation for France; in addition he sought to elucidate the origin of prismatic basalt, which he immediately presumed to be of the same origin everywhere, and therefore volcanic in Ireland if volcanic in France.

Desmarest's excursion of 1763 permitted him little more than a strong suspicion of his conclusion. But the next year he returned to Auvergne, this time with a cartographer, François Pasumot, and an artist, with a plan to map the volcanic regions, aided and protected by the Intendant, Simon-Charles-Sébastien Bernard de Ballainvilliers.[23] In 1765, Desmarest delivered a preliminary report of his findings to the Academy of Sciences, before departing for a year-long tour of Switzerland and

[21] Evidence of this astonishing fact is found in a letter to Grosley of September 7, 1761, wherein Desmarest stated that he had seen the sea the previous day "for the first time." He marveled at the "sublime" spectacle, and was baptized by his companions "according to the custom." (Bibliothèque Nationale, F.F., N.A., MS 803, fol. 133–134.) It is interesting that Desmarest devoted a substantial section of his *Dissertation* to the power of ocean currents and shore waves without ever having seen the sea.

[22] Sergei Ivanovich Tomkeieff, "The Basalt Lavas of the Giant's Causeway District of Northern Ireland," *Bulletin volcanologique, VI* (1940), 89–143.

[23] [L. Welter], "Carte des volcans d'Auvergne préparée par Pasumot," *Bulletin historique et scientifique de l'Auvergne, LXIV* (1944), 215–216.

Italy in the company of the young Duke Louis-Alexandre de La Rochefoucauld. Pasumot continued his mapping while Desmarest and La Rochefoucauld inspected the volcanic phenomena of Tuscany, Umbria, Latium, Campania, and the Euganean Hills in the district of Padua and Vicenza. Desmarest rejoined Pasumot in 1766, and they completed a geological map of the southern, Mont-Doré district that was published with Desmarest's report to the Academy of 1771. This article is generally taken as the first published announcement that basalt was volcanic. However, the discovery came to be known a good deal in advance of 1771, in part through Desmarest's unpublished announcement of 1765, which was probably discussed by members of the Academy, but also through the subsequent publication of the substance of this report with the plates of the *Encyclopédie* in 1768.[24] The existence of this report, published six years before the fuller memoir of the Academy, appears not to have been known to any of Desmarest's biographers, but it is not clear whether or not it was equally ignored by his contemporaries. It may well have been, for it is found as an explanation for one of the plates, unsigned; but at the very least it was read by Rudolf Erich Raspe, who acknowledged it in his letter establishing the igneous nature of German basalts.[25] Proof that word of Desmarest's 1765 report was spread in France is found in a 1766 memoir of Jacques Montet, who referred to the discovery in his account of volcanic mountains near Montpellier.[26]

The brief article in the *Encyclopédie* was long enough to do little more than describe the types of prismatic basalt found and to declare their certain association with volcanic flows. Desmarest was concerned there only with the configuration of basalt columns as he found them at the extremities of flows. In his two long memoirs on the Auvergne basalts, however, he turned to a historical task, that of reconstructing the former configuration of volcanic flows from the evidence at hand. These memoirs were presented to the Academy of Sciences in 1771 and 1775, respectively, and represented his fullest and most important contribu-

24 "Basalte d'Auvergne," *Recueil de planches, sur les sciences, les arts libéraux, et les arts méchaniques avec leur explication,* Vol. VI (Paris: Briasson, David, Le Breton, 1768), pp. 3–4 of section "Histoire naturelle, règne minéral." The two plates that this article accompanies depict a butte of articulated basalt columns upon which the château of La Tour d'Auvergne once stood (plate VII, section "Histoire naturelle, minéralogie, 6e collec., volcans"), and a mass of basalt in "boule" or ellipsoidal form, located by the château of Pereneire, near Saint-Sandoux (plate VIII).

25 "A Letter from Mr. R. E. Raspe, F.R.S., to M. Maty, M.D. Sec. R.S., Containing a Short Account of Some Basalt Hills in Hessia," *Philosophical Transactions, LXI* (1771), 580–583.

26 "Mémoire sur un grand nombre de volcans éteints, qu'on trouve dans le Bas-Languedoc," *Mémoires de l'Académie Royale des Sciences, 1760* (1766), 466–476. Although Montet's memoir is published with the proceedings of 1760, it was read in 1766.

tions to volcanic geology.[27] It was in the historical approach that Desmarest brought his earlier geological ideas to bear on the study of volcanic phenomena, and the result was an analysis of the Auvergne terrain that was original and effective. Applying the same principles of degradation and river-valley carving that he had begun to develop in the Champagne and Gironde regions, Desmarest was able to envision the former unity of lava structures now separated into distinct entities. He observed certain columnar basalt groups having no apparent connection with lava flows, resting in isolated patches, often perched atop hills and peaks whose bases were manifestly nonvolcanic. But being convinced now of the volcanic origin of all such basalts, without exception, he concluded that except in cases where they were distinct eruptions themselves, they must have been originally joined with neighboring lava flows, a connection demonstrable by identity in number and mineralogical type of corresponding layers. He ascribed the separation of such masses to "obvious alterations in the primitive disposition of the flows, either by the destruction of subsequent eruptions or, chiefly, by the degradation of water. . . ."[28] He was distinctly conscious of the power in connecting a current destructive agent with the reconstruction of older conditions.

> Water from rain and melted snow appeared to me as agents quite capable of working a large part of these degradations. Prisms, detaching easily from one another, can collapse by the removal of unsteady and light materials that serve as their base; for burned earth, pumice, scoriae, and porous lavas yield with the greatest ease to the action of the least trickle of water; and this play of water, whose last traces and most recent marks I had examined, once recognized, helped me, in my thought, to replace all the removed materials into their primitive state, and to restore without effort the ancient configuration of the terrain, as it was after the volcanic eruptions.[29]

27 The 1771 paper, read May 11, was published in expanded form in three parts, the first two together, under the title "Mémoire sur l'origine & la nature du basalte à grandes colonnes polygones, déterminées par l'histoire naturelle de cette pierre, observée en Auvergne," *Mémoires de l'Académie Royale des Sciences, 1771* (1774), 705–775, and the third, "Mémoire sur le basalte, troisième partie, où l'on traite du basalte des anciens, & où l'on expose l'histoire naturelle des différentes espèces de pierres auxquelles on a donné, en différens temps, le nom de basalte," in *Mémoires, 1773* (1777), 599–670. The memoir that he read on St. Martin's Day (November 11) 1775 was published in abbreviated form as "Extrait d'un mémoire sur la détermination de quelques époques de la nature par les produits des volcans, & sur l'usage de ces époques dans l'étude des volcans," *Observations sur la physique, sur l'histoire naturelle et sur les arts, XIII* (1779), 115–126. Much later, on 1 *prairial an* XII (May 20, 1804), he delivered essentially the same paper to the Academy: "Mémoire sur la détermination de trois époques de la nature par les produits des volcans, et sur l'usage qu'on peut faire de ces époques dans l'études des volcans," *Mémoires de la classe des sciences mathématiques et physiques, VI* (1806), 219–289.

28 "Mémoire sur l'origine & la nature du basalte," p. 720.

29 *Ibid.*, p. 738.

Desmarest's disposition to work from the present backward is illustrated in his comparison of different effects in different lava flows.

I noticed that the deepening of these valleys was in proportion to the antiquity of these flows and to the abundance of the streams' waters; that the streams that separated the different parts of the old flows had achieved the excavation of very deep cuts between them, whereas they were only rough-hewn and studded with falls and cascades when they were found cutting between parts of the more modern flows.[30]

His observations on the relative degree of degradation of lava flows of varying age were generalized into a principle of uniform action, which he enunciated in direct connection with the history of Auvergne terrain.

The results of the last operations are simpler and less modified by the changes which the primitive forms daily undergo, and that one recognizes there the agents more easily when the traces of their march are more manifest; moreover, this primitive state is a standard for comparison, to be kept continually before the eyes of the observer who would judge of the extent and progress of the successive alterations.[31]

Among the general conclusions Desmarest drew from his research in the Auvergne volcanoes, some of the most interesting include this link he saw between the process of degradation and uniform change in time. His proposition that "the same causes or parallel causes have equally altered, in the same place, different primitive conditions"[32] classed these causes among the "order of the most ordinary phenomena."[33] His method of research, he believed, showed that "Nature has been subject to the same order in the most remote ages as in the most recent times."[34]

Hooykaas rightly points out that although Desmarest claimed the uniformity of natural processes as a result of his investigation, he actually argued in a circle, as the assumption that nature proceeds uniformly was implicit in his method.[35] This circularity is illustrated in the passage already quoted (see note 30), in which Desmarest recognized a correlation between the age of a lava flow and its degree of degradation. For how did he determine the age of a lava flow? On this question he was not entirely explicit; but aside from cases in which he could assign relative

30 *Ibid.*
31 "Extrait d'un mémoire sur la détermination de quelques époques," pp. 117–118. This translation is taken from R. Hooykaas, *Natural Law and Divine Miracle: The Principle of Uniformity in Geology, Biology and Theology* (Leiden: E. J. Brill, 1963), p. 16.
32 "Mémoire sur l'origine & la nature du basalte," p. 739.
33 *Ibid.*, p. 737.
34 Extrait d'un mémoire sur la détermination de quelques époques," p. 117.
35 Hooykaas, *Natural Law*, p. 17.

ages from the circumstance that one flow partially covers another, he seems simply to have presumed that the more thoroughly destroyed the flow appeared, the older it was. There is no need to criticize this presumption, which is of a kind often employed by geologists, for it worked well, and the phenomena described by Desmarest fit neatly and harmoniously into the scheme based on the assumption. Nevertheless, Desmarest was either unwilling or unable to declare the uniformitarian principle, such as he used it, a necessary part of his theoretical framework, rather than a result of a series of demonstrations.

Although Desmarest used uniformitarian arguments, it should be recognized that there were limits to his use of the concept. The idea that nature acts uniformly was generally confined to "ordinary" phenomena, a category that included subaerial denudation but not volcanic activity. He held to the "accidental" nature of volcanoes for most if not all of his years. "I comprehended from the beginning," he wrote in 1806, "that, volcanic eruptions being accidents among ordinary phenomena of nature, the return of their attacks had not been subject to any fixed period."[36] In Desmarest's view, even though volcanoes had been the cause of substantial change in the landscape, they could not be relied on to account for regular effects of great consequence in earth history. It is a mistake to attribute to him the conviction that the eruptivity of volcanoes is subject to a uniformitarian order.[37]

This limitation on the power of volcanoes as agents of geological change is clearly expressed in Desmarest's last and largest geological work, the compendium *Géographie physique* of the series *Encyclopédie méthodique*.[38] Here he offered his opinions on a very wide variety of geological subjects, amplifying his ideas on questions discussed in his earlier works, and providing an unusual record of his thought, not only on the subjects listed alphabetically in Volumes II through IV, but also on the 40 geological thinkers dealt with in the first volume.[39] His ideas

36 "Mémoire sur la détermination de trois époques de la nature," p. 221. Desmarest made an almost identical statement in the 1779 version.

37 Hooykaas errs slightly in emphasizing Desmarest's uniformitarianism in the study of volcanic history; he translates the passage quoted on page 349 *supra* (note 34) as follows: "Nature, in the eruptions of volcanoes, has been subject to the same order in the most remote ages as in the most recent times." However, the phrase "in the eruptions of volcanoes" is not present in the 1779 version, having been added only in 1806.

38 *Encyclopédie méthodique, ou par ordre de matières: Par une société de gens de lettres, de savans et d'artistes: Géographie physique* (5 vols.; Paris: Chez H. Agasse, *an* III [1794/95]–1828).

39 He refused to write on living geologists, with two exceptions: Hutton and Pallas. Of the remaining 38, most lived in the seventeenth or eighteenth centuries. The fifth volume, published long after Desmarest's death, was completed by Bory de Saint-Vincent, Doin, Ferry, and Huot, and the contents appear to be more their doing than Desmarest's.

expressed here exclude the possibility that volcanoes or subterranean heat have operated throughout the earth's history. Following the accepted division of strata into three principal epochs, he explained his belief that the oldest, granitic formations (the *ancienne terre*) were not produced by fire or heat, as experiments showed that granite vitrifies when heated sufficiently.[40] Desmarest was following the lead of Guillaume-François Rouelle, whose courses he had taken and whose reflections on the history of the earth he greatly admired.[41] In Rouelle's opinion, the primitive core of the globe was crystallized out of an aqueous fluid.[42] Granite could not be formed from fusion, which results only in a homogeneous mass without distinct crystals of feldspar, quartz, or mica. Desmarest regarded granite as the primary stuff of lava, without supposing, however, that the *ancienne terre* necessarily represents the first and oldest stage of the earth's history — in fact, he speculated, without ever committing himself to an estimate of the earth's age, that a virtually endless succession of epochs and destructions may have preceded the three periods for which the evidence is clear.[43] There were then no signs of volcanic activity in the *ancienne terre*, but only in the secondary (*moyenne*) and tertiary (*nouvelle*) strata.[44]

This was consistent with Desmarest's conviction that the source of volcanic heat lay in subterranean masses of burning coal, which he preferred to the "fermentation" of pyrites favored by others. The massive and unstratified primary deposits did not contain significant collections of organic remains, and could not be presumed to have possessed coal beds adequate to produce volcanoes.[45] But later deposits had abundant coal, and with water this was enough to produce the required effects.

I admit that combustible materials alone are not capable of producing all the effects of volcanoes; but that should not cast any doubt on the necessity of these materials to serve as fuel for volcanic fire. Expanding water in

40 *Géographie physique*, Vol. II, p. 502 ("Ancienne terre").

41 That Desmarest attended Rouelle's public lectures in the early 1760's is inferred from two letters he wrote to Grosley, one [1761] in the Bibliothèque Nationale, F.F., N.A., MS 803, fol. 131–132, and the other [December 1762] in "Lettres inédites de Grosley et de quelques-uns de ses amis," ed. Truelle Saint-Evron and Albert Babeau, *Société Académique de l'Aube: Collection de documents inédits relatifs à la ville de Troyes et à la Champagne méridionale, I* (1878), 346–349.

42 *Géographie physique*, Vol. I, pp. 409–431 ("Rouelle").

43 *Ibid.*, Vol. II, p. 554 ("Anecdotes de la nature et de l'histoire de la terre").

44 *Ibid.*, Vol. II, p. 308 ("Alpes"); Vol. II, p. 502 ("Ancienne terre"). At one point Desmarest dismissed underground fire even from the secondary strata; *Ibid.*, Vol. III, p. 508 ("Couches de la terre").

45 *Ibid.*, Vol. I, p. 268 ("Lazzaro-Moro"); Vol. III, p. 366 ("Charbon de terre"). Desmarest set forth this opinion as early as 1754, in a note in his edition of Francis Hauksbee's *Expériences physico-méchaniques sur différens sujets, et principalement sur la lumière et l'électricité, produites par le frottement des corps* (2 vols.; Paris: Chez la Veuve Cavelier & Fils, 1754), Vol. II, p. 553.

the next instance seems to be the active force capable of carrying out the acts of eruption, and of producing the disturbance accompanying them; but it is no less true that the condition of all the products of subterranean fire shows that coal is the only material suitable not only to achieve their fusion, but also, by its immense quantity and its disposition in the bowels of the earth, to supply the subterranean fires. . . .[46]

Furthermore, volcanic remains show remnants of the combustion of coal.

Moreover, if one examines the vicinity of the craters and funnels by which flame and smoke surge out in frightful clouds, it will be seen that these craters and funnels are covered with immense masses of slag perfectly similar to the residue of the combustion of coal. . . .[47]

Coal was always Desmarest's preference for the cause of volcanic heat; he never accommodated himself to the opinion that related volcanic material to the earth's innate heat.[48]

Nor would Desmarest accept volcanic activity as a plausible cause of the uplift of mountains, a problem for which he was unable to offer a suitable mechanism. Recognizing that mountains and continents have incontestably risen, for whatever reason, he essayed an explanation only for their subsequent denudation. He rejected Anton Lazzaro Moro's suggestion that subterranean fire had caused mountains to rise early in the earth's history, by referring to the availability of coal only in the latter stages of the earth's development;[49] to the more ambitious system of Hutton, calling for both the consolidation of strata and mountain uplift by subterranean heat, he responded similarly.[50] As long as real and literal fires were the only source he could imagine for volcanic heat, and as long as he dismissed other sources of internal heat, it was difficult for him to think otherwise.

In short, Desmarest was no vulcanist. Instead, he was a neptunist, in the broad sense that he viewed water, rather than heat, as the main agent of geological change and the chief medium of creation of terrestrial features. This scientist, who had done much to extend the scope of volcanic action in geology, nevertheless thought volcanoes to be of far more limited importance in the earth's crust than water. Signs of his neptunist frame of mind are to be found in all of his geological works, dating back to his earliest publication, and it is unnecessary to document here the constancy of his attitudes. Suffice it to note that one of his later

46 *Géographie physique*, Vol. IV, p. 164 ("Feu des volcans").
47 *Ibid.*, Vol. IV, p. 163 ("Feu des volcans").
48 *Ibid.*, Vol. I, pp. 382–409 ("Romé de l'Isle"), and Vol. III, pp. 344–349 ("Chaleur du globe").
49 *Ibid.*, Vol. I, p. 268 ("Lazzaro-Moro").
50 *Ibid.*, Vol. I, p. 749 ("Hutton").

papers deals with prismatic columns formed like those of basalt, but from aqueous clay.[51] Far from regarding lavas as the sole material capable of this behavior, he was pleased to be able to extend the process of formation of prisms by contraction to aqueous as well as igneous substances. Desmarest's neptunism is not altogether surprising, as others prominent in volcanic geology of the time held similar opinions, but it calls for a clarification of the so-called vulcanist-neptunist dispute and of the specific nature of the controversy over the origin of basalt. The task of tracing the basalt controversy through the last four decades of the eighteenth century cannot be done here, but a few remarks and tentative conclusions can perhaps be offered.[52]

First of all, Dr. Rappaport is quite correct in indicating that a geologist's position on the basalt controversy is not a reliable index for classification in the vulcanist-neptunist dispute.[53] Many geologists not ordinarily reckoned among the friends of vulcanism spoke in favor of basalt's igneous nature. The neptunist Guettard, long a defender of the creation of basalt "in the humid way," came over to Desmarest's point of view late in his career; his change of mind on this subject is implied, as Dr. Rappaport has suspected, in his *Mémoires sur la minéralogie du Dauphiné*.[54] Jean-André Deluc, chosen by Professor Gillispie as a paragon of neptunism, nevertheless described basalts as volcanic as early as 1778.[55] Jean-Claude de Lamétherie, whose theory of geology owed much to Werner, attributed basalt to an igneous cause no later than 1795.[56] Furthermore, some geologists aside from Desmarest who devoted

[51] "Mémoire sur les prismes qui se trouvent dans les couches horizontales de plâtre et de marnes des environs de Paris, et sur leur analogie avec prismes du basalte," *Mémoires de la classe des sciences mathématiques et physique, IV* (*an XI* [1803]), 219–231; "Seconde mémoire sur la constitution physiques des couches de la colline de Montmartre et des autres collines correspondantes," *Ibid.*, *V* (*an* XII [1804]), 16–54.

[52] On the basalt controversy, see Otfried Wagenbreth, "Abraham Gottlob Werner und der Höhepunkt des Neptunistenstreites um 1790," *Freiberger Forschungshefte* (D 11), (Berlin: Akademie-Verlag, 1955), pp. 183–241; and Arnold von Lasaulx, "Der Streit über die Entstehung des Basaltes," *Sammlung gemeinverständlicher wissenschaftlicher Vorträge*, ser. 4, LXXVI (1869).

[53] Rhoda Rappaport, "Problems and Sources in the History of Geology, 1749–1810," *History of Science, III* (1964), 66.

[54] (2 vols.; Paris: De l'Imprimerie de Clousier, 1779). See Vol. 1, pp. 128–154.

[55] *Lettres physiques et morales sur l'histoire de la terre et de l'homme: Adressées à la reine de la Grande Bretagne* (5 vols. in 6; La Haye and Paris: De Tune and Veuve Duchesne, 1779–1780), Vol. II, 478–481; Vol. IV, pp. 146–161; *Geological Travels in Some Parts of France, Switzerland, and Germany* (2 vols.; London: Printed for F. C. and J. Rivington, 1813), esp. Vol. I, p. 232; Vol. II, pp. 264, 284. Gillispie deals at length with Deluc in *Genesis and Geology: A Study in the Relations of Scientific Thought, Natural Theology, and Social Opinion in Great Britain, 1790–1850* (Cambridge: Harvard University Press, 1951; reprinted Harper Torchbook, 1959), pp. 56–66.

[56] *Théorie de la terre* (3 vols.; Paris: Chez Maradan, 1795), Vol. II, pp. 47, 55–61.

substantial parts of their attention to volcanic activity nevertheless refused to behave as though a final choice had to be made between fire and water. Dolomieu and de Saussure fall into this class. It is not surprising, though, that there is confusion in the relation between the basalt controversy and the vulcanist-neptunist debate, as the participants themselves, and interested bystanders, often failed to distinguish between them. There are signs that the term *vulcanist* was first applied as an indication of belief only in the volcanic origin of basalt, and if this is so, then its quite natural subsequent extension to a more general geological doctrine has exacted its cost in historical clarity.[57] A few historians have since attempted to remedy the situation by employing the term *plutonism* for vulcanism of the variety requiring for heat a prime place in geological change, but this is difficult to carry through, and perhaps of doubtful desirability, since the principals in question frequently called themselves vulcanists.[58]

It is not unlikely that the stances of individual geologists regarding the basalt controversy were substantially influenced by their methodological predispositions. Perhaps the historian would find it useful to consider the opinions of the antagonists in the light of the various modes of research they most favored. Opposition to the igneous origin of basalt seems to have come particularly from mineralogists — aside from Werner, Bergman, Wallerius, Klaproth, and Romé de l'Isle were prominent mineralogists opposed to Desmarest on this count. On the other hand, support for the igneous view came especially from geological travelers who observed columnar basalt in its natural location and were less apt to study rocks from the standpoint of chemistry and mineralogy than from a dynamic and physiographical framework. The Auvergne basalts, in particular, were capable of convincing visitors, and Desmarest himself said that he might not have made his discovery anywhere else.

[57] *Vulcanist* and *neptunist* are used in a somewhat narrow sense by Saussure, "Observations sur les collines volcaniques du Brisgaw," *Journal de physique, de chimie et d'histoire naturelle, XLIV* (*an* II [1794]), 356, and by Faujas de Saint-Fond, "Essai de classification des produits volcaniques, ou prodrome de leur arrangement méthodique," *Annales du Muséum National d'Histoire Naturelle, III* (1804), 85–86. In the second edition of his *Théorie de la terre* (5 vols.; Paris: Chez Maradan, 1797), Lamétherie used the term *vulcanistes* in specific reference to believers in the igneous origin of basalt (Vol. II, p. 57).

[58] Geikie is perhaps the most notable example of a historian distinguishing vulcanists from plutonists. Frank Dawson Adams, *The Birth and Development of the Geological Sciences* (Baltimore: Williams and Wilkins, 1938; reprinted Dover, 1954), used the term *plutonist*, but then incorrectly summoned Desmarest to the anti-neptunist cause (p. 245).

Clarification of the role played by the basalt controversy in the development of plutonism will have to await research in the geological thought of a number of eighteenth-century scientists whose work has not been examined adequately. Too many geologists of this period have been understood only through the distorted pictures given us by other scientists in funerary orations; the alleged antagonism between Desmarest and Werner, for example, derives mostly from Cuvier's artistic sense, for in their rivalry he found a convenient common theme for their eulogies, which he delivered on the same day. Modern scholarship will be able to eliminate misrepresentations such as this, and perhaps will provide new categories through which the development of eighteenth-century geology can be better understood.

To make a clear distinction between the basalt controversy on the one hand and the vulcanist-neptunist dispute on the other is not to say that they were unrelated, but only that it is plain that many who believed in the volcanic origin of basalt would have found it necessary to change some of their basic presuppositions before they could have become vulcanists of the Huttonian mold. In Desmarest's case, at least, it was impossible to accept the action of volcanoes throughout all time or to treat them as geological agents of the highest importance because of his particular idea about their cause. A thoroughgoing vulcanist, on the contrary, required a vaster and more constant source of heat, and this could only come from the interior of the globe. Perhaps the radical vulcanists learned something from the adherents of "central heat" — Dortous de Mairan, Buffon, and Bailly. But it seems true, too, that vulcanists were capable eclectics, able to select useful ideas and information from their neptunist colleagues without being required to accept all neptunist explanations. Desmarest opened the way for later investigators in Auvergne without convincing them either that lava derives from granite or that volcanoes are burned-out vents from coal beds. And he was not the only neptunist who contributed to the development of vulcanism.

In conclusion, it can be said that Desmarest's geology is somewhat more difficult to characterize in the framework of eighteenth-century geology in general than has sometimes been thought, and that this perhaps reflects on the depth of our understanding of geological thought in this period. In particular, it may be suggested that vulcanism and neptunism should not be regarded as altogether hostile schools, but that valuable elements of one may have been derived from the other. The career of Nicolas Desmarest is an example of how some vulcanist concerns could play an important part in a geological outlook that always favored the primacy of neptunist agencies. In addition, the study of Desmarest's geology shows that actualistic ideas were present in his mind as early as the 1760's, even before his first visit to Auvergne, and that this actualism was strongly associated with that part of his geology that was not specifically volcanic; in fact, for him, volcanic activity lay beyond the limits of uniformitarianism. When one reflects that the fusion of plutonism with uniformitarian thinking helped geology take a new turn at the end of the eighteenth century, it seems pertinent to ask what alterations the two concepts needed to undergo in order to make this fruitful union possible.

The Development of the Geological Sciences in the U.S.S.R. from Ancient Times to the Middle of the Nineteenth Century

V. V. Tikhomirov

The Soviet Union occupies vast expanses of territory in both Europe and Asia. It is inhabited by a great number of peoples related from ancient times by trading and cultural contacts. These ties contributed to a mutual exchange of handicraft products, literature, and technical and scientific discoveries.

Rudimentary geological knowledge, including primitive ideas of the properties of minerals and ores as well as methods for the search for the most important minerals, originated with the people who inhabited this part of Eurasia in very ancient times.

Part 1: To the Nineteenth Century

ANCIENT PERIOD The development of mining served as the main impulse for the origin of geological concepts. The first societies widely engaged in mining were the earliest cultures — Chinese, Indians, Babylonians, Egyptians, Greeks. On the territory of the Soviet Union, also, mining began early. In different regions — the Ukraine, the Urals, Transcaucasia, Kazakhstan, Central Asia, the Altai mountains, and the Transbaikal area — traces of old mines, "chud diggings," have been found. Some of them date back to the Stone Age. In the Ukraine, for instance, numerous archaeological finds have been described that include various articles made of pyrophyllite. Primary deposits of pyrophyllite are known in the Volyn area. The processing of this comparatively soft stone began, apparently, as early as the Paleolithic and became widely developed in the Neolithic era. During this period there were real workshops that produced a variety of articles of pyrophyllite.

Translated by T. A. Sofiano, researcher of the Geological Institute of the Academy of Sciences of the U.S.S.R.

Three and a half thousand years ago, copper was smelted from malachite in the Ural area. Gold and copper ores were mined on an extensive scale in Northern Kazakhstan 1500–700 B.C. Production of copper ore from Kedabek, Kafan, and other deposits had started in Transcaucasia during the Bronze Age. At the end of the Bronze Age, about one thousand years before our era, silver-lead ores were mined in Central Asia. There are large quarries of the period extant, with labyrinthine underground passages and enormous worked-out chambers (Kanimansur, Kanigut, etc.) with total volumes of excavation amounting to tens of thousands of cubic meters.

At the beginning of the Iron Age, the tribes that inhabited the forest regions of the middle and northern parts of the Russian plain learned to use meadow, bog, and lacustrine brown iron ores occurring at the surface of the earth. The main sites of production for these ores were the basin of the Upper Dnieper, a vast area from the source of the Volga up to Onega Lake, and the territory between the present cities of Gorky and Kirov, the mid-course of the Oka.

In ancient Armenia, furnaces for smelting iron (Metsamor) that date back to 2,000 years before our era have been found, and there are special installations for the separation of the different ores that were mined in the region.

In Azerbaidzhan (Mingechaur) pottery has been found—vessels of Eneolithic type—that is referred to the third millennium before our era. The Nakhichevan salt mines were developed during the Neolithic, indicated by finds of dioritic stone hammers in abandoned excavations.

During the Bronze Age, industrial minerals were widely used for ornaments. In the burial grounds of Transcaucasia dated at the end of the first millennium before our era, there are articles made of native sulfur. Gypsum was used in Transcaucasia in very remote times, both as a building material and for the manufacture of ornaments. Crimean marmorized limestone was used for buildings in Chersonese and Panticapea (Bosporus) 500–400 B.C.

There are references to precious stones of "Piarmia" (Permian Priuralie) in old Scandinavian legends. Apparently, gem stones were produced there in sufficient quantity to be shipped far beyond the Ural area.

There are many more examples of the use of minerals. What has been said gives every reason to assume that even in ancient times the population inhabiting the territory of the USSR included competent men who knew certain properties of minerals — hardness, friability, malleability, etc., in other words, physical features — that were taken into account in processing and using these minerals.

MIDDLE AGES During the middle ages mining was widely developed on the territory of the USSR.

In Kiev, Russia, and in the lands of Great Novgorod, the production of iron articles flourished, still using the near-surface iron ores of sedimentary origin; pottery was popular; rock was quarried in great amounts for building, especially for the erection of churches. In Ovruch (Volhynia), in the tenth century, Proterozoic quartzites were used for the production of spindles — small stone discs used for spinning, in widespread use in all the lands of old Russia.

The artisanship of the tenth to sixteenth centuries used pyrophyllite from Volhynia, amber from the Dnieper, amethysts from Kandalaksha Bay. From very ancient times on, Russia used colorless transparent mica for windows. Such mica occurred in the lands around the Solovetzky monastery and on the Kandalaksha coast of the White Sea. Later, it not only was delivered to many settlements of the Moscow hegemony, but became an item of export to other countries, where it got the name of Russian or Moscow glass. This name was so generally accepted that at the beginning of the nineteenth century colorless transparent mica received the mineralogical name *muscovite*, from the word "musca," as Moscow was called by the Italians in former times.

Beginning with the seventh century, Transcaucasia produced oil, which was exported into the countries of the Near East, and silver-lead, copper, and iron ores were also mined there.

Articles were fabricated from iron mined in Central Asia. Silver, gold, lead, and mercury were mined; silver objects were shipped to distant areas, even to Central Europe.

The wide development of handicrafts helped the formation of scientific concepts. During this period prominent scientists appeared in Central Asia and Transcaucasia. Judging by those of their papers that reached us, some of their ideas on natural history belong to the sphere of the geological sciences. An important place in the early stages of this period belongs to the Armenian thinker of the seventh century, Ananiĭ Shirakatsi,[1] who thought, like ancient philosophers, that the natural environment consists of a combination of four basic elements —

[1] A. G. Abramian, *The Lunar Cycle Tables of Ananii Shirakatsi* (Erevan: Erevan State Univ. Press, 1962) (in Russian).

earth, water, air, and fire. Everything that exists is in a state of constant motion and change. He emphasized the destruction of things of the past and present and their replacement by new things. Contrary to the then popular view of a flat (patelliform) or quadrangular shape of the earth, Shirakatsi thought that the world was of an egg shape, with the earth corresponding to the yolk, the surrounding air being the white, and the sky corresponding to the egg shell.

A number of papers by Shirakatsi that reached us are devoted to astronomy; we do not know of any papers directly concerned with geology or mineralogy.

An outstanding place in medieval science is occupied by Abu Ali Ibn-Sina (about 980–1037), widely known under the name Avicenna. He was born and spent most of his life within what is now Tadzhikistan.

Ibn-Sina was a naturalist who combined a materialistic approach to nature with theology and mysticism. This did not prevent him, however, from voicing thoughts that were very advanced for his time. His geological ideas were formulated in a special treatise that became a part of his fundamental work *Book of Cure*, written in 1021 to 1023. A part of this treatise was translated into Latin and published in Venice in 1608.[2] In full, the book (*Kitâb-Al-Shifâ*) was republished in various European countries more than 30 times.

Ibn-Sina was interested in the problem of mountain building; he studied the composition and properties of minerals and meteorites. His classification of rocks and minerals was of a very high standard for his time. He distinguished stones, fusible matter (metals), sulphuric combustible matter, and salts. This classification was widely known and persisted to the second half of the eighteenth century. Among other scientists, A. G. Werner, professor at the Freiberg Mining Academy, used it as a basis in his textbook on mineralogy.

Ibn-Sina described the alteration of soft clay into a stone and thought that loose rocks could become lithified as result of the same process as the precipitation of salt from solution; however, he did not take into consideration any cementing factors and assumed that the induration of a rock happens under the influence of some hypothetical (special) force. Having noticed the presence of fossils in a mass of stratified rock, Ibn-Sina came to the conclusion that at some earlier

[2] Avicenna, *De congelatione et conglutinatione lapidum. Being sections of the Kitâb Al-Shifâ.* Trans. and ed. E. J. Holmyard and D. C. Mandeville (Paris: Paul Genthner, 1927).

time, the place had been covered by the sea. He indicated that beds that formed a mountain did not all originate simultaneously. In his opinion, the process of mountain building was the result of tectonic factors (strong earthquakes), as well as erosion, which washes out soft rocks, leaving rocky cliffs and mountain slopes. He also pointed out that exogenetic factors operate slowly, which means that the formation of a complex dissected relief demands a very long time. The world, in Ibn-Sina's opinion, was just as eternal as God. During the long period of the existence of the earth, the land and the sea changed places repeatedly. In this way Ibn-Sina established the existence of transgressions and regressions.

Though Ibn-Sina admitted the existence of God, his checking by experience and practice of theoretical deductions regarding nature made him disagree with Islamic dogmas, which caused the reactionary Moslem clergy to persecute him and burn his papers.

Ibn-Sina's contemporary, the Choresmian (Uzbek) scientist Biruni (about 972–1048), left a number of valuable geographical and mineralogical papers. He was the first among the scientists of the Middle East who, in his astronomical papers, assumed the revolution of the earth around the sun, even if he repeated the assertions of the Koran that the earth is the center of the universe. Contrary to opinions predominant at that time, he regarded the earth as a sphere and computed its size, obtaining figures similar to the present data.

Biruni expressed his thoughts on geology in a number of special papers. The most important of them is *Collected Information on Precious Minerals*, written about 1048 and later distributed in several handwritten copies, as well as in the form of translations of certain parts of the book into European languages.[3]

The treatise consists of two parts: Part 1 deals with precious stones and Part 2 with metals. The author gave a description of about a hundred minerals known at that time, of their varieties, and of rocks, quoting up to 300 synonyms of their names. Giving the details of the characteristics of each, he supplied information on color, hardness, methods of production, deposits, cost of precious stones, etc. Of special importance were the specific gravity determinations made by the author. For this purpose, Biruni designed a special device and weighed the water displaced by a definite amount of the substance. In this way, he deter-

[3] al-Biruni, abu-al-Rayhān Muhammad ibn-Ahmad, "Collected information on the nature of precious minerals" (1048), an Arabic manuscript on the subject of precious stones.

mined the specific gravities of 18 minerals, obtaining figures very close to the present values and within limits determined by isomorphic admixtures.

In Europe the determination of the specific gravity of minerals with a precision equal to Biruni's computations was not done until the eighteenth century. By introducing into mineralogy the figures for specific gravities, Biruni began the transformation of this branch of science from a descriptive into an exact science, using numerical constants as a means of investigation. On the basis of similarity in certain features (specific gravity, hardness, etc.), Biruni distinguished several groups of allied minerals. He included rock crystal, amethyst, carnelian, and onyx in the quartz group. In the corundum group, Biruni included ruby, sapphire, and leucosapphire. In this case, he scientifically confirmed the working jeweler's conclusion of the mineralogical unity of these groups.

Biruni noticed that some minerals contain inclusions of a liquid, and he took this for proof of the aqueous origin of crystals. In describing minerals and ores, Biruni noted which rocks usually accompany them in natural conditions, thus providing very useful indices for search and prospecting.

Biruni sharply criticized and ridiculed opinions that were very popular at that time about the mystical powers of precious stones. Stating that in the interiors of the earth some minerals are growing while others are being destroyed, Biruni expressed the idea of a continuous evolution of nature.

In his treatise, Biruni described deposits of minerals in the entire world known to him, including China, India, Ceylon, Byzantium, Egypt, the territory along the Baltic Sea, and the Mozambique straits. He treated mineral deposits then known in Central Asia with special detail.

For some time, Biruni's treatise remained a most important reference book on precious stones and was often quoted by the most prominent scientists.

In a number of Biruni's other papers there are ideas that characterize him as a naturalist with a very broad mind and great erudition. He wrote, for instance, that artesian wells operate on the principle of difference of hydrostatic level. Remains of fish and shells found on land he regarded as proof of a previous presence of the sea and assumed that from time to time the land and sea exchanged places.

He studied the geological process of delta formation at the mouth of the Ganges and established that in the remote past the sea had covered the territory of the great Indian rivers.

His papers of paleogeographical content include researches on the position of the ancient bed of the Amu-Daria.

The Azerbaidzhan mathematician-astronomer Mukhamed Nasireddin (1201–1274), also known under the name Tusi, was the author of a great number of papers on various problems in natural history, including a mineralogical treatise "Dzhavakhirname," which means "Fundamentals of Minerals." This paper describes 34 minerals: diamond, emerald, ruby, spinel, turquoise, lazurite, agate, jasper, and so on. A detailed enumeration of their physical properties is given: color, luster, hardness, specific gravity, transparency, friability, etc. The author also gave the principal methods for determining minerals. Except for the mineralogical treatises by Ibn-Sina and Biruni, the paper by Tusi was the only one of the period that did not simply repeat the data of his predecessors, but also gave additional scientific information.

In this way, out of the general background of the lull in creative thought typical of the Middle Ages, in the Near and Middle East there was a development of some geology, particularly mineralogy.

SIXTEENTH AND SEVENTEENTH CENTURIES The development of handicraft and mining in Russia was interrupted by the Tartar-Mongolian invasion and the subjugation of the native Slavic population. This lasted from the thirteenth through the fifteenth centuries. After its liberation, the country resumed its development. Substantial attenion was paid to the search for minerals and to their use. In 1584, the *Kamenny Prikaz* was founded in the state of Moscow — an agency for the output and processing of natural building materials. The production of iron ore was increased; the mining of sideritic ores was added to the output of bog ores that had been mined from ancient times.

In 1637, the *Siberian Prikaz* was established to direct extensive work on a search for deposits of mica and various ores in the Urals and Siberia. Large and long expeditions were organized throughout the seventeenth century with prospect parties of people in government employ; as a result, deposits of iron and other ores were found in many parts of the Urals and Siberia. Encouraging decrees of the czar attracted a great number of private prospectors to the development of subterranean treasures.

As indicated by old manuscript sources, prospectors were already familiar with indices for the search of iron, copper, silver, and tin ores. This made their work not accidental but purposeful.

In addition to the Siberian *Prikaz*, other departments were also engaged in a search for mineral resources during the second half of the seventeenth century. For instance, the *prikaz* of "secret affairs" sent expeditions to different parts of the country. Minerals and rocks of any special interest found by these prospectors were delivered to the "minor goods (governmental) chamber" organized at the *prikaz.* Here, quite a large number of samples were accumulated representing all sorts of ores and minerals from various deposits in Russia. This collection served as a prototype for future geological museums.

REGINNING OF THE EIGHTEENTH CENTURY A significant advance in geological research and in the development of theoretical geology took place in Russia during the eighteenth century. Many useful innovations were introduced into mining and science during the reign of Peter the Great, who ordered in 1700 the creation of a *Prikaz Rudokopnykh Del* (Administration of Mining). Numerous experienced and qualified people were invited to take part in reconnaissance and prospecting. Information about finds of minerals began to appear in the first printed Russian newspaper: "Annals of military and other affairs worthy of knowledge and memory, occurring in the Moscow state and adjacent lands."

The discovery of new deposits became so important that the carrying out of a death sentence was postponed if the criminal stated that he had information about previously unknown mineral deposits.

In 1719, after a number of reforms in the mining service, a Berg-Collegium was created, in other words, a Board of Mining, with a group of "ore reporters" — ore prospectors.

In 1723, an expedition directed by G. G. Kapustin discovered coal deposits in the Donetz Basin. Outcrops of coal were established also in the Kuznetsk area, and shows of oil were reported in Bashkiria and on Ukhta. In the Urals, deposits of magnetic and brown iron ore, asbestos, and gemstones were found. Under Peter the Great, the Urals became the main center of the mining industry. During this time, a number of pig iron and copper smelting plants were built there.

Along with successes in the mining industry, important steps were also undertaken to develop science and disseminate specialized knowledge. The Academy of Sciences was founded in 1725 in Petersburg, where important research began along the most varied lines, including mineralogy, the study of ore deposits, and other branches of geology.

The training of qualified employees, mining specialists, took place in special schools. The first schools of this kind were opened in 1721 at the Kungursk, Uktussk, and Alapaevsk plants in the Urals, and by 1737 similar schools existed at all the large government plants of the Urals.

The founder of these schools was a prominent figure in the mining industry and a versatile scientist, V. N. Tatishchev (1686–1750). He was the author of several articles published from 1720 to 1730 in Russia and Sweden; in these he discussed the nature of fossils and proved their organic origin, despite the then prevailing view that they were sports of nature. His articles on the mammoth were especially interesting.[4] At this time it was believed that mammoths had been brought to Siberia from India during the flood or that they were the remains of fighting elephants. Tatishchev stated correctly that the mammoth inhabited Siberia at a time when the climate there was much warmer, thus assuming the possibility of physico-geographical changes in environment with time.

At about the same time, another Russian scientist, G. V. Richman (1711–1753) was writing about the constant change undergone by the earth's surface.[5] He saw the main cause of such changes in the scouring action of water flowing according to the law of gravity from elevations towards the depressions. The summits of mountains were destroyed first, because they were more subject to the effect of rain and wind than the other parts of the land; the areas of lower relief were, in the meantime, filled in by the disintegration products of the elevations.

He spoke about the destructive and creative activity of rivers, which deposit in their mouths the material they have washed down from upstream, and pointed out that it is possible by the geological sequence to judge of recurrent advances and retreats of the water. Richman also described a number of minerals.

In the 1830's, several expeditions went to Siberia to get some idea about this land, which was still virtually unknown. Valuable geological data was collected by S. P. Krasheninnikov (1711–1755), who spent five years on Kamchatka. He gave the first geological-geophysical account of this peninsula, listing the rocks and minerals found there, with descriptions. His observations of three volcanoes — Tolbachikha, Avacha and Kliuchevskoi — and of seismic phenomena, were particularly interesting. The two-volume monograph by Krasheninnikov, "Description of the Land of Kamchatka,[6] attracted much attention and was published abroad in Russian, French, English, Dutch, and German. Despite these separate scientifically interesting ideas, geology in Russia

[4] V. N. Tatishchev, *Mamontova cost, n.e. ossa subterranea, fossilia, ingentia, ignot animaliae Siberia adeferri coepta* (Stockholm, 1725), and "Of the bones of the animal, which the Russians call mamont," *Acta germanica . . .*, London, 1743.

[5] G. V. Richman, "On remarkable changes, to which the surface of the Earth is subjected from time to time," *Primechaniia na Vedomosti,* part 89–96 (1739) (in Russian).

[6] S. P. Krasheninnikov, *Description of the Land of Kamchatka* (St. Petersburg: Academy of Sciences, 1755) (in Russian).

at the beginning of the eighteenth century was, generally speaking, still at a rather low level, and its formation as an independent branch of natural history took place in the middle of that century, thanks to the papers and researches of the outstanding Russian scientist and encyclopedist, M. V. Lomonosov.

MIDDLE OF THE EIGHTEENTH CENTURY Mikhail Vasilievich Lomonosov (1711–1765) received his first mining education in Germany; he was a student of H. Wolf in Marburg and J. Henckel in Freiberg. A further amplification and accumulation of his scientific knowledge took place in Russia. In 1763, Lomonosov published a most remarkable geological paper, "On the Earth's Strata"[7] as a "supplement" to a treatise on "Metallurgy." An earlier paper written in 1757 "Oratio de generatione metallorum a terrae motu,"[8] was also very important.

An important principle of these papers is the idea of the continuous evolution of the earth. Lomonosov severely criticized scientists who asserted that the observable environment was unchanging and was found now just as it was made by the Creator. He argued that the earth's relief and accumulations of ores and minerals originated gradually and were constantly changing. He stressed that mountains rose in places that were formerly covered by the sea, while older lands had sunk under the waters of the sea.

An important feature of the geological papers of Lomonsov was a perfectly intentional and skillful use of the method of interpreting events of the geological past by comparing them with present geological processes; that is, the actualistic method. A masterly application of this method enabled Lomonosov, for instance to approach the problem of peat, coal, and amber formation correctly, proving their vegetative origin conclusively, and to voice a very daring opinion for that time on the organic genesis of oil as a distillation product of plant remains under the effect of internal heat in the depths of the earth. Though using the actualistic method, Lomonosov was very far from regarding the nature of present geological events as absolutely identical with phenomena of the remote past.

Lomonosov distinctly realized the extremely great length of geological time (obviously more than several hundreds of thousands of

[7] M. V. Lomonosov, *On the Earth's Strata*, Second Supplement to *First Principles of Metallurgy or the Science of Ore Bodies*, first edition (St. Petersburg: Academy of Sciences, 1763); *The Collected Works of Lomonosov* (Leningrad, 1934), Vol. VII, pp. 168–271 (in Russian).

[8] Lomonosov, *On the Generation of Metals by Earthquakes* (St. Petersburg: Academy of Sciences, 1757) (in Latin).

years) and unhesitatingly rejected attempts to fit the history of the earth into the Procrustean bed of biblical chronology.

Lomonosov was the first to divide geological processes into external (the activity of wind, rain, rivers, sea, waves, floods) and internal (fast and slow oscillations of the earth's surface, volcanism). He correctly assigned a decisive role in the shaping of the face of the earth to internal abyssal processes.

Referring to floods and "deluges," Lomonosov correctly indicated that they could be caused by two factors: abundant rains or an incursion of the sea occurring as the result of a slow, almost imperceptible sinking of the land.

Although contemporary opinion held that the earth's crust was stationary, Lomonosov distinguished several types of tectonic movements. Besides rapid displacements during catastrophic earthquakes, he spoke of slow and gradual "imperceptible" movements. His papers give, in fact, the first classification of tectonic movements, not very different from the present interpretation. Lomonosov was the first to formulate precisely the principle of conjugation between uplifts and subsidences. He wrote that depressions were formed simultaneously with mountains — either owing to a partial stoping and collapse along fractures originating during the uplift, or as the result of a sinking of strata into cavities formed in the interior of the earth. He pointed out that uplifts and subsidences of the earth's surface are a cause of the folding and faulting of beds.

Lomonosov tried to determine the thickness of the crust and the depth at which underground heat is located, using for this purpose data on the distribution and nature of earthquakes and on the size of orogenic uplifts. The figures he obtained for the thickness of the crust are similar to present data, although in some details his computations were erroneous.

Lomonosov was the first to record that disturbances in the occurrence of seams and the formation of ore veins take place repeatedly, and he indicated certain features as evidence. He said that "volcanoes are nothing but pipes or vents, through which surplus heat emerges" and that the products of their activity covers vast areas.

In his papers, Lomonosov gave much space to problems of the genesis of rocks. He showed that under the effect of certain processes, massive forms are subjected to disintegration, leading to the formation of sand. At the same time, he wrote about the various ways in which sedimentary rocks originate: consolidation under the influence of temperature and pressure; cementation; deposition of new beds as the result of precipitation of water-transported material; precipitation of amorphous bodies from solution; and crystallization of dissolved matter. However, according to Lomonosov, this list did not exhaust the conditions under which rocks could be formed, and he specifically said that there could be other ways.

The premises worked out by Lomonosov on the role of natural waters and organisms in the formation of certain minerals, ores, and rocks were also new to geology. He indicated that aqueous solutions, as well as plants and animals, play a definite role in the cycle of materials in nature. Special attention was given by Lomonosov to chemical processes, both at the surface and in the interior of the earth, that determine various geological phenomena, including ore formation.

Lomonosov was extensively concerned with mineralogy. He was the first to show that minerals of ore veins are grouped into natural associations and, consequently, that the discovery of one mineral can serve as an indication of the possible presence of another or others included in the same association. Later, at the very end of the eighteenth and the beginning of the nineteenth centuries, these ideas of Lomonosov were developed by the outstanding Russian mineralogist V. M. Severgin (1765–1826), who devoted himself to the study of the joint occurrence of minerals related by conditions of their formation. Later, this concept was termed *paragenesis* by J. F. Breithaupt. Lomonosov's work on the structure of crystals was a great scientific achievement for that period. In his paper "Discourses on the hardness and liquidity of bodies"[9] he described two systems of packing of spheres (positions of atoms) to form the crystals of both compact and rarefied minerals. In another paper,[10] Lomonosov suggested a crystal structure plan for saltpeter as a closest packing of globular particles, as a result of which crystals are formed with angles of 120°. For a cubic crystal of rock salt, he placed the "corpuscles" (atoms) in a pattern, with the lines connecting their centers forming quadrangles. In this way, forestalling Western European crystallographers of the end of the eighteenth century, Lomonosov in 1750 supplied grounds for a crystal structure theory and specifically spoke of the constancy of interfacial angles.

Lomonosov's contributions to geology are enormous. He was the

9 Lomonosov, "Discourses on the Solidity and Liquidity of Bodies," presented on Sept. 6, 1760, in *The Complete Collection of the Works of Lomonosov* (Moscow-Leningrad, 1952), Vol. 3, *Papers on Physics 1753–1765*, pp. 378–409 (in Russian).
10 Lomonosov, "On the Origin and Nature of Saltpeter" (St. Petersburg: Acad. of Sciences, 1749). This paper was presented before the British Academy of Sciences (in Russian).

first scientist who understood geology as the science of the development of the globe with due consideration of all the exogenetic and endogenetic processes.

His deductions were fully based on the facts he observed, and he never resorted to celestial or mystical forces to explain natural phenomena. He interpreted geological processes as controlled by the laws of physics and chemistry; he was one of the first to use the microscope for the study of geological materials.

Abroad, Lomonosov's researches in geology and in other branches of science became known while this great encyclopedist was still alive. In many countries, full or summarized translations of his papers were published, as well as their reviews.

From a scientific point of view, Lomonosov's papers rate considerably above the works of the majority of his contemporaries and in many respects forestalled discoveries of later geologists. This explains why many deep thoughts voiced by Lomonosov in his geological papers did not immediately find acceptance even in his own country.

Lomonosov's contemporary, the German geologist J. G. Lehmann (1700–1767), who moved to Russia, made interesting statements on the age differences of rocks, which was one of the first steps in stratigraphy.

Speaking in 1762 at a meeting of the Academy of Sciences in Petersburg,[11] Lehmann indicated the possibility of dividing all the formations that make up the crust into two groups: "primary" existing from the creation of the world; and "mountains of the second order"—which originated after the deluge. Primary mountains are rich in metal, do not contain fossils, and are characterized by a regularity of occurrence so that they can be fitted into three belts ("tracts"). In their distribution, the ore veins are governed by these mountain tracts, each direction being typified by particular minerals. The Primary mountains are the result of a recrystallization of matter from the world ocean; the Secondary are products of earthquakes and are affected by the subterranean fire. In these valuable statements are not only the initial elements of stratigraphy, but also the first concept of the existence of ore belts. In this way, in the middle of the eighteenth century, which with full justification can be called the Lomonosov period, theoretical geology in Russia made great progress, placing it in the first ranks of world science.

11 J. G. Lehmann, "An attempt at a General Orography, or Description, of the Main Mountains Extending over the Globe," presented on Sept. 23, 1762 (St. Petersburg: Acad. of Sciences, 1763) (in Russian).

SECOND HALF OF THE EIGHTEENTH CENTURY On Lomonosov's initiative, complex regional studies were begun on the territory of Russia, in the process of which great attention was paid to geology. The Academy of Sciences organized extensive expeditions, directed mostly by Lomonosov's pupils. Some of the young Western European geologists who came to Russia at that time also, to a certain extent, felt the influence of Lomonosov's ideas. One of them, P. S. Pallas (1741–1811) mainly on the basis of his investigations in the Urals, established that the core of mountain ranges usually consists of granite, against which, along the flanks of elevations, are inclined, first, masses of slate and then marine limestone sediments; marginal foothills consist of less consolidated reddish sandstones, marls and clays, containing the remains of terrestrial animals and plants.[12] Pallas thought that the origin of such mountains proceeded in the following way: the oldest rocks are granites that rose as primeval islands from a primary sea; together with barren slates they form the Primary formation; the slates are younger than the granites because they were formed as the result of a disintegration of the granites. The limestones, which contain fossils, are still younger and form a Secondary formation. Unconsolidated rocks of the foothills distinguished as a Tertiary formation are youngest, their accumulation resulting in a filling in of depressions with sediments. The uplift of the mountain range intensified by volcanic processes was accompanied by a retreat of the sea. This caused the inclination of the beds, steepest with the oldest rocks. This fact, well known to the geologists of our time, was not generally recognized in the middle of the eighteenth century, inasmuch as, owing to an absence of geological maps and sections of folded areas, it was very difficult then to establish any regularity in the change of the inclination of beds. These ideas of Pallas were a continuation of Lomonosov's ideas, which stressed that "the nearer the seams are to the ore mountains, the steeper the angle they make with the horizon!"[13]

Pallas' papers describing regularities in the structure and formation of mountains played a major role in the further development of theoretical geology. According to Cuvier, they substantially influenced many geologists of that time, including such people as Werner at Freiberg and the first investigator of the Alps, H. B. de Saussure.

From the time of the diluvialists of the end of the seventeenth cen-

[12] P. S. Pallas, "Observations sur la formation des montagnes et les changements arrivés au globe, particulièrement à l'égard de l'Empire russe," *Acta Acad. Petropoli*, t. 1, (1777), 1778.

[13] Lomonosov, *On the Earth's Strata*, p. 117.

tury and later, during the period of ascendancy of neptunism, it was generally recognized that erratic boulders strewn over the territory of Northern and Central Europe were brought down from Scandinavia by powerful currents of water. For a long time these ideas hindered a correct establishment of Quaternary history — a period of continental glaciation. And yet, one of the Russian travelers, I. I. Lepëkhin (1740–1802) in 1780 correctly approached the solution of the problem of the forces that could transport gigantic stone blocks over considerable distances.[14] Openly objecting to the idea of a universal deluge, he wrote about the role of ice in the dissemination of boulders of alien rocks in the north of the European part of Russia.

During the second half of the eighteenth century, great attention was paid in Russia to regional mineralogy. The foundation of these researches was laid by Lomonosov, who worked out a project involving wide masses of the population in the search for minerals. Considerable credit for the creation of a topographical mineralogy of Russia belongs to V. M. Severgin, whose work is discussed in the next section of this paper, which deals with the early nineteenth century.

Much was done for the study of Russian mineralogy by E. G. Laxman (1737–1796), a member of the Petersburg Academy. From 1778 to 1796 he investigated the northwestern regions of European Russia from the southern coast of Finland to the Onega river. He also visited Moldavia and traveled through Siberia, where he made geological observations at many points along the extensive route from Barnaul to Okhotsk. Laxman discovered a number of new minerals.

Geological researches of the eighteenth century were accompanied by the first elements of geological mapping. In Russia, plans and maps with indications of pits, ore outcrops, deposits of building materials, and springs appear in vast numbers, beginning with those of the 1730's. A manuscript geological (lithological) map of the eastern Transbaikal area compiled by Dorofei Lebedev and Mikhail Ivanov between 1789 and 1794 is an indication of the level attained by Russian geology by the end of the eighteenth century. In its precision, this map compares favorably with many maps compiled even in the nineteenth century. This map was not an exception, however; during the period between 1798 and 1804, P. K. Frolov (1775–1839) was in charge of the compilation of maps of the courses of the rivers Angara, Verkhniaia Tunguska, and a part of the Yenissei, as well as the stretch of Irtysh between the

14 I. I. Lepëkhin, *Continuation of the Diary of Ivan Lepëkhin, Academican and Doctor of Medicine, during his Travels across Various Provinces of the Russian State in 1771* (St. Petersburg: Acad. of Sciences, 1780) (in Russian).

mouths of the Naryn and Shulba. On these maps, different colors are used to designate the occurrence of granite, limestone, sandstone, and other rocks.

Expeditions in the second half of the eighteenth century embraced nearly the entire territory of the Russian Empire of that period — from the Kola peninsula to the Caucasus, from Bielorussia to the Sea of Okhotsk, from the lower reaches of the Yenissei to Kiakhta. This was really the heroic period in Russian natural history, and although geological observations were only a part of the program of expeditionary studies, they proved to be most fruitful.

An abundance of empirical material was accumulated, which, together with the theoretical researches of M. V. Lomonosov and his school, served as a reliable basis for the further development of Russian geology during the first half of the nineteenth century.

Part II: First Half of the Nineteenth Century

The development of Russian industry in the eighteenth century created a demand for mineral resources that resulted in the expansion of serious geological prospecting. The famous academic expeditions of this period investigated for the first time tremendous, previously unknown expanses in the northeastern part of European Russia, the Caucasus, and Siberia. The discovery of numerous new deposits of minerals contributed to the accumulation of a vast amount of various geological data. There was also a substantial increase in the number of geologists with specialized training when, in 1773, the Petersburg Mining School (later Petersburg Mining Institute) opened. By the beginning of the nineteenth century, Russia had many people engaged in research, accumulating extensive data that needed scientific interpretation and summary.

The first third of the nineteenth century is often called the Heroic Period of geological history because of the amazing progress in most branches of this science. In Russia this process developed along the same lines as in Western Europe, but with certain specific features determined by social and economic factors, as well as by the peculiarities of the geological structure of Russia.

The limited scope of this paper does not permit discussion of all divisions of geological science; for this reason only general geology, including stratigraphy, tectonics, and paleogeography, will be covered. No special mention will be made of branches of geology dealing with natural stones (mineralogy, petrography, etc.). The most important findings of Russian geologists during the first half of the nineteenth century will be reviewed and summarized.

PALEONTOLOGY The discovery of the paleontological method, which immediately demonstrated its exceptional value, focused a great deal of interest on the study of fossils. Special courses in paleontology were introduced in a number of universities and institutes about 1820. Papers by Y. G. Zembnitsky, A. M. Karpinsky, and others popularized the achievements of foreign paleontology and contributed to the application of the paleontological method in the stratigraphic researches of Russian mining engineers. All this had an important positive effect on geological literature, for more and more paleontological papers appeared, contributed by many different authors.

A discussion or enumeration of these works would require an excessive amount of space; therefore, only the most significant will be mentioned. G. I. Fisher produced the first major collection of paleontological monographs in Russia.[15] It was devoted to the description of his fossil collections from the Moscow area, and although it contained some errors in the relative ages of these fossils, it is still of interest and use.

An earlier paper had been published describing fossils collected around Petersburg.[16] Its author, C. I. Pander, continued his research, and in the 1850's he contributed a number of papers on the fossil fish of Silurian and Devonian deposits, in addition to studies of conodonts — problematical remains that he assumed to be the teeth of fish. Pander not only described these fossils but supplied theoretical discussions of the potential effects of facies conditions upon organisms. His comparative study of different regions led him to notice regularities in organic remains contained in the mass of sedimentary rocks as well as their transition from older to younger deposits.

C. F. Rouillier's paleozoological hypotheses, based on the analysis of extensive fossil collections, were of great theoretical importance.[17] Rouillier established the existence of paleozoogeographical provinces that characterized old climatic features of various regions in the geological past. He demonstrated that every organism displayed some signs of transformation and that changes in environments caused changes in external appearance and internal structure. His assertion that an organism is affected not only by climate but also by its ecological relationships to other organisms anticipated paleoecology — a science that developed about half a century later. Also interested in the complex problem

15 G. I. Fisher, Oryctographie du gouvernement de Moscou (Moscow, 1837).

16 C. I. Pander, *Beiträge zur Geognosie des Russischen Reiches* (St. Petersburg: K. Kray, 1830).

17 C. F. Rouillier, "Doubts in zoology as a Science," *Otechestvennye Zapiski, 19* (1841), part 2 (in Russian).

of biological species, Rouillier thought as a real evolutionist. He pointed out that there are no distinct boundaries for any species, that a gradual transition from one to another could be observed. Such transitions occur, he believed, because there is nothing immutable in nature — everything changes.

The development of paleobotany was advanced by C. E. Mercklin's investigations[18] of the anatomy of fossil plants. He gathered many samples of fossilized wood, forming a world-famous collection of prepared mounts that is still recognized as valuable and important.

The abundance of paleontological material accumulated by the middle of the nineteenth century inspired E. I. Eichwald to compile a summary of the paleontology of Russia; the result was a three-volume monograph published both in French and Russian.[19] This impressive book contained more than 2,500 pages of text and several atlases illustrating more than 200 fossil organisms. A number of species first described in *Paleontology of Russia* are currently being used as index forms, and the pictures in the monograph are considered a valuable aid in distinguishing fossil remains. Despite certain errors and defects, Eichwald's fundamental summary was a significant contribution to paleontology; it indicated, furthermore, the mature state of the Russian geological school.

Russian paleontologists were eminent in the first half of the nineteenth century. In addition to many descriptions of previously unknown forms, they proposed a number of theoretical concepts. Lomonosov's theory of development, incorporating the idea of transformism, occupied a firm theoretical position. During the second quarter of the century, when the concept of creation dominated the West, many prominent naturalists in Russia discussed continuous changes in organic material from simple to complex forms.

STRATIGRAPHY The increase in paleontological research encouraged stratigraphic study. Attention was directed to Paleozoic deposits. Papers by Pander, Eichwald, and others were directly relevant to the study of strata. By the end of the 1830's, British geologists had divided the Lower Paleozoic into Cambrian and Silurian systems. D. I. Sokolov, who analyzed a great number of Western European and Russian sections given in the geological literature, suggested dividing these deposits into

[18] C. E. Mercklin, *Palaeodendrologikon Rossicum, Vergleichende anatomisch-mikroskopische Untersuchungen fossiler* . . . (St. Petersburg, 1855).

[19] E. I. Eichwald, *Lethaeo Rossica; ou, Paléontologie de la Russie* (Stuttgart, 1852–1868), Vols. 1–3.

three independent systems.[20] A similar thought (the Ordovician) voiced in 1879 by Lapworth was only formally accepted in our own time by the XXI Session of the International Geological Congress in 1960.

A. D. Ozersky carried out a very important study of Silurian deposits in the Baltic provinces and proposed a detailed subdivision of these deposits.[21] He outlined characteristic fossils for the majority of stages and horizons. His paper was significant in that Baltic Silurian systems differ substantially from classical sequences in England. F. B. Schmidt continued the study of the Silurian; he published a paper that supplied crucial details to Ozersky's, and worked out a modern stratigraphic scheme.[22]

Outstanding among several papers on Devonian deposits were investigations by Pander. He attempted to justify a tripartite division of this system into lower and upper sandstone and middle limestones. Precisely this division is accepted in modern stratigraphy. Unlike Devonian deposits in England, where beds with marine invertebrate fauna (Devonshire) are distant from analogous formations with freshwater fish fauna (Scotland), in northeastern Russia there are joint occurrences of fish, brachipods, and fossil molluscs. Pander was the first to appraise the significance of this fact, and confidently proposed a correlation of Devonian deposits, despite their different facies. Murchison, in the third edition of his *Siluria*,[23] included excerpts from Pander's letters to support his claim that synchronous British Devonian deposits belonged to different facies.

G. Grewingk devoted much time and attention to a detailed examination of the Russian Devonian, citing the data of Pander and others on the stratigraphy of these deposits.[24] Grewingk was especially concerned with Carboniferous deposits within the Donetz Basin, around Moscow, and in the northwestern regions. Contrary to the theory of L. Elie de Beaumont, he demonstrated the Paleozoic age of the Donetz coal, estab-

20 D. I. Sokolov, *A Course in Geognosy*, Vols. 1–3 (St. Petersburg. Prats & Co., 1839) (in Russian).

21 A. D. Ozersky, "Geognostic Sketch of Northwestern Esthland (Estonia)," *Gornyi Zh.*, 1844, Nos. 5 & 6 (in Russian).

22 F. B. Schmidt, "Untersuchungen über die silurische Formation von Estland, Nord-Livland und Oesel," *Archiv für die Naturkunde Liv-, Est- und Kurlands*, Ser. 1, Vol. 2, 1858, Lief. 1 1–248; "Nachträge und Berichtungen zu den Untersuchungen über die silurische Formation von Estland, Nord-Livland und Oesel," ibid., Ser. 1, Vol. 2, 1859, Lief. 2, 465–474; Beitrag zur Geologie der Insel Gotland nebst einigen Bemerkungen über die untersilurische Formation des Festlandes von Schweden und die Heimat der norddeutschen silurischen Geschiebe," ibid., Vol. 2, 1859, Lief. 1, 403–465.

23 R. I. Murchison, *Siluria* (3rd ed.; London: John Murray, 1859).

24 C. Grewingk, "Geologie von Liv- und Kurland mit Inbegriff einiger angrenzender Gebiete," *Archiv Naturkunde Liv-, Est-, Kurlands, 2* (1861), Lief. 3.

lishing a three-part division of the Carboniferous in the Moscow region.

At this time, Russian geologists proposed that the lower part of the Carboniferous deposits could be distinguished by the presence of *Productus giganteus*, while the top horizons contained *Spirifer moscowensis*. These statements were confirmed by von Buch.[25]

The controversy over the size of the Donetz Basin and its Carboniferous deposits was important. Outstanding Western European geologists, including Murchison, asserted that only the lower part of the sequence was developed there, i.e., Carboniferous limestone, while the top part — Carboniferous sandstone — which contained the main coal reserves in the West, was absent in Southern Russia. Sokolov argued, on the contrary, that a complete Carboniferous sequence was present in the Donetz Basin and that coal reserves were, consequently, extremely abundant. Sokolov's assertion was confirmed during the second half of the nineteenth century.

Russian interest in the study of Permian deposits was of special importance. In 1831, before these deposits were distinguished as an independent system, Sokolov correctly correlated a thick mass of sandstones in the Western Urals with the red sandstones of Germany, which were classified as upper Carboniferous. Later concurring with A. Burat, who distinguished an independent "terrain" in the roof of the Paleozoic, Sokolov convincingly demonstrated the existence of radical differences in the terrigenous rock mass in the territory beyond the Volga from its underlying and overlying strata. Early in 1841, in an explanatory note to his map of European Russia, G. P. Helmersen called these sandstones Permian;[26] at the end of the year, Murchison applied this label.

A study of the fossils in the terrigeneous rock mass of the Ural area by Russian and foreign paleontologists enabled Murchison to list 188 species, 168 of which were characteristic of Permian deposits only; the remaining 20 were also known in the Carboniferous.[27]

The history of the establishment and study of the Permian is connected with joint Russian and foreign research on the territory between the Volga and the Urals.

Triassic deposits were comparatively rare in European Russia. They were established conditionally only after the mid-nineteenth cen-

25 L. von Buch, "Addendum to the Classification of Mountain Formations in Russia," translated from German by Ensign Afaifer, *Gornyi Zh.* (1842), No. 10 (in Russian).

26 G. P. Helmersen, "Explanatory Notes to a General Map of Rock Formations in European Russia," *Gornyi Zh.* (1841), No. 4 (in Russian).

27 R. I. Murchison, E. de Verneuil, and A. Keyserling, *The geology of Russia in Europe and the Ural mountains* (London-Paris, 1845), Vols. 1, 2.

tury by J. B. Auerbach, who discovered Triassic marine fossils in the deposits found in the Bolshoe Bogdo mountain area.[28]

Jurassic deposits were known in Russia for quite some time. Information about them had appeared in the early papers of Fisher and Eichwald, although their papers contained a number of mistakes. Rouillier's investigations of the Central Russian Jurassic were much better substantiated.[29] He separated these deposits into three divisions, indicating key forms for them and establishing the presence of a particular fauna in their top horizons. The study of this last division (later referred to as the Volgian stage) established the existence of a boreal province. Some fossils from the mid-Russian Jurassic were sent to von Buch, who confirmed Rouillier's designation of them as Callovian and Oxfordian.

Jurassic deposits were also discovered and studied in other regions of Russia as well: the Western Urals, the Polar Urals, Southern Russia, and the Caucasus.

Cretaceous deposits were well known and described in the mid-eighteenth century. They were most thoroughly studied in the territories around the upper reaches of the Volga, where in the 1820's and 1830's P. M. Yazykov conducted his investigations.[30] He suggested a stratigraphic subdivision of Cretaceous deposits in the Simbirsk area on the basis of paleontological and lithological data. His three-part division of Upper Cretaceous deposits proved to be so adequate that it has survived to the present. Another scheme, suggested by A. D. Orbigny ten years later for the Upper Cretaceous in France, was identical with Yazykov's.[31] However, the later plan received scientific recognition and the earlier did not, probably due to the fact that Yazykov did not name the stages he proposed.

In the 1840's and 1850's, the Cretaceous deposits of Southern Russia and the Urals underwent detailed investigation. Their development within European Russia was established, and the stratigraphy of the

28 J. B. Auerbach, "Report on Geognostic Studies in the Mountain Areas of Big Bogdo, Chapchachi and Beś-Chokho and in the Zhiguli Mountains," *Vestnik Russkogo Geograficheskogo Obschestra,* part 12, sec. 6 (1854), supplement (in Russian).

29 C. F. Rouillier, "Naturhistorische Notiz über die Umgegend von Moskau," Moscow, *Soc. Nat. Bull. XVII,* 3, (1844), 625–635; "Explication de la coupe géologique des environs de Moscou," Moscow, *Soc. Nat. Bull. XIX,* 2 (1846), 359–485; "Über die Fauna des Moskauer Gouvernements und ihre Veränderungen . . . ," *Erman, Archiv Russ., V* (1847), 443–482.

30 P. M. Yazykov, "A brief Review of Cretaceous Formation of the Simbirsk Province," *Gornyĭ Zh.* (1832), No. 5 (in Russian) *Cf.* Jasikoff, P.

31 A. D. Orbigny, "Sur l'ensemble des mollusques gastéropodes des terrains crétacés," *Paris Soc. Geol. Bull., XIV* (1842–1843), 460–485.

Volga and Trancaucasia areas were paleontologically substantiated. A detailed survey of these deposits was made, including descriptions of many fossils belonging to this age. Papers by Russian geologists, particularly those of H. W. Abich and E. I. Eichwald, became widely known abroad, contributing to the world fund of geological literature.

Cenozoic deposits in Russia were frequently studied. Worthy of attention are papers by A. L. Andrejeowsky on Tertiary deposits of Podolia, H. W. Abich on the Neogene of Transcaucasia, E. I. Eichwald on the Pliocene of Eastern Azerbaidzhan, and G. P. Helmersen on the Oligocene and Miocene of the Ural area.[32] Data on the distribution and character of Tertiary deposits in Russia were summarized in textbooks by Sokolov and Eichwald.[33] Sokolov, discussing the physicopetrographic conditions of that time, thought it necessary to separate the lower part of the system from the upper, thus anticipating the present subdivision into independent Paleogene and Neogene systems.

Deposits near the Recent were found occurring above the Tertiary; these attracted the attention of Russian researchers. This interest was intensified with the discovery of gold placers in the Urals in 1814. Earlier, in 1798, Severgin had used the expression "mountains of the fourth formation." In 1827, I. I. Eichfeld assigned Quaternary deposits a place of equal importance with all the older systems.[34] In his textbook,[35] Sokolov preceded many of his contemporaries in writing that the deposits could be united under the name Quaternary.

Scientists were very interested in the erratic boulders widely distributed on the Russian plains. Severgin was probably the first to voice the opinion that they were brought from Scandinavia by moving ice.[36] The theory of the direction of movement of these boulders was con

[32] Andrejeowsky, "Remarques sur l'ouvrage de M. Frederic DuBois de Montpéreux, ayant pour titre: Conchyologie fossile, ou aperçu géognostique des formations du plateau Volhynie-Podolien, Berlin, 1830," *Bull. Soc. des natur. de Moscou* (1832), t.4, No. 3, 513–558; "Liste des fossiles tertiares de la Podolie russe," in Deshayes, "Determination rectifiée d'une collection des fossiles tertiaires de la Podolie russe," *Bull. Soc. géol. de France 6* (1835), 321–322. H. Abich, "Vergleichende geologische Grundzüge der Kaukasischen, Armenischen und Nordpersischen Gebirge. Prodromus einer Geologie der Kaukasischen Länder," *Mem. Acad. Sci. de St. Pétersbourg, sci. math. et physiques*, ser. 6 (1859), 359–534, 8 ill. E. I. Eichwald, "Fauna Caspio-Caucasica nonnullis observationibus novis illustravit," *Nouv. mém. Soc. des natur. de Moscou* (1841 [1842]) t. 7 (13), 1–4, I–V, 1–290, 1–2.

[33] Sokolov, *A Course in Geognosy;* Eichwald, *Geognosy, predominantly as applied to Russia* (St. Petersburg, 1846), 572 pp., 2 maps (in Russian).

[34] I. I. Eichfeld, "An Orographic View of Valakhia, Moldavia and Bessarabia," *Gornyi Zh.* (1827), Nos. 5 & 6 (in Russian).

[35] Sokolov, *A Course in Geognosy.*

[36] V. M. Severgin, "Comments on the Probable Age and Origin of Various Mountain Ranges in Russia," *Umozritelnye issledovania imperatorsroi S. Peterburgskoi Akademenii Nauk*, Vol. 4 (1815) (in Russian).

firmed by Razumovsky, who established two displacement phases each differing in the direction of movement.[37]

K. E. von Baer interpreted striae and grooves observed on the surface of limestone and other rocks as traces left by floating ice blocks in which the boulders had been frozen.[38] With retreat of the shallow sea which had covered European Russia during the Quaternary, the boulders were released from the melting ice and strewn over the plain.

Russian geologists studied the stratigraphy of all deposits from the Lower Paleozoic to the Recent, achieving recognition and esteem equal to that of the best scientists in other countries.

TECTONICS In the study of the structures of the crust and its movements, Russian geologists were strongly influenced by ideas prominent in the West — at first, neptunism, and later, the elevation crater hypothesis. During the height of popularity of any particular theory, however, Russians were, as a rule, cautious in their acceptance. They usually pointed out discrepancies between field observations and fundamental points of such hypotheses. For instance, if von Buch's theory stipulated the presence of volcanic rocks in certain elevations, and if they were not there, the author writing and describing the origin of the elevations would indicate this absence.

Careful study and summary of empirical data permitted Russia's geologists to make important deductions, such as Lomonosov's statement that beds in the axial part of a range should be more steeply inclined than beds on the periphery, a theory developed during the last third of the eighteenth century by Pallas and reiterated by Sokolov in his 1839 textbook.

During the 1830's, the concept of crustal movements as a cause of folding and mountain building was extremely popular among Russian geologists. Interest in tectonic processes and study of their specific features persisted through the following years. While studying the peculiarities of tectonic processes within a single zone of the crust, A. D. Ozersky noticed that positive movements were followed by negative movements; and named this phenomenon "oscillating movements." The problem of the alternation of ascending and descending movements was discussed in great detail by G. E. Shchurovsky. In an 1856 paper he expressed the opinion that the energy of the earth's interior was

[37] G. P. Razumovsky, *A Geognostic View of the Northern Part of Europe and Russia in Particular* (St. Petersburg, 1816) (in Russian).

[38] K. E. von Baer, "Nachricht von der Wanderung eines sehr grossen Granitblockes über den Finnischen Meerbusen nach Hochland," *Bull. Acad. Sci. de St. Petersburg, V* (1839), 154–158.

manifested in two ways: as volcanic activity or in the form of slow uplifts and subsidences of the crust.[39]

These views sharply contradicted the hypothesis of elevation craters, which stated that uplifts take place virtually instantaneously. Consequently, Russian geologists quickly gave up von Buch's ideas. Such a departure from von Buch was also indicated in Eichwald's 1846 textbook, in which he pointed out that the formation of a mountain range does not happen at once but in several stages.

While in geological papers of the 1840's frequent mention was made of the relative time of folding and the stage of uplift, such statements were rare before the development of stratigraphic research.

Studies of folding resulted in the discovery of cleavage first noticed by E. P. Kovalevsky, who indicated that jointing does not coincide with bedding originates in intense deformation.[40]

By the middle of the nineteenth century, some characteristic features of tectonic processes were established. G. D. Romanovsky, in trying to formulate these, discussed the gradual formation of mountain ranges, and indicated that no general law for the distribution of mountains was then established.[41] He also indicated that he felt ridges were not solely the result of lateral pressure.

A number of statements of the problems of tectonics showed the considerable interest in crustal movements shared by Russian scientists; many of their observations were original and were significant contributions to the newly developing science.

LITHOLOGY AND PALEOGEOGRAPHY At that time there were in Russia only a few studies that would now be included in lithology. They were primarily concerned with different building materials, although special attention was paid to physical properties. Problems of the genesis of certain sedimentary rocks (limestones, clays, etc.) were considered only incidentally. Investigations that approached what we would call lithology included the compilation of soil maps for European Russia, which appeared in several editions during the 1850's.

Coincident with the study of rocks and strata, Russian geologists were engaged in examinations of the chemical composition of lake and sea water. H. W. Abich suggested studying the composition of shells

[39] G. E. Shchurovsky, *Oscillating Motion of the European Continent during Historical and Nearly-Historical Times* (Moscow, 1856) (in Russian).

[40] E. P. Kovalevsky, "Geognostic Sketch of the Donetz Ridge," *Gornyĭ Zh.* (1829), Nos. 1, 2, & 3 (in Russian).

[41] G. D. Romanovsky, "On the Origin and Inner Structure of Mountain Ridges," *Gornyi Zh.* (1854), No. 4 (in Russian).

belonging to Recent molluscs in seas of different compositions.[42] He felt that by analyzing the shells of fossil animals in this way it would be possible to form an idea of the composition of the water during the time when such organisms lived.

In piedmont regions, especially along the Ural mountains, there were extensive studies of river alluvium, its mineralogical composition, the size of fragments, and amount of rolling. The results of these studies produced confidence and stimulated the search for primary gold deposits and other important minerals.

Geologists were familiar with the laws of clastic particle transportation by water, which took into account shape, size, and specific gravity of fragments, as well as the sequence of precipitation from suspension (heavier particles first, then lighter ones). Observing these regularities, Eichfeld made the first successful attempt in world geological literature to decipher the genesis of flysch rocks in the Eastern Carpathians.[43]

From a study of the conditions of formation of sedimentary rocks, Russian geologists tried to distinguish mechanical sediments from chemical. They were not always successful in these endeavors, however, as Sokolov pointed out in his 1839 textbook, for very often there was a mixture of both types.

The important role of organic remains in the formation of sedimentary rocks was stressed, and serious attempts were made to reconstruct the paleogeographical conditions of past periods. Y. G. Zembnitsky indicated that it was possible to judge the conditions in which now extinct organisms had lived by the external features of fossilized shells.[44] From the nature of fossil fauna and flora, geologists tried to form an idea of paleoclimates. In the 1830's, by deciphering data of geological sequences for paleogeographical reconstructions, they learned that the appearance of a bed of sandstone in a calcareous or argillaceous sequence indicated the formation of an uplift subjected to a washout. Describing the specific features of various geological eras, Sokolov noted picturesque characteristics of the distribution of land and sea, as well as the nature of the plant and animal kingdoms in large diverse territories — the Caucasus, European Russia, the Fore-Urals, etc. Frequently, Russian geologists made attempts to reconstruct paleogeo-

42 Abich, "Vergleichende chemische Untersuchungen der Wasser des Kaspischen Meeres, Urmia und Van-See's," *Mém. Acad. Sci. de St. Pétersbourg, sci. math. et physiques,* sér. 6 (1856), Pt. 1, t. 7 (9), 1–57, 1 Karte, 1 Tafel.

43 Eichfeld, "An Orographic View."

44 Zembnitzky, *Conchyliology, or presentation of data on shells and animals that produce them* (St. Petersburg, 1831), VIII+254+6 pp., 5 ill. (in Russian).

graphical conditions of the Carboniferous period of the Donetz and Moscow Basins, even though the alternation of freshwater and marine sediments were interpreted differently. One approximate solution of the problems of physicogeological conditions of the Carboniferous was C. F. Rouillier's. He wrote that in place of present coal deposits there had previously been peat bogs covered by the sea from time to time.[45] In another paper he analyzed paleontological remains of Jurassic deposits and confidently discussed the different temperatures of the northern and southern zones of the Jurassic sea of Middle Russia.

H. A. Trautschold outlined a trend in the changes of land and sea contours in the Russian plain from one era to another.[46] Consequently, by the middle of the nineteenth century, it was possible not only to give paleogeographical sketches of the distribution of seas during different geological periods but also to outline a theory of geological developments on the Russian platform from the Cambrian to the Quaternary.

AREAL MAPS AND REVIEWS During the second half of the eighteenth century and the first half of the nineteenth century, maps began to appear in Russia that were prototypes of geological maps. Before this time, the maps should properly be called lithological-petrographic maps. The maps of the 1830's were geological in the true sense, for in 1834 a systematic geological survey of mining districts was begun. When the mapping of separate portions was completed, a summary areal map of the territory was compiled and its description prepared. The volume of material grew so rapidly that by 1839 a compilation of a summary geological review of the entire European part of the country was started. The first experiment of this kind was a map and brief description compiled by G. P. Helmersen.[47] One should note, however, that three decades earlier another much more extensive summary had been written by Severgin,[48] although it was not accompanied by any cartographic material. Severgin had collected all existing information on rocks, mineral resources, and mineral springs in each province of Russia, yet his efforts did not provide any definite idea of the geological structure of the country. Its purpose was much more modest: to establish local

45 Rouillier, "Coal Formation," *Vestnik Estestvennykh Nauk, 4* (1857), Nos. 3 & 4 (in Russian).

46 H. A. Trautschold, "Bemerkungen über die stratigraphischen Verhältnisse des Gouvernements Kaluga," *Bull. Soc. des natur. de Moscou* (1860), t. 33, Pt. 2, No. 2, 589–600.

47 Helmersen, "Explanatory Notes."

48 Severgin, *An Attempt at Mineralogical description of the Russian State* (St. Petersburg, 1809), Part 2 (in Russian).

resources of mineral raw materials so that industry would be spared the problems of distant transportation. Although it was schematic and included only eight stratigraphic categories, Helmersen's map supplied information about the development of rocks by geological ages in European Russia. It outlined the fundamental scheme of the geological structure for this enormous territory.

Nearly simultaneous with Helmersen's, another map was compiled by A. K. Meyendorf. It was a bit more detailed than the preceding one, including some territories adjacent to Russia in Scandinavia and Eastern Europe, but it was not really superior in quality.

Summarization of this great increase in research data was needed. An invitation to visit Russia was extended to R. I. Murchison, who at that time was recognized as an outstanding specialist. For two years Murchison and Verneuil traveled with two Russians, A. A. Keyserling and N. I. Koksharov, visiting the mining districts of European Russia, inspecting characteristic sequences, and studying collections of fossils, geological maps, and other primary material accumulated during many years by Russian mining engineers. The assistance and benevolent attitude of high Russian authorities in the geological and mining industry enabled Murchison and his colleagues to prepare a remarkable monograph on the geology of European Russia and the Urals.[49] It was the crowning achievement of contemporary geological knowledge of that territory.

The above-mentioned reviews (most of which appeared between 1840 and 1845) provided reliable information on the geological structure of specific areas and indicated regions that required more study.

THEORETICAL PROBLEMS Genetic theoretical principles at the beginning of the nineteenth century were rather contradictory. In the first quarter of the century, materialist ideas such as the Kant-Laplace theory were forbidden and persecuted by both the clergy and official science. Later, under the pressure of irrefutable facts, obsolete doctrines were gradually discarded and more progressive viewpoints triumphed.

The philosophical principles and cosmogenic ideas that predominated in each period generated various hypotheses that were applied to purely geological problems. For example, neptunist theory developed among Russian geologists during the first quarter of the century, becoming a great impediment to scientific progress. Although its errors became more and more obvious, few geologists dared to criticize or amend it. Those who did were deeply involved in practical investiga-

49 Murchison, *et al.*, *The geology of Russia.*

tions, and their corrections were connected with problems of magmatism, crustal movements, and geological chronology.

By the end of the 1820's, when the vulcanist hypothesis began to be popular, it was favorably received in Russia; within a comparatively short period (5 to 8 years), an overwhelming majority of Russians gave up neptunist principles and became followers of the new theory. This contributed to the rapid development of theoretical geology. Since the vulcanist hypothesis was progressive as compared with neptunism, it helped to formulate ideas of developmental processes in nature, the duration of geological time, the relation between magmatism and tectonics, and the phenomenon of metamorphism.

Catastrophism, which spread over Western Europe during the same years, was greeted with reserve in Russia. Although the literature contained the terminology (e.g., "revolutions"), this was not meant literally for geological events; often, the descriptions of geological processes were widely popularized views of the development of nature that had originated at the time of Lomonosov. A criticism of catastrophism that appeared in the Russian press in the 1830's completely undermined the position of this theory.

Charles Lyell's outstanding monograph in the 1830's[50] marked the beginning of an era in which uniformitarianism flourished. Yet even before this the opinion was voiced by Dvigubsky that there was no difference in principle between the present and the remote past. Sokolov maintained, furthermore, that a continuous transformation was in process resulting from the accumulation of minor deposits. The present was therefore the key to the interpretation of the past. Such ideas prepared Russian geologists (morally) to reject concepts, which were dominant at the beginning of the century, of the immutability of the surrounding world and the impossibility of knowing the geological past.

At the end of the 1820's there was a marked transition from general discourse on the development of nature to empirical research on transmutation phenomena in organisms. Indirect evidence of an alternation of physicogeographical conditions of the surface of the earth was collected. The most progressive scientists of the time firmly expressed original ideas on the unity of the universe and its operating forces, on causal relations and interdependence among all observed phenomena, on the origin of organic life in inorganic material, and on continuing genetic processes in nature. In biological research, Russian naturalists tried to account for the influence of the environment on organisms to

[50] C. Lyell, *Principles of Geology* (London: John Murray, 1830–1833), Vols. 1–3.

show that their morphology was changed by their living conditions, that their habitat affected the functioning of their organs.

Lyell wrote about constant changes of the earth's surface, but for a long time he did not recognize change in the organic world. Russian scientists, on the other hand, particularly C. F. Rouillier, asserted the constant transformation of flora and fauna since very earliest times. Development and evolution were intimately intertwined with uniformitarianism in the papers of many Russian geologists long before the classic paper by Lyell.

During the second quarter of the nineteenth century, Russians began to use a comparative historical approach to research more widely, its integral part being the actualistic method. This method was used formally when uniformitarianism was first introduced; later, however, refutation of the concept of absolute identity between present and past necessitated corrections in actualistic comparisons. In the middle of the nineteenth century, it was not clear what sort of corrections should have been applied, for there were too few empirical data from which to deduce them. This explains the primitive and often mistaken nature of paleogeographical and other hypotheses that originated in the analysis of geological sequence. Analogous difficulties were experienced by naturalists in other countries, not only during these years but also much later.

This sketch of the development of Russian geology during the first half of the nineteenth century indicates that natural history was closely related to the advanced ideas of Western European scientists, although Russian geologists and biologists sometimes foreshadowed them. Within the limits of this communication it was impossible to substantiate certain premises by excerpts from original sources. Such data can be found in detail in two previously published books by the author.[51]

[51] V. V. Tikhomirov, *Geology in Russia in the First Half of the 19th Century* (Moscow. Academy of Sciences, 1960, 1963) Vols. I, II (in Russian).

Changing Ideas About Moving Shorelines

Eugène Wegmann

As soon as records of former sea levels were transmitted from one generation to another, the displacements of the shorelines and the question of how and why land became sea and sea became land began to puzzle mankind. Several interpretations were proposed, some of them probably as early as in prehistorical times. These first explanations are characterized by *mythological patterns.* Different tribes reached such explanations according to their philosophy and their representations of the world. Many of these mythologies seem to have originated independently, but upon establishment of reciprocal contacts some of them became interrelated by partial exchanges.

Introduction: from Mythology to Science

In personalized pictures of the world like those of mythology, the reasons for the changing environments are explained by divine motivations. According to the optimistic or pessimistic philosophy of the tribes, different trends became apparent. The immediate causes of shore displacements were explained in different ways by several kinds of observations. For instance, observations of the results of heavy rains and periods of drought in lakes and rivers suggested the models of the great floods, whereas the awareness of the changing shores and beaches by alluvial deposition led to another explanation. A third category of mechanisms was afforded by the concept of a personalized earth, which was not a dead object, but had its own movements and motivations. Therefore, crustal movements were easy to understand.

In this way, some of the basic patterns of explanations appeared very early, probably in prehistorical times. The concepts of these relationships can be expressed both in mythological and scientific language. The structure of these explanations belongs to archetypical types penetrating first the mythological layer and then the scientific superstructure. They represent, therefore, fundamental ways of thinking, and reveal some of the first approaches of the human mind to this problem. These ways were used and reused later, also.

The displacements of the shorelines were first considered as a consequence of changing sea level related to climatic fluctuations controlled by divine forces. The description was, of course, given in the manner of the time; the translation in modern language is given in order to facilitate comparisons. The last step, which linked the causal succession to divinities, disappeared during the rationalistic approach, especially during the Age of Enlightenment, but some reasonings of the former stage continued to control scientific evolution up to recent times. The names of some principles are often more or less personalized, even in recent papers, and these simplified concepts are used by some authors, reminiscent of the mythological style. A deepened awareness of the evolution and operation of scientific hypotheses and theories can be obtained by analyzing the anatomy and development of this type of human activity. In order to do so, it is necessary to recognize the main components and their relationships, to study their evolution (as in comparative anatomy) and their modes of connection, and to follow the different patterns from mythological ages through the transitional stages into the scientific era. Through such a process one becomes aware that many approaches are rooted in archetypical thinking, and that these archetypical frameworks are adorned by decorative figures taken from the general views of the different epochs. Those who are interested in knowing *how* theories work, have to analyze the sources of old and new learning.

The problems of the moving shorelines, of transgressions and regressions, examined under a different viewpoint remain fundamental questions, even in modern geology. The topic is sometimes discussed in a rather confused manner, because many authors do not distinguish the displacements of the shorelines from the supposed causes — the oscillations of sea level, or the movements of the crust. They do not distinguish observable changes from the presumed immediate causes. Within their minds the phenomena and their causes are so closely connected that the displacements of the shorelines and the changing sea level, or the moving crust, are, so to speak, synonyms. Students of such textbooks are therefore greatly astonished if someone distinguishes between the observable phenomena (the displacements of the shorelines) and the hypothetical causes. The difference between these two elements is not always realized. It is superfluous to mention modern examples of this tendency, but predecessors occur in the history of science, because most of these approaches have been used earlier. In a historical perspective, these late simplifiers appear as descendants of the pioneers who opened the ways in such areas. It is interesting to observe that the most modern solutions coexist in present-day literature with the most archaic types of explanation, as in today's flora and fauna some species of long extinct genera live together with most recent types.

The mythological approach will not be discussed any further. The literature relating and comparing the variations of traditions and legends concerning the great flood is very extensive and cannot be abstracted here.

The scientific approach began in Fennoscandia (Finland and Scandinavia or the Baltic Shield). After an early start, the main stages of development were first reached in these countries, and the recent flourishing of ideas also came largely from this area. The history of this part of the world is therefore an outstanding example of a problem studied and restudied by a regional scientific community. The recent results have not yet penetrated the international scientific consciousness. History shows how, starting from simple cause-effect chains, one complication calling another, the entire question became a network of interacting factors. In this example, the main lines of evolution and the different stages appear much clearer than in any other of the history of science, and therefore they represent an ideal case-history.

Facts concerning moving shorelines at first attracted scarcely any scientific attention outside Fennoscandia. Later, in the first half of the nineteenth century, the facts became known to international scientific circles, mostly through traveling geologists. However these phenomena were at first often considered as curiosities, not only because they occurred in a distant country, but also because their place was not yet ready in the picture of geological knowledge. As the general understanding of the earth's evolution developed, and the number of observations requiring marine inundation of many land areas increased, the importance of the studies on moving shorelines was recognized. Eventually, these investigations moved into a central position within geology and became interrelated with many other fields of the earth sciences.

The problem of moving shorelines also represents an important aspect of the interface between terrestrial phenomena on different scales and the attempts of human understanding. It is characteristic of the way in which many generations tried to build an understandable picture of a puzzling phenomenon, and to determine its place and its dimensions in space and time, and in the picture of the world.

First Approach in Fennoscandia

Observations on the displacements of the shorelines in Fennoscandia were recorded as early as the beginning of historical times; the code of laws of the Vikings contains rules about the ownership of recently emerged land lots.

Scientific research began in 1694, when Urban Hjärne, one of the first great geniuses of Sweden (he was the chief of the medical corps, president of the board of mines, member of many political and economical authorities, an innovator in science, literature, and social organization) sent a questionnaire to prominent people in the kingdom asking a series of questions, among which were some about the displacements of the shorelines. His correspondents were invited to give their opinion about the reality of these displacements, if the land was growing or the water diminishing, if rocks became visible where none had existed before, if they knew waterways that in recent times had become not navigable. They had to add their views about the causes: Were the displacements due to changing water level, to alluvial deposition, or to the rising earth's crust? The answers were published in 1702 and 1706.

It is remarkable that Hjärne considered several possible explanations: changing sea level, alluvial deposition, and deformation of the earth's crust. In his commentaries, he discussed some of the possible mechanisms. He considered the hypothesis of a change in the Danish Straits providing an easier outlet to the waters of the Baltic sea. He declared he was puzzled by the fact that the waters of the Baltic Sea seemed to diminish, while inundations and advances of the sea were reported from the countries around the North Sea; but he emphasized that there were only assumptions, and that he would accept the views of those who were better informed. He distinguished between local, regional, and general phenomena, and between immediate and ultimate causes, and proposed an order in which the questions could be solved. This open-minded discussion contrasts sharply with the simplified solutions given for about two centuries, and still presented in some recent papers. This is one of the very rare examples in which research began by multiple explanations but was followed by a succession of one-way hypotheses.

The subsequent development proceeded quite differently, being controlled by the general tendency of science during that period. Nature had to be described and classified, phenomena interconnected, and general rules discovered. It was believed that these rules had to be simple, because science was to provide the fundamental requirement of a unified picture of the world. This need had formerly been satisfied by monotheistic religions. The unifying tendencies still predominated. Scientific explanation had to establish the relationships of causality, so that *one* cause was followed by *one* effect. Chains of causality were set up by concatenating such causal links, and it was believed that these chains converged towards an ultimate cause, as in religion. It was hoped that in this way the general laws regulating the world would be found. Diversifying tendencies were a minority, but they occasionally appeared.

Unicausal explanations thus correspond to an important stage of evolution of many hypotheses and theories, but they can also characterize the scientific climate of an epoch.

Early Problems of the Changing Sea Level

The problem of the changing sea level started its unicausal stage with two well-known scientists, Celsius and Linné.

It is understandable that the movements were first considered as belonging to the mobile element, water; and once this assumption was admitted, the reference mark, the earth's crust, had to be stable in order to measure the changing levels.

Celsius (1701–1744) was well acquainted with the physiography of the Swedish coasts. He had learned, during his numerous travels, the traditions about new lands, emerging meadows, rocks appearing in navigable channels during the last generations, and other similar changes. He became convinced of the reality of the phenomena and began to wonder how the rate of the falling sea level could be estimated or measured. He devised an interesting method to appreciate the secular differences of water level—to ask seal hunters about the elevation of the stones used as resting places by the seals. Computing his data, mostly from Northern Sweden, he estimated the diminution of the water level to be about 4½ feet during the past 100 years. This estimation became known as the "Celsius-value." It was afterwards reduced to 4 feet, and played an important role during later times. Celsius also provided the basis for future research by engraving marks on the coastal rocks indicating the water level of his time. This was a very important contribution, because his example was followed in many places, so that a great number of such records exist now around the Baltic Sea, sometimes several subsequent inscriptions on the same rock.

Linné had a very different approach to the problem (Hagberg, 1939). During his extended travels he had certainly also heard about the former extent of the sea, and he had the opportunity to visit many of the typical localities. These observations fitted well into his picture of the history of nature, which was somewhat different from the orthodox doctrines, as pointed out by some of his contemporaries. Linné imagined that the whole earth was initially covered by water, with the exception of one small island. Every original species (Linné distinguished between original and subsequent species; see Hagberg, 1939, p. 257) was represented on this island, the hermaphrodites by one individual and the dioecious by couples. Dry land emerged from the decreasing ocean and was populated by the expanding flora and fauna. The theory was explained in his *Oratio de Telluris Habitabilis Incremento* (1743). This was the theory of "the diminution of the waters," an idea that preoccupied several scientists of that epoch (as, for instance, Newton). The span of time allotted to these events was short, since the world was created, according to Bishop Usher, in the year 4004 B.C. This time limit gave a very different perspective to natural science than was given later on, when it became more and more probable that the earth existed before the historical epoch, and that this earlier period was immensely long; for some scientists, such as Hutton and Playfair, it was boundless.

Eventually, Celsius and Linné published together, in Holland, a volume in Latin for the international public, in which the hypothesis of the diminution of the waters was presented.

This theory played an important role during the eighteenth century, not only in natural history, but also in historical works and theological discussions. For instance, the number of Celsius was used by the Swedish historiograph Olof Dahlin when he tried to reconstruct the earliest shape of the kingdom of Sweden by the most modern methods of his time. Extrapolating the secular diminution of the water level, he found that the country had begun as a tiny archipelago, where only the highest mountains emerged, and that it existed only during a third of the time since the creation of the world. This interpretation gave rise to scandal, and the opposition proposed a vote of censure, because, as they believed, the dignity of the kingdom had been challenged; but this action had no consequences. This was probably one of the first applications of geological methods to historiography.

The essence of the theory is the simultaneous change of the water level susceptible to be recorded everywhere along the shores of the oceans. Complements and additions to this framework were supplied by other fields of the earth sciences as they evolved, and these additions were tentatively inserted at different places. Consequently, not only was the internal structure of the theory changed by these new articulations, but also its significance within the general picture of the earth and its history; and this history became, in turn, a different one. The most important change was the concept of the geological time scale, different from the historical one. Within this changing environment of geological knowledge and concepts, the evolution of ideas changed its direction. Some of the main episodes will be sketched in this paper.

The concept of Celsius and Linné was not the only one of its epoch. Another hypothesis was also based on worldwide displacements of the shorelines, but they did not have the same direction everywhere. Emmanuel Swedenborg (1688–1772) considered the sinking water level in northern latitudes as the consequence of a change in the rotational speed of the earth. In other words, with increasing speed the hydrospheroid became more flattened than the geoid; waters collected into a protuberance around the equator and diminished at higher latitudes. Astronomical solutions were therefore not absent from the panorama of eighteenth-century science. This type of interpretation had the advantage that no "refuge" had to be found for the waters during the decrease of the ocean level. The additional question of the destiny of the disappearing waters received different answers in the course of time, from Newton to McLaren and Suess.

Criticisms of the theory of the "diminution of the waters" were possible at the following different levels: the observational basis, the internal organization of the hypothesis, the place in which the observations were inserted, the external relationships within a more general picture of the world, and the place it had to occupy in such a system. The theory was therefore not unchallenged. The opposition came from several different quarters, because the whole construction offered, even at that time, several weak points, and the opponents did not fail to find some of them and tried to refute them with the means of their time.

Most of the counterarguments were collected in a book published in 1755 by Johannes Browallius (1707–1755), Bishop of Åbo-Turkku. The different facets of the theory were closely discussed in the light of the Bible, and of classical and modern authors, but also disproved by interesting observations, mostly made by some friends of the author. For instance, Jakob Gadolin (1719–1802) had measured the height above sea level of the old castle of Åbo in 1751. If the Celsius-value were right, during several centuries one would have had to wade through the water in order to reach the entrance of the castle, a situation improbable from a practical point of view. P. A. Gadd (1727–1797), another professor at the Academy of Åbo, was the creator of the method of *dendrochronology*. He had counted the tree rings of a series of pines growing near the shores of the sea, and found ages between 232 and 310 years. Were the Celsius-value right, one had to believe that these pines had been growing below the sea for centuries.

After this attack, the theory lost many of its former followers. The presidential address for 1765 to the Royal Academy of Sweden by Bengt Ferner was centered on the theory of the "diminution of the waters," still being discussed in scientific circles. He mentioned the possibility of earth movements, but his important contribution was the distinction between the present and the past; i.e., if the shorelines are moving in our time, it is not sure that they have moved in the same way in the past.

Crustal Movements

In the same year, 1765, the problem took a new turn. Efraim Otto Runeberg (1722–1770), the director of the cadastral survey of Finland, often had to allot parcels emerged from the sea, and in this way he had collected an abundance of observations, especially in the gulf of Bothnia. His attention was focused on the fact that the gains of land were more pronounced in the northern countries with a cold climate. He described the mechanisms of the thick winter ice and their evolution playing an important role in the accumulation of loose materials on the beaches. Another important factor was the movements of the rocky basement. Runeberg tried to approach the nature of the subterranean forces by the study of its appearances. He laid stress on the concept that nature works by the same means on large and small scales and that the great movements of the earth's crust are due to the same causes as the small ones recognizable in northern Finland. He mentioned that joints play a role in these deformations, but that the same forces that are suddenly released in an earthquake can raise the rocky basement slowly, so that some areas emerge from the hydrospheroid while others subside and are covered by the waters. He emphasized that such movements are not always accompanied by cataclysms. It is remarkable that this idea was put forward in 1765, four years before Cuvier (1769–1832), the father of the cataclysmal theory, was born.

Runeberg was therefore a pioneer of uniformitarianism and living tectonics; he had an inkling of these chapters of structural geology 20 years before the *Theory of the Earth* (1785) of Hutton (1726–1797).

His work contains very advanced geological ideas interwoven with the views typical for his epoch, all of them expressed in an original style. At that time, earth sciences did not provide a framework of knowledge and a network of concepts in which the ideas and observations of Runeberg could have been inserted, thereby attaining international recognition. His work is a remarkable one, but his observations and ideas were presented too early.

Torbern Bergman, one of the great Swedish scientists, stated in 1766 that the fundamental changes admitted by Runeberg were impossible without catastrophic consequences; the earthquake of 1755, which destroyed Lisbon, was still in everybody's mind. Runeberg replied in 1769, but in vain, because it was apparently too difficult for his contemporaries to imagine, that the rocky ground was not stable. The thought that benchmarks were moving day and night under the houses, the woods, the fields, and the lakes, the fundamental discoveries and interpretations of Runeberg, did not have the necessary theoretical background, and therefore his contemporaries did not understand their importance. A series of advances in knowledge had to take place before earlier observations could be utilized. This is one of the frequent examples in the history of science in which *many excellent observations are ignored by most of the scientists as long as no niche is prepared for them in the framework of the generally accepted hypotheses.*

The period from Urban Hjärne to Runeberg can be considered as a first stage: the studies were limited to Fennoscandia; the problem was recognized and observations were made, partly in order to support different hypotheses, partly to establish the dimensions of the phenomenon. The question, at least in the beginning, was a part of the history of the creation and its development. Three main explanations were proposed: change of sea level, deformation of the earth's crust, and alluvial deposition. The causes were general, regional, or local. The results were inserted in different pictures of the world, but geological knowledge was not sufficiently developed to permit the recognition of their general importance. Another reason for the confinement of the discoveries was probably the fact that most papers were published in Swedish, and, in the case of Runeberg, by an outsider.

Connections Between International and Fennoscandian Scientific Developments

As mentioned previously, a series of advances in knowledge had to take place before the importance of earlier results could be recognized. Some of them will be mentioned here.

An important point of the changing background was the *extension of the chronological scale beyond the historical epoch*, and the vanishing time limits. The problem, seen at this new angle, appeared in a very different perspective, as compared with the times of Celsius and Linné. The difference between historical and geological times, the inequality of their dimensions, and the difference of their observational bases appeared during the second half of the eighteenth century, at first indistinctly, then more and more clearly. The development of geological chronology has been reviewed by several authors and will not be discussed here.

The significance and the importance of the moving shorelines for geological history could not be recognized until evidences showing that dry land had been repeatedly covered by the sea became so numerous as to be a rule, and until the events, later called *transgressions* and *regressions*, were ascertained in so many places that Lamarck, in his *Hydrogeology* (1802), could assume that every spot on earth had been repeatedly below the sea.

The importance of the studies was increased by the principle of *uniformitarianism*, which says that "the present is the key to the past." It was therefore important to have the opportunities to observe and describe examples moving before one's eyes in order to understand the records of the past.

The *time-space dimensions* of the Fennoscandic regression, with the exception of the Celsius-value, were very little known. Measures of the areal extent, the height, time, and rates became subjects of research in the second half of the nineteenth century, but satisfactory approximations were only available in our century.

Several distinctive features characterize developments in the nineteenth century and its continuation: research branched out into two main directions; the geological data were observed and interpreted in order to reconstruct the prehistorical evolution; and the measurements of the dimensions of actual and historical movements were recorded. The immediate causes of the moving shorelines were considered by some scientists as more important than the ultimate causes.

Fennoscandian research became interconnected with the international scientific movement, especially with the emerging geological centers of Western Europe. Numerous well-known geologists visited Fennoscandia. They became witnesses of the phenomena, and made them known to an international public. Consequently, observations and interpretations of the moving shorelines became more familiar and were used as important arguments in geological discussions. The list of foreign visitors is a long one; therefore, only some of the most striking examples can be mentioned here.

One of the first eminent travellers was Leopold von Buch (1774–1853) who visited Fennoscandia in the years 1806 to 1808. His penetrating mind discovered the main problems very soon. In September 1807, during his stay at Torneå-Tornio, in Northern Finland, he noticed that there was hardly a spot that did not show that the sea had recently withdrawn. "The phenomenon is curious, remarkable, surprising! How many questions crowd into one's mind! It is certain that the sea-level cannot drop; this is decidedly impossible, because of the equilibrium of the sea. But, because the lowering [of the water level] cannot be doubted, there remains, as we can see now, no other way out of this than the conviction that the whole of Sweden is slowly rising. . . ." (von Buch, Vol. II, pp. 503–504). The problem was clearly recognized, and the proposed solution, the crustal deformation, fitted well into the views of Leopold von Buch.

Another famous traveller was Charles Lyell (1797–1875), the herald of uniformitarianism and the author of the *Principles of Geology* (1830). The first edition of this work contains five pages on the "Celsian controversy with its contestants." The author takes the side of the theory of the oscillating sea level. His visit in Sweden, in 1834, "had three main aspects: namely (1) geologically recent, relative elevation; (2) its actual continuance till today; and (3) its localization, pointing to uneven earth-movement". (Bailey, 1962, p. 113). The observations and conclusions published in one paper (1835) were included in the fourth edition of the *Principles* (1835). Lyell emphasized that the rising was more intense in the north than in the south, and he remarked that the crust was deformed without seisms.

The coasts of the Arctic Ocean were surveyed by the French geologist and mineralogist Auguste Bravais (1811–1863) during the years 1838 and 1839. He observed and measured the raised beaches in the Altenfjord in northern Norway. The same continuous shorelines are situated at a higher level in the interior of the fjord than towards the ocean. These observations were later questioned by Eduard Suess (1888, pp. 441–443), but they were corroborated by the careful measurements of Väinö Tanner (1930). The observations of Bravais attracted the attention of Elie de Beaumont (1842). They seemed to fit his ideas about the shrinking earth and the tangential movements in folded mountains. Bravais believed that the tilting movement discovered in the Altenfjord was the continuation of the diastrophism that had created the mountains of Norway. The chronological limits of the tilting were not yet known at that time.

Uniformitarianism emphasized the similarity between older crustal movements recorded in folded rocks and unconformities, conglomerates, and changing facies, and recent, still active movement. By comparison, time scales of three orders of magnitude became increasingly apparent (1) time scale of geology; (2) time scale of historical chronology; and (3) shorter periodicities and annual recurrences. The relationships of movements on these scales were displayed in the borderland to which the study of moving shorelines belonged and provided several important results. The distinction between different time scales opened several perspectives: geological movements were continuous during long periods, and therefore were slow and mostly not catastrophic. The present-day stage is not exceptional and not the final goal of the evolution, but only one point in a long time-space development. However, these conclusions were only slowly entering the consciousness of many geologists, and it is interesting to follow this penetration.

Discontinuous and rapid deformations had not lost their supporters, and these views reappeared in different forms from time to time. The ideas of Hutton and Playfair were, at the beginning of the last century, somewhat overshadowed in many circles by the cataclysmal theory of Cuvier, and in our century they appeared again as Stille's "orogenic phases." The antagonism between continuous and discontinuous movements of different magnitudes (Wegmann, 1950), shows how the atmosphere of general views imprinted its marks on every generation. The return of antagonistic concepts maintained a tension field, shed new light on known facts, and stimulated further research.

Not only visitors came to Fennoscandia; new ideas also passed the borders and had to be tested. They transformed the outlines and the structure of the theoretical framework in a new manner, and therefore changed the value of former observations, brightening up some of them, putting others in the shadow. Some of these structural changes will be mentioned here.

The first of these ideas was the *glacial theory*. This new interpretation discovered by Venetz and de Charpentier in western Switzerland was brought to Fennoscandia by Agassiz. The studies of glacial and postglacial deposits soon demonstrated that the recorded phases of movement happened essentially after the melting of the ice cap. Its position in the geological time scale was thus established as postglacial.

Another concept was closely connected with the existence of continental glaciers, as shown by McLaren's review in 1842 of Agassiz' work on the glaciers for the *American Journal of Science*. In a footnote he proposed the idea that the level of the oceans had to be lowered if great masses of water were accumulated on the continents as ice caps, and that the sea level rose when these quantities of water returned to the oceans during the melting of the ice. This concept was soon accepted, because it provided a simple principle for the parallelization of all recent terraces around the oceans and, through the network of streams, in the interior of the continents. The theory was easy to understand but difficult to test; and furthermore, the balance of water was hard to establish. The story of the "diminution of the waters" took in this way a new form, in which the problem of the "refuge" of the waters was explained. This concept later received the name *glacio-eustacy.*

The deformationists were also supported by new concepts. Thomas F. Jamieson (1865, 1882) presented the idea that crustal areas loaded by thick ice caps were pressed down and rose again when the glaciers melted away. This principle, fertile for both geology and geophysics, is known as the *glacio-isostatic principle.*

Another, more general interpretation of the multiple shorelines on the ocean coasts was proposed by Eduard Suess (1831–1914) in the second volume of his fundamental work *The Face of the Earth* (1888). His interpretation served as a base for many discussions during the following years, and is still the standard explanation in many textbooks. It is necessary, therefore, to sketch briefly his ideas. He gave a new name to the positive and negative oscillations recorded simultaneously at every sea coast, calling them *eustatic movements* (Suess, 1888, p. 680). The euphoniousness of this name impressed many geologists and contributed much to its spreading, as is usual in such cases.

The oscillations formed an essential part of his general picture. As a consequence of the contraction of the earth, vast portions of the crust collapsed and the waters followed, filling the depressions and forming the ocean basins (Suess, 1888, p. 680). But although the subsidences of the lithosphere were limited, the level of *all oceans* was lowered. The collapses occurred at discrete intervals, beating time for geological history. The formation of the ocean basins controlled the negative eustatic movements. The "diminution of the waters" was thus ultimately explained by the shrinking earth.

Suess described in his *Memoirs* (1916, p. 139) some of the stages of evolution of the idea, and he said that it was clear for him at the start "how this idea touched upon the fundamentals of the predominant ideas in geology" (Suess, 1916, p. 139). He studied during the summer of 1861 the sequences of Tertiary beds of the northern basin of Vienna. He was again struck by this thought, when in 1869, assisting at the opening of the Suez Canal, he studied the younger formations of this area. He noted that these beds were strictly horizontal, so that the isthmus could not have been raised; therefore, the sea level must have been lowered (Suess, 1916, pp. 208–209). At that moment he planned to make a comparison with the raised beaches in northern Europe, but he had to wait until 1885 to realize this project. In that year he visited northern Norway (Suess, 1916, p. 365 *et seq.*), but his manner of traveling did not allow him to make systematic measurements by levelings along the series of raised beaches.

Suess's main interest was the great transgressions and regressions of the geological past. He distinguished several orders of magnitude, and explained that similar phenomena at different scales can have different causes and a different mechanism. The similarity is sometimes only feigned because of the use of the same descriptive term for features of different dimensions. Suess was a man of vast erudition, and he was also well acquainted with the history of scientific research in Scandinavia. His chapters on these matters are classic models (Suess, 1888, Chap. 1, 8–10, 12). The problems of the Baltic Sea and the raised beaches of Norway are treated separately.

The raised beaches of northern Norway, considered as tilted by Bravais, did not fit the picture drawn by Suess. He therefore mentioned other sources, questioning the exactitude of Bravais's measures. After having described similar cases from other countries and other continents, he concluded that the doctrine of secular deformations of the continents was not able to explain the submersions and emersions of the lands. The changes were too extensive and too uniform to be caused by earth movements (Suess, 1888, p. 684).

The case of the Baltic Sea was treated in another manner. This interior sea was considered as a special kind of lake, controlled by climatic changes. The discussions about climatic oscillations are very old; they belong to the mythological, the transitional, and the scientific periods, and have often been renewed. However, it was difficult to find precise observations to support the different theories, because regular meteorological measurements started too late. The theories were mostly based on pessimistic or optimistic traditions. It is not appropriate to give here a summary of the different phases and the multiple levels of these discussions. In order to corroborate his point of view, Suess reproduced the curves of the water gauges around the Baltic Sea from 1873 to 1875 (Suess, 1888 Pl. IX), concluding that there was no discernible tendency of a rising coast and that the oscillations were due to climatic fluctuations.

The celebrity and the authority of the great Viennese geologist and the synthesis of his extensive knowledge made a great impression on his contemporaries, so that the following generations, including people who had never heard his name, were affected by his intellectual position. The methods based on the eustatic principle are still used and described in many textbooks.

Suess continued the series of *unicausal explanations* started by Celsius and Linné. The phases of collapses of the ocean basins were for him the *only* causes, and many chapters of his book are written in order to show the inadequacy of other factors. This unicausal manner, initiated for the same problem by Celsius and Linné, is still found in many modern papers. It is probably a legacy from Cartesian philosophy, and from the epoch of the Enlightenment.

The movements of the shorelines had to be due to *one* cause, for instance, the changing sea level, which, in turn, was imputed either to true eustatic oscillations caused by the shrinking earth or to glacio-eustacy related to climatic changes; or the displacements were considered as effects of crustal deformations, from which the concatenations were followed to further causes, but the chain of causality had to be single-tracked. Many pages were used to show that no other cause was operating and that all others could not be considered.

In the background of these discussions about general problems, a new approach was developing. Two paths may be distinguished, each one ramifying into a multitude of branches: first, we see the unparalleled development of the Quaternary and Postquaternary geology of Fennoscandia; second, we have the isolation and the analysis of the different factors composing the secular displacements of the Baltic seashores.

Toward the New Leading Position of Fennoscandian Geological Research

For a long time, marine fossils had been found in the interior of the land, but their stratigraphical relationships could only be established during our century. The oldest member of the series known at that time were the marine beds with *Yoldia arctica*, overlain by the deposits of the Ancylus Lake, and, as the youngest member of the fossil sequence, the sediments of the Litorina Sea. An earlier stage below these series was discovered by Henrik Munthe in 1910; it was called the Baltic Ice Lake, because its deposits were contemporaneous with the melting period of the ice cap. The systematic studies of the fossil water sheets, their relationships with the glacial deposits, and their present situation were synthesized in 1896 by Gerard De Geer. Already this comprehensive view established several important points: the remains of the Yoldia Sea are more widespread than those of the Litorina Sea, and they occupy higher levels today. All fossiliferous beds reach their highest elevation in the areas around the Gulf of Bothnia.

Several new methods and techniques were invented and developed in order to obtain more data about the postglacial period. They became so closely interwoven in the new pattern of this branch of science that its aspect and reasoning process were radically changed. It is henceforth impossible to imagine this field without these methods and their results. The most important are discussed here.

The *absolute chronology by counting the annual varves*, a tremendous work, was started and brought to a successful conclusion by Gerard De Geer (1858–1943) (1905, 1910, 1940) and his school, surveying the enormous expanse from southern to northern Sweden. A similar chronology was established by Matti Sauramo (1889–1958) for Finland. These studies made it possible, long before the invention of radiochronological methods, to determine the age of formations and events, and this with only a small margin of error. The chronology begins with the melting of the ice cap. The beginning of the time scale is thus not the same at every place. These studies provide, also, the rate of melting from year to year and represent a paleoclimatological record.

Another fundamental method is *palynological analysis* created by the Norwegian Blytt and the Swede Sernander. The method and its results have developed into a vast branch of science. Von Post in Sweden and Sauramo and Auer in Finland have adapted this knowledge to the problem of the raised beaches. This kind of research is not only used for the establishment of chronological correlation, but also provides fundamental data on the spreading and development of the land vegetation.

A third part of the new research activities consists of *micropaleontological studies*, especially of *diatoms*, started by Cleve-Euler and her followers. These fossils make it possible, in many cases, to distinguish the nature of the former biotopes and the salinity, temperatures, and size of the basins (pool, pond, small or extended lake, etc).

The study of *peat bogs* (Anderson, 1898) with their different horizons and their interbedded layers of minerogenetic deposits has provided a wealth of data about the changing level of the seas and the lakes, the tilting movements, and many other topics (Auer, 1924). The peat bogs now drowned below the present sea level have been the subject of many discussions.

Archaeological explorations were long ago closely related to the studies of postglacial formations. The exchange of information between archaeology and geology has played an important role for the chronological correlations.

Not only was the structural pattern of the knowledge about postglacial evolution altogether changed by these new methods and their results, but also the general direction, the goals, and the purposes became fundamentally different. For a long time the main goal was to find the one unique cause for the events, excluding every other cause. An important part of the papers was devoted to the defense of the proposed cause and to the annihilation of the arguments supporting opposing views. This style of presentation is still encountered where unicausal thinking survives. This orientation was partly changed under the influence of the great masters of Fennoscandian geology. Their first goal was to reconstruct a time-space picture of the events, a film, so to speak, of what really happened. Every episode had to be established as precisely as possible. Every detail had to be placed in time-space coordinates in order to approach the pattern of the real sequence of events as they had taken place, and not as they should have happened had they been the effects of one or the other of the assumed causes.

This new approach produced many observations that, at first, seemed sometimes contradictory and could not be explained by unicausal concepts without suppressing important points. It became very difficult, for example, to coordinate closely spaced stratigraphical sections and to understand the reason for such differences. Applying orthodox eustasy or pure deformationism resulted in very strange pictures, but they were sometimes accepted by the supporters of unicausal doctrines. The need for uniform causal chains helped many to accept strange conclusions and still represents a palliative.

On the other side, it was hoped that time-space pictures of the events, once established, would become the base for better explanations. Many scientists recognized the hopelessness of the earlier style of explanation.

The New Categories of Synthetical Patterns

A new synthetical pattern was created by Wilhelm Ramsay (1865–1928), one of the famous geologists of Finland, and his school. Ramsay did not seek "the real cause"; he abandoned the unicausal pattern, considering the displacements of the shorelines as the result of the interference between the oscillations of the sea level and the deformations of the earth's crust; i.e., as the result of two independent variables. Such a bicausal concept is more complicated, because a multitude of methods and techniques of analysis have to be created in order to approach, by degrees, the sequence of the events. This is for a great part the achievement of the great generation of Finnish Quaternary geologists — Ramsay (1896, 1898, 1917, 1920, 1921, 1924, 1926), Sauramo (1889–1958) (1918, 1923, 1929, 1934, 1939a, 1939b, 1944, 1947, 1954, 1955), Väinö Tanner (1881–1948) (1929, 1930, 1938), Auer (1924, 1957, 1959, 1963, 1964), and many others. Their work is continued by younger generations, from Hyyppä (1963) to Donner (1965) and Ignatius (1958). They have contributed to shape the new pattern of this branch of science. These methods created in Finland are now spreading over Sweden, Norway, Denmark, and the countries east of the Baltic Sea.

New tools, such as diagrams of different kinds, isobase maps for synchronous sea levels, and many other helpful methods and techniques, have been developed and are constantly being improved. One of the main difficulties is the fact that both the sea level and the crust are moving; therefore, no stable reference mark is given. The two variables are independent. The careful analysis of the interference makes it possible to determine the components by successive approximations, and to disentangle the complicated sequence of transgressions and regressions recorded at many places and not interpretable in another way.

The main lines of the methods and techniques have been presented briefly in several papers (Ramsay, 1924; Sauramo, 1929; Tanner, 1938; Donner, 1965; Wegmann, 1967). They will not be discussed again here, although they are fundamental parts of the present-day structure of this branch. The picture resulting from these studies is not only a detailed time-space analysis of the interfering movements and their dimensions, but also, a stupendous fresco of the physiographical and biogeographical evolution, of the land climates and hydroclimates, of the spreading of faunas and floras, and of the changing conditions for human settlement. The picture represents the most elaborate model of modern stratigraphy. The network of connections between the observations and the interpretations grows denser every year. The velocity of movements and displacements and the changing rates at each place can be established within narrow time limits. Also, for living tectonics, this example is one of the most detailed.

The series of records that can be interpreted with today's methods breaks off at the moment when a given locality was covered by the ice cap, or begins when this ice cap disappeared. This unknown period includes, with the exception of the marginal zones, the main part of the Quaternary epoch. Further developments will perhaps supply these studies with means to extend investigations into this still empty portion of the picture, but for the moment we have one part of the sequence provided with a wealth of details and another one nearly devoid of the main trends. The first is the postglacial period; the second corresponds to the glacial epoch. Dynamic interpretations, as many geophysical models, are therefore based on an incomplete picture. In most cases, the lines of this part of the picture are provided by assumptions and extrapolations, which have no observational base but only an assumed starting stage and an assumed dynamic principle.

As in many areas of science, the step from the unicausal to the bicausal concept is important because it implies fundamental structural changes. This step represents a real renewal, but the bicausal stage is only the preparation for an advance toward the multicausal explanations.

Väinö Auer (born 1895), a former student of Ramsay, transplanted the Finnish methods and concepts to southern Argentina and Patagonia and continued his investigations there for 30 years, adding new ways of investigation, such as the important *tephrachronological methods* (Auer, 1959). The comparison with the Fennoscandian pattern brought interesting new results (Auer, 1964). The graph representing the changing sea levels since late Quaternary times shows not only rising movements and stops, but real oscillations with depression and culmination points. These oscillations are indeed influenced by the growing and diminishing ice caps, as Auer has shown, but other factors appear also in the diagrams. The shape of the curve is the result of the interference between the ice-water balance and some other, not yet analyzed factors. One of them is probably the deformation of the oceanic receptacle (Auer, 1957). The concept of the eustatic oscillations (Suess, 1888) is therefore introduced again, but this time not as a unicausal explanation excluding other factors, but as one of the interfering determinants. The way to multicausal analysis is thus open.

It is impossible to present here the multiple relationships between the studies on the displacements of the shorelines both for the sea and the lakes within their tilting basins and the other problems of the physiographical, geochemical, and biological evolutions. The main lines are sketched in another paper (Wegmann, 1967). The changing history of the lakes has played an important role, because these water sheets record the movements in the interior of the land (Auer, 1924; Sauramo and Auer, 1928, Sirén, 1951, 1964; Kääriäinen, 1953, 1963; Tanner 1938, pp. 634–645; Wegmann, 1967). The limnological and paleolimnological studies are connected with nearly all aspects of this vast problem, both for the geological past and for the determinations of secular movements.

Measuring Secular Displacements

The measurement of the rates of secular displacements developed in a somewhat different way from the methods used in geological approach, but a careful examination shows some parallel features. The main stages will be briefly sketched here.

For a long time, some scientists considered the secular movements as a fact, whereas others, less in touch with nature but most erudite, questioned the reality of the phenomenon. If the movement was real, it had to be measured. This is already shown by the attempts of Celsius (p. XX). His bench marks and those of his followers were important achievements, reminding many visitors to the coastal area of the problem, so that it would not fall into oblivion.

From the middle of the last century, the pilot stations in Sweden and in Finland had to measure and to record the daily water level. The quality and homogeneity of these observations were not quite satisfactory in the beginning but were improved with time. Another difficulty soon became apparent, when avalanches of figures reached the offices.

A. E. Nordenskiöld (1832–1901), the first seafarer who sailed through the Northeastern Seaway, adapted and organized the calculations of the annual averages. In order to show the efficiency of his methods, he published a comprehensive study of the observations made in Stockholm between 1774 and 1852 (Nordenskiöld, 1858).

The comparative study of series of measurements from different localities was, at the start, somewhat disappointing, because the averages were not, as expected, located on a straight line but were scattered over a large zone (Fig. 1). For such a station as Rönnskär in the Gulf of Bothnia, jumps of more than 10 centimeters and sometimes up to 16 centimeters from one year to another were not uncommon, so that the averages seemed to support the opinion of Eduard Suess (1888, pl. IX, where some of these series are reproduced), that the crustal deformations could not be proved. In the same year, 1888, Leonhard Holmström stated the following in the Swedish Academy of Science: The question of the moving shorelines has recently become more interesting because the hypothesis of the oscillating sea level is more and more accepted, whereas the views of John Playfair and Leopold von Buch and their followers about crustal deformations have lost nearly all ground.

Deformations were thus in the minority, but the turn was not far away. The countercurrent started with the first map of secular isobases drawn by the Austrian geographer Robert Sieger (1864–1926), in 1893.

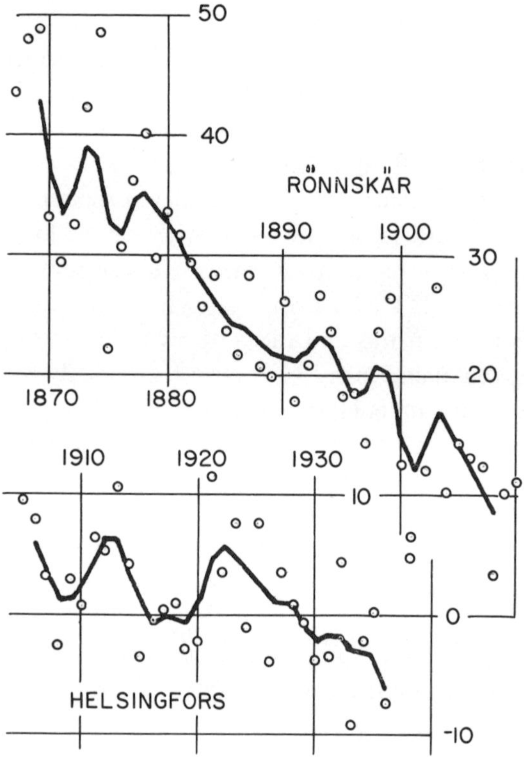

FIG. 1. *Mean annual sea levels at Rönnskär, Gulf of Bothnia, and Helsingfors (Renquist, 1948).*

This work is based on careful and detailed historical studies, on critical analysis of the data, and on many original observations. Sieger used the observations on the Baltic Sea as well as on the Great Lakes. He tried to eliminate the meteorological factors, but the information available at his time was not adequate. His map shows the more intense rising around the Gulf of Bothnia. The principle of the secular isobase maps proved to be advantageous. New maps of this kind were later published, based on observations between 1889 and 1912 around the Baltic Sea (Renquist and Blomquist, 1914), but the meteorological factors were still scattering the annual averages. This is the reason that Rolf Witting (1879–1944) subjected them to a close examination, considering the real water level as a multicausal system. He limited his first studies to a period of 15 years, but tried to detect and analyze every disturbing factor: the direction, intensities, and persistence of the winds; the distribution of barometric pressures; the inflow and the changing distribution of fresh water; the balance of water flowing through the Danish Straits; the evaporation; and other minor factors. The influence of each factor was estimated for every station for the 15 years, and eliminated in order to isolate the secular displacement and to chart the results on a series of diagrams and new isobasic maps (Fig. 2). One of these maps contains the whole series of measures, other maps represent only a fraction of the time, down to 3 years. For the shorter intervals, the difference between the different maps can be very important. The comparison of maps presenting longer intervals shows fewer differences. Despite this careful analysis, the diagrams representing the sea level at one locality did not produce a straight line. This feature has been explained in several ways; unknown factors could be hidden by the dispersion of the averages, but one can also admit, with Witting, that the majority of the disturbing factors have been eliminated by the operations and that the real movement is revealed. In other words, the displacement is really discontinuous in space and in time. A model consisting of a mosaic of independently movable blocks resting on a more or less mobile cushion would produce movements of this kind.

Later maps based on longer series by Witting (1928, 1945) and by Bergsten (1939) have not produced essential changes of the isobases. The limit of efficiency of the hydrographical methods is probably reached by these maps (Lisitzin, 1963, 1964).

The secular isobases based on observations at the seashore have been made more complete by limnological measurements in the interior of the land. The movements of the lakes attracted the attention of Rudbeck, the professor of Linné. It was known by tradition that the meadows and woods on the low parts of southern shores on the Great Finnish Lakes had a tendency to become swampy, and the vegetation was drowned, whereas marshy areas on the northern shores were drying up. This is the consequence of the tilting movement of the basin started in Quaternary times and continuing through the postglacial epoch up to the present. The effects of these movements have been studied both for the geological past and for the historical past. The consequences of the tilting were sometimes spectacular. Former fjords opening toward the north, the Gulf of Bothnia, have been changed into lakes by the threshold between the Gulf and the basins. The Great Lakes previously had their outlets toward the north but were, in relatively recent time, tipped over to the south. These southern outlets of the Great Lakes did not have sufficient time, a few thousand years, to reach their profile of equilibrium. They still show rapids and falls, and these features are responsible for the location of the industries of Finland. The watersheds have been displaced during a very short time from southern to central and northern Finland. Parts of the lakes have been emptied, sometimes

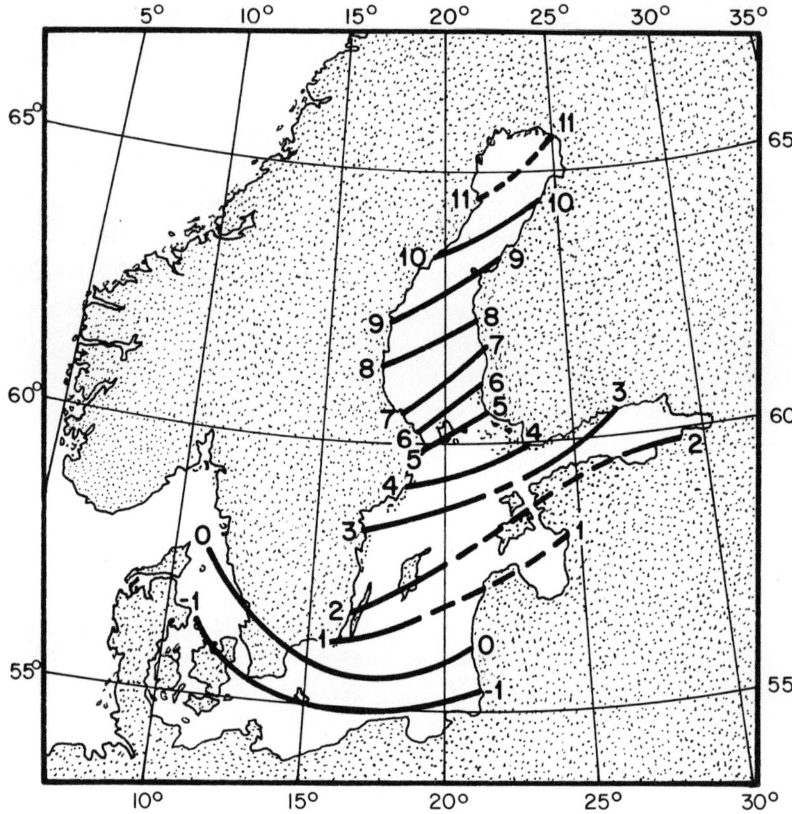

FIG. 2. *Isobases of recent movements (after Witting, 1943).*

by catastrophes, and great areas of the lake bottom were dried up. During other periods, woods, peat bogs, and other vegetable associations were drowned, so that the stratigraphy is somewhat complicated. The history of these lakes is not only important for the local geology, but it is also interesting from a general point of view because it provides carefully studied examples of stratigraphical correlations in tilting basins (Auer, 1924; Sauramo and Auer, 1928; Tanner, 1938, p. 634 *et seq.*). In other areas the gradient of the drainage is diminishing; water is not sufficiently drained off, so that wide areas have become marshy and some districts are up to 50 percent covered by peat bogs where there were much fewer formerly. They provide examples, still observable in the making, for the growth of caustobioliths favored by tilting ground, as one finds them from the Carboniferous up to recent times in the coal-bearing stratigraphical series.

The crustal warping can also be ascertained by geodetical methods. They have led Bravais (1840, p. 691) to the hypothesis of tilting movements. This branch of research has been renewed with the methods of modern geodesy and geophysics (Hela, 1953; Kääriäinen, 1953, 1963; Kukkamäki, 1963; Niskanen, 1943, 1949). Additional new methods are introduced as technology develops (Simonsen 1946, 1949, 1960), so that several new facets can be controlled, and new data will be secured in the near future (Aulanko, 1966), especially on the way detailed displacements take place.

Conclusion

The two branches of the studies of moving shorelines — the geological evolution and the measurement of secular movement — originally started from the same phenomena, but since approximately the middle of the last century, they have evolved in different directions. Scientists in touch with nature considered the displacements, long ago, as a fact to be studied, others, more influenced by theoretical views, doubted its reality. A first antagonism was thus centered about the existence of the problem. Further questions pertained to the causes. They were considered the main object of the study. During this stage, detailed investigations advanced only slowly. The complexity of nature had to be discovered.

Three main stages are easily distinguished: the first, from 1694 to about 1800, was characterized by studies limited to Fennoscandian scientific circles. During a second stage, new ideas permeated from outside and were discussed. The third stage corresponds to the flourishing of Fennoscandian Quaternary and Postquaternary geology and to a fundamentally new orientation. The energies were devoted to gathering the real sequence of phenomena, to describing them in detail and to locating them as exactly as possible in a time-space grid. The main goal was to establish the sequence of the events as they appeared in the record, not as they could be reconstructed from presumed causes and their mechanism. A new, successful way was found.

Unicausal explanations were recognized as inadequate for organizing the abundant observations flowing in from the older methods of investigation and also from the many newly created fields. Unicausal interpretation led to contradictions; if not, important results had to be eliminated. *Bicausal and multicausal analysis* not only developed a multitude of original methods, impossible in the unicausal conceptual environment, but they also created series of new models matching successfully the wealth of data and giving better and better synthetical pictures.

Looking back at the two leaves of the diptych — the geological studies on the one side and the measurements of secular movements on the other — attention is first attracted by the differences of perspective, of means, and of methods; but a more careful examination reveals some similarities in the change of style during the development. A main trend, hidden by a multitude of epiphenomena, appears and marks the transition from unicausal to bicausal and to multicausal mental environment. A similar change is found in many fields of science if the evolution is analyzed, not only in geology but in many other disciplines as well. This transition can assume different forms in various areas. It is an interesting problem to find out the conditions controlling the differences in the transition and their relationships with the structure of the discipline.

The concept of the evolution stages (unicausal, bicausal, multicausal) should not be applied in a schematical way, but should be used only as a general leading idea in order to establish the different trends of evolution in various fields. In this way the individual characteristics of every field will appear more distinctly, and it will be possible to undertake a further step: the comparative study of evolution. These trends of evolution and transition can be compared and related to the conditions controlling the modes of transition, and they will provide some important results for the survey of the winding path of knowledge between the Scylla of the unifying tendencies and the Charybdis of the diversifying tendencies.

Acknowledgements

The author is indebted to Professor Cecil J. Schneer for giving encouragement and to Professor Albert V. Carozzi for critical reading of the manuscript.

References

Agassiz, Louis (1837). "Des glaciers, des moraines et des blocs erratiques." *Actes Soc. Helvét. Sc. Nat. 22-me Session, Neuchâtel, 1837*. Pp. V–XXXII.

Andersson, Gunnar (1898). "Studier öfver Finlands torfmossar och fossila kvartärflora." *Fennia, 15* (1898), 210 p., 21 fig., 4 pl.

Auer, Väinö (1924). "Die postglaziale Geschichte des Vanajavesisees." *Bull. Comm. Géol. de Finl., 69* (1924), 134 p., 10 fig., 10 pl.

——— (1957). *Ueber die eustatische Bewegung des Meeresspiegels in Feuerland-Patagonien und anschliessende Fragen.* Stuttgart: Lautensach-Festschr., 1957. Pp. 407–418, 2 fig.

——— (1959). "Wie die auf vulkanische Ausbrüche gegründete geologische Zeitrechnung entstand." *Sitz.-Ber. Finn. Akad. der Wiss. (Festvortrag)*, 1958 (1959), pp. 37–50.

——— (1964). "Finländsk geologisk forskning i Sydamerika." *Nordenskiöld-Samfundets Tidsskr., 24* (1964), 3–20, 12 fig.

——— (1964). "Late-glacial and post-glacial shore-line displacements in South America as established by tephra-chronology, compared with displacements of the Baltic shore-lines." *Fennia, 89* (1964), 51–55.

Aulanko (1966). "Proceedings of the Second International Symposium on Recent Crustal Movements, Aulanko, Finland, August 3–7, 1965." *Annales Acad. Sc. Fennicae, 90* (1966), 498 p., figs., pls.

Bailey, Edward (1962). "Charles Lyell." *British Men of Science.* Garden City, N.Y.: Doubleday, 1962. X + 214 p., 24 pl., 20 fig.

de Beaumont, Elie (1842). "Rapport sur un mémoire de M. A. Bravais relatif aux lignes d'ancien niveau de la mer dans le Finmark." *C. R. Acad. Sci. Paris, 15* (1842), 817–849.

Bergsten, Folke (1939). "Den nutida landhöjningen vid Sveriges kustar." *Ymer, 59* (1939), 131–149, 6 fig.

Berzelius, Jakob (1847). "On the polished rocks of Sweden. Extract from a letter to Professor Leonhard from Jakob Berzelius dated Stockholm, Jan. 12 1847." *Quart. J. Geol. Soc. London, 3* (1847), 76.

Bravais, Auguste (1840). "Sur les lignes d'ancien niveau de la mer dans le Finmark." *C. R. Acad. Sc. Paris, 10* (1846), 691.

Buch, Leopold von (1810). *Reise durch Norwegen und Lappland.* 2 vols.; Berlin, G. C. Nauck, 1810.

De Geer, Gerard (1888, 1890). "Om Skandinaviens nivåförändringar under kvartärperioden." *Geol. För. Förh., 10* (1888), 366–379, 1 pl.; *12* (1890), 61–111.

——— (1894). "Om strandliniens förskjutning vid våra insjöar." *Sveriges Geol. Unders., 141* (1894), 1–15.

——— (1894). "Om kvartära nivåförändringar vid Finska Viken." *Sver. Geol. Unders., 141* (1894), 1–17.

——— (1905). "Bidrag till istidens kronologi och klimatlära." *Geol. För. Förh., 27* (1905), 219–222.

——— (1910). "A geochronology of the last 12,000 years." *C. R. XII Session Congr. géol. Internat., Stockholm, 1910.* Pp. 241–253, 2 pl.

——— (1940). "Geochronologia Suecica, Principles." *Kgl. Svenska Vet.-Akad. 18* (1940), 367 p., 65 fig., 90 pl.

Donner, J. J. (1965). "Shore-line diagrams in Finnish Quaternary research." *Baltica, 2* (1965), 11–20, 5 fig.

Hagberg, Knut (1939). *Carl Linnaeus.* Stockholm, 1939. 342 p., 14 pl.

Hela, Ilmo (1953). "A study of land upheaval at the Finnish coast." *Fennia* (1953), 38 p., 14 fig.

Holmström, Leonhard (1888). "Om strandlinjernas förskjutning vid Sveriges kuster." *Kgl. Vet.-Akad. Handl., 22* (1888), 1–99.

Hyyppä, Esa (1963). "On the Late-Quaternary History of the Baltic Sea." *Fennia, 89* (1963), 37–48, 1 pl., 4 fig.

Jamieson, Thomas F. (1865). "On the History of the Last Geological Changes in Scotland." *Quart. Geol. Soc. London, 21* (1865), 161–203, 10 fig.

——— (1882). "On the Cause of Depression and Re-elevation of the Land During the Glacial Period." *Geol. Mag. N. S., 9* (1882), 400–407, 457–466.

Ignatius, Heikki (1958). "On the Rate of Sedimentation in the Baltic Sea." *Bull. Comm. Géol. Finl., 180* (1958), 135–142, 2 fig.

Kääriänen, Erkki (1953). "On the Uplift of the Earth's crust in Finland." *Fennia, 77* (1953), 106 p., 17 fig.

——— (1963). "Land Uplift in Finland Computed by the Aid of Precise Levelings." *Fennia, 89* (1963), 15–18, 1 fig., 1 pl.

Keilhau, A. M. (1838). "Undersögelser om hvorvidt i Norge, saaledes som i Sverrig, findes tegn til en Fremstigning af Landjorden i den nyere og nyeste geologiske Tid." *Nyt Mag. f. Naturv., 1* (1838), 105–254.

Kukkamäki, T. J. (1955). "The Land Uplift in Finland Determined with Two Levellings as Well as with Water Level Observations." *Proc. Sect. of Lev. Rome Sept. 1954.* Copenhagen, 1955. Pp. 27–29, 1 fig.

——— (1963). "Symposium on Recent Crustal Movements in Finland with Bibliography." *Fennia, 89* (1963), 69 p.

Lamarck, J. B. (1964). *Hydrogeology.* Trans. by Albert V. Carozzi. Urbana, Ill.: University of Illinois Press, 1964. VII + 152 p., 4 fig., 1 pl. (Original edition Paris, 1802. 186 p.)

Lisitzin, Eugénie (1963). "Land Uplift as Sea-Level Problem." *Fennia, 89* (1963), 7–10, 1 fig.

——— (1964). "Contribution to the Knowledge of Land Uplift Along the Finnish Coast." *Fennia, 89* (1964), 1–22, 3 fig.

Lyell, Charles (1837). *Principles of Geology, being an inquiry how far the former changes of the Earth's surface are referable to causes now in operation.* 5th ed.; London, 1837. 4 vol.

McLaren, Charles (1842). "The Glacial Theory of Prof. Agassiz." *American J. of Sci., 42* (1842), 346–365, 9 fig.

Niskanen, E. (1943). "On the Deformation of the Earth's Crust Under the Weight of a Glacial Ice-Load and Related Phenomena." *Ann. Ac. Sc. Fennicae, 17* (1943), 1–59, 2 fig.

——— (1949). "On the Elastic Resistance of the Earth's Crust." *Ann. Ac. Sc. Fennicae, 21* (1949), 22 p., 3 fig.

Ramsay, Wilhelm (1896). "Till frågan om det senglaciala hafvets utbredning i södra Finland." *Bull. Comm. Géol. de Finl., 3* (1896), 1–44, 1 pl.

——— (1898). "Ueber die geologische Entwicklung der Halbinsel Kola in der Quartärzeit." *Fennia, 16* (1898), VI + 152 p., 5 pl., 1 map.

——— (1917). "De s.k. marina gränserna i södra Finland." *Fennia, 40* (1917), 8 p., 1 fig.

——— (1920). "Litorina gränsen i det sydliga Finland." *Geol. För. Förh. 42* (1920), 243–262, 2 fig.

——— (1921). "Strandlinjer i södra Finland." *Geol. För. Förh., 43* (1921), 495–497.

——— (1924). "On Relations Between Crustal Movements and Variations of Sea-Level During Late-Quaternary Time, Especially in Fennoscandia." *Fennia, 44* (1924), 39 p., 10 fig. (Also, *Bull. Comm. Géol. Finl., 66*).

——— (1926). "Dränkta länder och kulturer." *Soc. Sc. Fenn. Arsbok-Vuosikirja, 4* (1926), 1–15, 8 fig.

——— (1927). "Nivåförändringar och stenaldersbosättning i det baltiska omradet." *Fennia, 47* (1927), 1–68.

——— (1931). "Changes of Sea-Level Resulting from Increase and Decrease of Glaciations." *Fennia, 52* (1931), 1–62, 3 fig.

Renquist, Henrik (1923). "Vasatraktens topografi, landhöjning och geografiska namn." *Fennia, 44* (1923), 1–10, 1 pl.

——— (1927). "Några gamla skrifteskartors vittnesbörd om landhöjning och abrasion." *Fennia, 47* (1927), 1–24, 2 fig.

——— (1948). *Landhöjningen vid våra kuster.* Helsingör: Skärgardsboken, 1948. Pp. 74–94.

Runeberg, Efraim Otto (1765). "Anmärkningar om några förändringar på jordytan i allmänhet och under de kalla klimat i synnerhet." *Kgl. Sv. Vet.-Akad. Handl., 26* (1765), 81–115.

——— (1769). "Förklaring på några omständigheter rörande fragan om vattuminskningen." *Kgl. Sv. Vet.-Akad. Handl., 30* (1769), 177–197.

Sauramo, Matti (1918). "Geochronologische Studien über die spätglaziale Zeit in Südfinnland." *Bull. Comm. Géol. de Finl., 50* (1918), 44 p., 5 fig., 4 pl.

——— (1928). "Der Anteil der eustatischen Komponente an den Niveauverschiebungen in Fennoscandia" *Fennia, 50* (1928), 10 p., 2 fig.

——— (1929). "The Quaternary Geology of Finland." *Bull. Comm. Géol. de Finl., 86* (1929), 110 p., 39 fig., 25 pl., 1 map.

——— (1939a). "Aus der Erforschungsgeschichte der Ostsee." *Schriften der Phys.-ökon. Ges. zu Königsberg, 71* (1939), 11–36, 9 fig.

——— (1939b). "The Mode of Land-Upheaval in Fennoscandia During Late-Quaternary Time." *Bull. Comm. Géol. de Finl., 125* (1939), 1–26, 9 fig., 1 pl.

——— (1944). "Der Mechanismus der Landhebung." *Sitz.-Ber. Finn. Akad. d. Wiss. 1943* (1944), 93–114, 4 fig.

——— (1947). "Waldgeschichtliche Gliederung des Spätglazials." *Sitz.-Ber. Finn. Akad. d. Wiss., 1946* (1947), 231–235, 2 fig.

——— (1954). "Das Rätsel des Ancylussees." *Geol. Rundschau, 42* (1954), 197–233, 17 fig.

——— (1955). "Land Uplift with Hinge-Lines in Fennoscandia." *Ann. Acad. Sc. Fenn., 44* (1955), 5–23, 11 fig.

Sauramo, Matti, and Auer, Väinö (1928). "On the Development of the Lake Höytiäinen in Carelia and Its Ancient Flora." *Bull. Comm. Géol. de Finl., 81* (1928), 1–40, 20 fig., 4 pl.

Sederholm, J. J. (1899). "Les Dépôts Quaternaires de Finlande." *Bull. Comm. Géol. de Finl., 10* (1899), 1–28, 2 fig., 1 map.

Sieger, Robert (1893). "Seenschwankungen und Strandverschiebungen in Skandinavien." *Zschr. f. Erdkunde* (Berlin), *28* (1893), 1–116, 393–488, 1 pl.

Simonsen, O. (1946). "Orienterende Bemærkninger om Nivellementsnul i Danmark." *Matem. Tidskrift, B* (1946), 53–60.

——— (1949). *Nivellements-Nul paa Sjælland, Mön og Lolland-Falster med særligt Henblik paa Köbenhavn og Frederiksberg 1845–1945.* Copenhagen: Thèse, (1949). 183 p., 19 fig., 5 pl.

——— (1960). "Report from the International Commission For European Levelling for the Period September 1957–July 1960." Submitted to the General Assembly of the International Union of Geodesy and Geophysics, Helsinki, 1950. Pp. 2–35, 3 tabl.

Sirēn, Allan (1951). "On Computing the Land-Uplift from the Lake Water Level Records in Finland." *Fennia, 73,* (1951), 181 p., 37 fig.

——— (1964). "Recent Land Uplift in Finland Computed from Lake Water Level Records." *Fennia,* 89, (1964), 11–13, 3 fig.

Suess, Eduard (1888). "Die Meere der Erde. In *Das Antlitz der Erde,* Bd. II. Vienna, 1888. VI + 704 p., 42 fig., 3 pl.

Tanner, Väinö (1929). "Några Synspunkter angående de senkvartära nivåförändringarna i Fennoscandia." *Medd. f. Dansk Geol. För.*, 7 (1929), 356–362.

—— (1930). "Studier över kvartärsystement i Fennoscandias nordliga delar. IV. Om nivaförändringarna og grunddragen av den geografiska udvecklingen efter istiden i Ishavsfinland samt om homotaxin i Fennoscandias marina avlagringar." *Bull.Comm.géol.de Finl. 88* (1930), II + 589 p., 84 fig., 4 pl.

—— (1938). "Die Oberflächengestaltung Finnlands. Eine übersichtliche Darstellung der Morphographie und Morphologie sowie der Morphogenie in chronologischer Beziehung." *Bidrag till Kännedom af Finlands Natur och Folk,utg. af Finska Vet.So., 86* (1938), 762 p., 520 fig.

Vogt, Thorolf (1930). "Strand-lines as interference phenomena." *Det kgl. Norske Vid.-Selsk.Forh., 3* (1930), 40 p.

Wegmann, C. E. (1950). "Diskontinuität und Kontinuität in der Erdgeschichte." *Geol.Rundschau, 58* (1950), 125–132, 1 fig.

—— (1955). "Tectonique vivante." *Geol.Rundschau, 43* (1955), 273–306.

—— (1957). "Tectonique vivante, dénudation et phénomènes connexes." *Revue de Géogr.Phys. et de Géol.Dynam., 2* (1957), 3–15, 4 fig.

—— (1958). "Das Erbe Werner's und Hutton's. *Vol. Commem. Serge von Bubnoff, Geologie* (Berlin), 1958, Pp.531–559.

—— (1961). "Anatomie comparée des hypothèses sur les plissements de couverture (le Jura plissé). *Bull.geol.Inst.Univ.Uppsala, 40* (1961), 169–182, 2 fig.

—— (1967). "Evolution des idées sur le déplacement des lignes de rivage. Origines en Fennoscandie." *Mémoires de la Soc.Vaud. d.Sc.Nat., 14* (1967), 129–191, 10 fig., 2 pl.

Witting, Rolf (1918). "Hafsytan, geoidytan och landhöjningen utmed Baltiska Hafvet och vid Nordsjön." *Fennia, 39* (1918), 346 p., 25 fig., 2 pls.

—— (1928). "Ueber den Ablauf der fennoskandischen postglazialen Landhebung." *Fennia, 50* (1928), 1–17.

—— (1945). "Landhöjningen utmed Baltiska Havet under åren 1898–1927." *Fennia, 68* (1945), 1–40, 8 fig.

Early Geological Observations in the American Midwest

George W. White

Geological observations in America were made along the Atlantic seaboard by some of the first explorers: Verrazano in 1524, Thomas Hariot in 1588, and Captain John Smith in 1612 (White, 1953). There was almost no travel by the English across the Appalachian Mountains until about 1700, although one Englishman may have reached the Ohio River as early as 1669 (Banta, 1949, p. 49). In the first half of the eighteenth century, more and more travelers and traders moved beyond the folded rocks of the mountains into western New York and western Pennsylvania and on into the eastern margin of the "Ohio Country." In the second half of the century the whole central interior was traversed by observing travelers — army officers and others — some of whom recorded observations that provided geological information about this part of the continent. This paper will attempt to summarize the geological observations in English to 1787.

This paper does not deal with the many observations by the French when they controlled the northern and central parts of North America (e.g., coal near Joliet, Illinois, was described as early as 1673). Nevertheless, one publication should be specifically mentioned, that of Jean Etienne Guettard (1715–1786), who presented to the Royal Academy in Paris in 1752 a memoir on the minerals and rocks of North America that was published in 1756. Guettard, who was never in North America, compiled a map from French publications and through personal correspondence. He had access to many French maps, one as early as 1703 (Tucker, 1942, Pl. 13, 15, 17), that showed mineral locations in the upper Mississippi Valley. The Guettard map is not only the earliest geological map of this continent but one of the oldest geological maps of any continent (Wells, 1959).

The final defeat of the French in North America and the peace of Versailles in 1763 opened the "Western Country" to legitimate and authorized exploration, travel, and trade. For over a century this region had been either French territory or under *de facto* French control, and the English on the Atlantic seaboard knew of it only vaguely through reports by the French or from oral reports of a few daring English frontiersmen who made quick and clandestine trips as far as the lower Ohio Valley. Immediately after 1763, several colonial Americans, army officers who had fought the French and their Indian allies, set out on exploratory and mapping expeditions in the newly acquired vast territory of the Mississippi-Great Lakes basin. The best known of these, Robert Rogers, Jonathan Carver, and Thomas Hutchins, produced books in the next 15 years that informed English-Americans about the "Western Country" and included geological information. Thereafter followed several other books with further information. It is these books that provided knowledge of the geography and information about drainage, topography, and geological features. These, in turn, provided a basis for the accounts of more scientific travelers, who, beginning with Volney in 1803, published organized accounts of the geology of the trans-Allegheny region.

Lewis Evans

Although Lewis Evans (1700–1756) in his travels reached only the eastern part of the Mississippi basin, his great map of 1755 included territory as far west as Illinois, and he deserves an honored place in the earliest history of geology of the Middle West (White, 1951). Lewis Evans was the great colonial cartographer, surveyor, protogeologist, engraver, and pamphleteer (Gipson, 1939). His map of the British colonies is almost a geological map, for he showed on it mountain ridges, the Allegheny front, the glaciated part of northern Ohio — "level rich land intermix'd with swamps and ponds" — the location of deposits of petroleum, coal, clay, freestone, salt, and "elephant bones." The map has an inset portion for Illinois and Indiana.

Lewis Evans' *Map of the Middle British Colonies* (1755) extended as far west as the Mississippi River, and the accompanying text, *An Analysis of the General Map*, contains some descriptive material of the Mississippi Valley, most of which he had secured from others. His great map was reprinted with or without credit more than a score of times through a period of more than 40 years. This map, and its accompanying text, is one of the great treasures of early American geology. The *Analysis* has many pages of organized geological and geomorphic description and some theoretical speculations of high merit on fossils, isostasy, erosion, peneplains, and Coastal Plain origin. Evans himself did not have personal knowledge of the country west of Pennsylvania, but he effectively synthesized the more important and accurate observations of others.

Evans' classification of physiographic provinces of the United States is the first such attempt; it is accurate and in broad outlines it is followed today. Although usually unacknowledged, his physiographic classification and descriptions provided the basis for classification by many later authors. His work was important and widely influential, and had he lived longer he might well have held a position in America comparable to that of the Englishman, William Smith, who is considered the "father of stratigraphy."

Peter Kalm

Although Peter Kalm (1716–1779), who traveled in the United States from 1749 to 1751, only touched the edge of the Mississippi Valley, he did get beyond the mountains into the flat-lying Paleozoic strata, and he knew that a very extensive plains region of horizontal beds extended far to the west. His description of Niagara Falls, published in 1751, (in Bartram, 1751) was a very early one in English.

Kalm's *Travels*, published in Swedish in 1753, was published in English in 1770–1771. His book contains more than 150 excellent geological observations and includes perceptive speculations about the origins of rocks and their structures. Although his remarks are almost entirely on the eastern part of the continent, he deserves at least mention here. His descriptions, not available in English until 1771, were widely quoted in Europe and in America were the source of information to many writers beginning with the geographer Morse (1789).

Major Robert Rogers

Major Robert Rogers (1731–1795) of Rogers' Rangers fame, was born in New England and early achieved brilliant success in the Seven Years' War and in the campaign against Quebec. Immediately afterward, he was sent to command the post at Detroit, from which he traveled even farther northwest. He soon went to England, where his book, *A Concise Account of North America*, was published in 1765. He returned to the Great Lakes region in 1766 and was commander at Mackinac when Jonathan Carver set out on his trip to the northwest. Rogers returned to England, where in London his career ranged from "checkered" to distinctly unhappy. The unfortunate circumstances of his later life are in sharp contrast to the brilliant military, geographic, and literary successes of his earlier career. Rogers' geological references are not collected in one section of his book, but are mentioned in connection with other matters. Rogers knew about the oil seeps in Oil Creek, Pennsylvania, referring to them as follows:

> . . . those remarkable springs, greatly esteemed by the Indians as a remedy for almost every disease; they called the oil-springs, on account of an oleous substance that issues forth with the water, and rises upon the surface of it. The Indians use these springs for consumptions, asthmas, and various internal disorders, by drinking the water; and for rheumatick pains, strains, dislocations, &c. by bathing, with great success. (p. 177)

Rogers (pp. 199–200) speaks of the territory that we now know as Middle West in a most favorable way — too optimistically, indeed, of the prospect of a great city at the mouth of the Ohio River; he could not know in 1765 that it would be the railroad nexus at the south end of Lake Michigan — Chicago — and not Cairo at the junctions of the rivers that would be one of the world's major cities.

Rogers' *Concise account* was the first book in English to give information from personal knowledge of the western country. It is particularly important because, scanty as the geological information was, it did provide a basis for elaboration by later writers. His book has always been highly prized as an item of early Americana, but it also deserves a place in the very early dawn stage of history of geology in the Mississippi basin.

Jonathan Carver

Jonathan Carver was born in 1732 in Connecticut. Becoming an Army officer at the age of 18, he fought in the Indian wars and was a Captain at the conclusion of the French and Indian War in 1763. He then set out on a long series of travels through the western and northwestern parts of the British dominions, reaching as far as the headwaters of the Mississippi in northern Minnesota in 1766 to 1768. In 1769 he went to England — as he says, "to communicate the discoveries I had made, and to render them beneficial to the kingdom." He was bitterly and justly disappointed by the lack of recognition of his contributions to the knowledge of the far frontier and of the Indian tribes with whom he had lived. He assembled his notes and maps for publication in 1778 as *Travels Through the Interior Parts of North America*. Although the book was successful, passing through three editions by 1781, Carver derived little profit from it, and the privations he endured (far worse in London than in the wilderness!) led to his early death in 1780. It is sad to realize that the present price of one copy of any of the editions of his book would have relieved for a long period the extreme hardships of his family in England and would have prevented Captain Carver's own death from actual destitution.

Carver's book describes the topography of the Great Lakes region, the Wisconsin River Valley, and the upper Mississippi Valley. South of the Wisconsin River he saw the region of the driftless area (pp. 47–48). Carver was one of the first of the English to record the changes in the level of the Great Lakes that had long been known to the French. His observations particularly referred to Mackinac, where he had stayed at the fort, which at that time was under the command of Major Robert Rogers. No doubt the two men discussed this phenomenon:

> I had like to have omitted a very extraordinary circumstance relative to these Straights. Acording to observations made by the French, whilst they were in possession of the fort, although there is no diurnal flood or ebb to be perceived in these waters, yet, from an exact attention to their state, a periodical alteration in them has been discovered. It was observed that they arose by gradual, but almost imperceptible degrees till they had reached the height of about three feet. This was accomplished in seven years and a half; and in the same space they as gently decreased, till they had reached their former situation; so that in fifteen years they had completed this inexplicable revolution. At the time I was there the truth of these observations could not be confirmed by the English, as they had then been only a few years in possession of the fort; but they all agreed that some alteration in the limits of the Straights was apparent. All these lakes are so affected by the winds, as sometimes to have the appearance of a tide, according as they happen to blow; but this is only temporary and partial (pp. 146–147).

Carver described the features of various lakes, particularly those of Lake Erie, which discharges through the Niagara River into Lake Ontario. His description of Niagara Falls is disappointingly brief, because he says it has been "so frequently described that I shall omit giving a particular description of them." (p. 69).

Carver's book, in the various editions, was widely used and quoted by those who came after him. It is useful for the knowledge of land routes and water routes for travel, and some nuggets of geology may be mined from it by close attention to references primarily about other subjects.

Thomas Hutchins

The little book, *A Topographic Description*, by Thomas Hutchins (1730–1789) was written in 1778, but it was based on personal knowledge he had gained during the preceding 25 or more years. The author was a military engineer, surveyor, and cartographer who had traveled as far as the Illinois country. His book was a source of information for many years and was often reprinted, with and without acknowledgment. Hutchins was not only a traveler in the central interior, but had also been a resident of the region in his younger days. He was the first to map accurately the Ohio River in 1766. His field notes on the courses of the river included many comments on the materials of the banks and channel and on the rocks along the shore (Bond, 1942). He accompanied Col. Bouquet's march against the Indians into central Ohio in 1764 and made a very accurate map of the route (Williams, 1960). Earlier, in 1762, he had traveled across Ohio to Sandusky and thence to Detroit, Mackinac, and Green Bay, returning across Illinois to the Wabash Valley and then to Portsmouth, Ohio (Bond, 1942, p. 14). He thus had the most extensive personal knowledge of anyone to that time when he wrote his book of 1778.

Hutchins' book contains many geological observations. He noted coal, salt springs, iron ore, and lead deposits along the Ohio and Scioto Rivers, and the great vertebrate fossils at "Big Bone Springs," Kentucky (pp. 10–11).

Hutchins' reference to a silver mine near the present location of Lafayette, Indiana, was repeated by other writers for many years. Hutchins was not referring to an actual opening from which ore was produced, but rather to a "deposit" and perhaps only to a single specimen — the word "mine" having had this meaning in earlier times. By the time he wrote, this use of the word was already obsolete. Perhaps the material he saw was marcasite, the light-colored iron sulfide. At any rate, Hutchins was quite correct, as he usually was, about the salt springs, coal, clay, and limestone in the Wabash Valley (p. 28).

Hutchins' intimate knowledge of the Ohio River and the details of the Mississippi upstream from the Ohio stemmed from his careful surveys, which he recorded in a beautifully detailed map of parts of these valleys. Whereas Rogers, the line officer, thought the location of the junction of the Ohio and Mississippi rivers would be the site of an important city, Hutchins, the engineer, readily recognized that the location was subject to severe floodings, and he knew, of course, that it would be unsuitable for a major city. Hutchins' description of the valleys, the origin of the flood plain material, and the rock walls of the valleys are those of an engineer with a keen eye for topography (pp. 33–36).

The large map accompanying the report is significant geologically because it shows "petroleum" at several places and has many other notes on coal, clay, freestone, and topography. It shows a one-half-mile outcrop ("mine") of coal on the Illinois River near the mouth of the Vermillion River. The Mississippi Valley part of the map, precisely the part of our present interest, is beautifully reproduced as Plate 29 in a recent atlas of the Illinois State Museum (Tucker, 1942).

The history of geology in the Mississippi Valley really begins with Thomas Hutchins. He was one of the earliest to observe the difference between the inclined strata of the Appalachian Mountains and the flat-lying beds that extend far to the west of the mountains. He had a keen eye for the kind of topography in the various parts of the country and the differences from place to place. He realized the country was soon to be settled and pointed out the location of the important mineral resources that would be needed by the settlers — salt, clay, fuel, and stone. His careful descriptions were to be widely quoted for several succeeding decades.

Johann David Schöpf

Another observer whose work dealt almost entirely with the eastern part of the United States, but who made some mention of the "western territory," was Dr. Johann Schöpf (1752–1800). During the Revolutionary War Schöpf, a scientist, scholar, and academician, was attached as a physician to the Hessian troops. Immediately after the war he set out on travels that took him through the eastern and southern part of the country and as far west as Pittsburgh, which brought him into the Mississippi Valley. Besides being a keen and competent observer, he was a lucid, compact writer, and in the years following his travels he published several books and many articles about the scientific features of America.

Schöpf's book on geology and minerals, *Beyträge zur Mineralogischen Kenntniss des Ostlichen Theils von Nordamerika und Seiner Gebürge* (1787), never translated into English, is actually the first book on the geology of the United States. It was followed the next year by a two-volume book on his travels in the United States (1788), and the two books form almost a single entity. The *Reise*, translated in 1911 as *Travels in the Confederation*, gives a day-to-day account of what he saw, and the *Beyträge* gives an organized summary of the geology of his reports.

Schöpf was very well acquainted with the European literature on mineralogy and with the American reports of Kalm, Catesby, Evans, Pownall, and others. From his own observations he confirmed Evans' division of the country into Coastal Plain, Piedmont, and the Mountains. He described the attitude and bedding of the strata at various places, and the relations of sand, clay, and fossiliferous limestone. He described and named various fossils. Although he gave most of his attention to the eastern part of the country, with which he was most familiar, he did describe the "western country" from his own knowledge as far as Pittsburgh and the upper Ohio Valley. His description of the coal and associated shale and limestone at Pittsburgh is accurate and is probably the earliest published description by a scientist of this, the world's greatest coal bed.

The concluding theoretical part of Schöpf's book is notably perceptive in its discussion of sedimentation in ancient seas, with their falling and rising levels, and of the very long time necessary for the extensive erosion of mountains and plateaus.

Conclusion

With Schöpf we come to the first book that contains an organized description of the geology of the United States and thus arrive at the end of the first chapter in the history of geological observations in the central interior. There were to be other incidental descriptions of the geology, some of considerable importance, such as those of Thomas Jefferson (1743–1806), who included some material on the central interior in his *Notes on the State of Virginia* (1788) and who is well known for his collection of vertebrate fossils (Martin, 1952). The book by Gilbert Imlay (1754–1828), *Topographical Description of the Western Territory*, first published in 1792, has some important geological notes. His contribution will be discussed in later papers.

At the beginning of the nineteenth century the important work of Contantin François Chasseboeuf, Comte de Volney, originally written in French in 1803 and translated in 1804 by Charles B. Brown as *View of the Climate and Soil of the United States*, contains important material on geology, geography, and climatology. His association with Thomas Jefferson has been discussed by Chinard (1923). He derived some of his geological information from the work of Mitchell (1798), Hutchins, Evans, Imlay, Morse (1789), and Alexander Mackenzie (1802). Volney's geological contributions have been discussed in a recent publication (White, 1968).

After 1787, especially after 1803, the history of geology in the "western country" becomes more and more that of organized scientific description and analysis, in contrast to the "observations" in the earlier works. A period of more extensive and lengthy reports extends from 1803 to about 1838, when the first state survey reports of some of the Mississippi Valley states appeared and ushered in the "modern" era of geological investigation and writing. These later men did not enter an unknown land — the pioneer observers had provided a basis of general knowledge for those who came after them.

References

Banta, R. E. (1949). *The Ohio.* New York: Rinehart and Company, Inc., 1949. 542 p.

Bartram, John (1751). *Observations on the inhabitants, climate, soil, rivers, productions, animals, and other matters worthy of notice. Made by Mr. John Bartram, in his travels from Pensilvania to Onandago. . . . To which is annex'd a curious account of the cataracts at Niagara by Mr. Peter Kalm, a Swedish gentleman who travelled there.* London: J. Whiston and B. White, 1751, 94 p., 1 pl.

Bond, B. W. (Ed.) (1942). *The courses of the Ohio River taken by Lt. T. Hutchins Anno 1776 and two accompanying Maps.* Cincinnati: Historical and Philosophical Society of Ohio, 1942. 85 p., 11 figs., 3 maps.

Carver, John (1778). *Travels through the interior parts of North America, in the years 1766, 1767, and 1768.* London: J. Walter, 1778. 544 p., 4 pl., 2 maps. (Many later editions.)

Chinard, Gilbert (1923). *Volney et l'Amerique d'apres des documents in-edits et sa correspondence avec Jefferson.* (Johns Hopkins Studies in Romance Literatures and Languages, Vol. 1) Baltimore: The Johns Hopkins Press, 207 p., port.

Evans, Lewis (1755). *Geographical, historical, political, philosophical and mechanical essays. The first, containing an analysis of a general map of the middle British colonies in America. . . .* Philadelphia: B. Franklin and D. Hall, 1755. 32 p. The large map was published separately. (Facsimile of text and map in Gipson, 1939.)

Gipson, L. H. (1939). *Lewis Evans, to which is added . . . facsimile of his . . . Analysis of a General Map . . . Also facsimiles of Evans' Maps.* Philadelphia: Historical Society of Pennsylvania, 1939. 219 p., 6 maps.

Guettard, Jean-Etienne (1752). "Memoire dans lequel on compare le Canada a la Suisse, par rapport a ses mineraux." *Histoire de L'Academie Royale des Sciences.* Paris: 1752. Pp. 189–220, 323–360, 524–538.

Howes, Wright (1962). *U. S. iana (1650–1950). A selective Bibliography.* New York: R. R. Bowker Company for the Newberry Library, 1962. 652 p.

Hutchins, Thomas (1778). *A topographical description of Virginia, Pennsylvania, Maryland, and North Carolina. . . .* London: J. Almon. ii + 67 p., 2 pl. (The map was issued separately.)

Imlay, Gilbert (1792) *A topographical description of the western territory of North America. . . .* London: J. Debrett, 1792. xv + 247 p. (Third edition, 1797. xii + 598 + xxvi p., 4 maps.)

Jefferson, Thomas (1788). *Notes on the State of Virginia.* Philadelphia: Prichard and Hall, 1788. 244 p. (First French edition, 1786; first London edition, 1787.)

Kalm, Peter (1770–1771). *Travels into North America.* Trans. J. R. Forster. Warrington-London: 1770–1771. 3 vol., 24, 400, 352; 8, 310, 14, map, 6 pl., (Another ed., 1772; 1938. Original Swedish 1753–1761; German and Dutch editions, partial French editions.)

Mackenzie, Alexander (1802). *Voyages from Montreal, on the River St. Laurence, to the Frozen and Pacific oceans; in the years 1789 and 1793. . . . With original notes by Bougainville, and Volney. . . .* London: T. Cadell, etc., 1802. 2 vols., xiv + 290; 332 p. (The first edition of 1801 did not have the "Notes.")

Martin, E. T. (1952). *Thomas Jefferson; Scientist.* New York: Henry Schuman, 1952. x + 289 p., 15 pl.

Merrill, G. P. (1924). *The first one hundred years of American geology.* New Haven: Yale University Press, 1924. xxi + 773 p., 130 figs., map; New York: Hafner, 1964.

Mitchill, S. L. (1798). "A sketch of the mineralogical and geological history of the State of New York. . . ." *Medical Repository, 1* (1798), 293–314, 445–452; *3* (1800), 325–335; *5* (1801), 212–214.

Morse, Jedidiah (1789). *The American geography; or, a view of the present situation of the United States of America. . . .* Elizabethtown: Shepard Kollock, 1789. 534 p., 2 maps. (Also many later editions.)

Rogers, Robert (1765). *A concise account of North America: . . . situation, extent, climate, soil . . . also of the interior or westerly parts of the country. . . .* London: J. Millan. 1765. vii + 264 p.

Schöpf, Johann David (1787). *Beyträge zur mineralogischen Kenntniss des Ostlichen Theils von Nordamerika und seiner Gebürge.* Erlangen: J. J. Palm, 1787. 12, 194, 2 p.

——— (1788). *Reise durch einige der mittlern und südlichen Vereinigten Nordamerikanischen Staaten. . . . 1783 und 1784.* Erlangen: J. J. Palm, 1788. 2 vol., 24, 644, 2, 40, 352 p., map.

——— (1911). *Travels in the Confederation, 1783–1784.* Translated and edited by A. J. Morrison. Philadelphia: William J. Campbell, 1911. 2 vol. x + 426, 344 p.

Tucker, S. J. (1942). *Indian villages of the Illinois Country,* Part I. Springfield, Ill.: Illinois State Museum, 1942. xiii + 18 p., 54 pl.

Volney, C. F. (1803). *Tableau du climat et du sol des Etats Unis d'Amérique. Suivi d'éclaircissemens sur la Floride, sur la colonie Française au Scioto, sur quelques colonies Canadiennes et sur les Sauvages. Enrichi de quatre planches gravées, dont deux cartes géographiques et une coupe figurée de la chute de Niagara.* Paris: Courcier, Dentu, 1803. 2 vol. xvi + 300, 532 p., 2 folding maps, 2 folding pl. A second state has vol. 2, 524, 8, 4 p. Large paper copies ("quarto"), 2 vol. in 1, are known. There are also later editions, including English, German, and Italian translations.

——— (1804). *View of the climate and soil of the United States of America: to which are annexed some accounts of Florida, the French colony on the Scioto, certain Canadian colonies, and the Savages or Natives: translated from the French of C. F. Volney.* London: J. Johnson, 1804. xxiv + vi + 503 p., 2 folding maps and 2 folding p.

——— (1804). *A view of the soil and climate of the United States of America . . . with supplementary remarks upon Florida; on the French Colonies, on the Mississippi and Ohio, and in Canada; and on the aboriginal tribes of America. Translated with occasional remarks by C. B. Brown.* Philadelphia: J. Conrad and Co., 1804. xxvii + 446 p., 2 maps, 2 pl. Facsimile reprint with introduction by G. W. White, New York: Hafner, 1968.

——— (1787). *Voyage en Syrie et en Egypte, pendent les années 1783, 1784, et 1785. . . .* Paris: Desenne, 1787. 2 vols. English translation 1787. (Many later editions.)

Wells, J. W. (1959). "Notes on the earliest geological maps of the United States, 1756–1832." *Jour. Washington Academy of Sciences, 49* (1959), 198–204.

White, G. W. (1951). "Lewis Evans' contributions to early American geology — 1743–1755." *Ill. Acad. Science Trans.*, *44* (1951), 152–158.

White, G. W. (1953). "Early American Geology. *Scientific Monthly*, *76* (1953), 134–141.

——— Editor's Introduction, in facsimile edition of Volney, 1968, q.v., pp. v–xvii.

——— and Slanker, B. O. (1962). *Early geology in the Mississippi Valley: an exhibition of selected works held in the University of Illinois Library, November, 1962.* Urbana, Ill.: University of Illinois Library 1962. 26 p.

Williams E. G. (Ed.) (1960). *The orderly book of Colonel Henry Bouquet's Expedition against the Indians. 1764.* Pittsburgh: Mayer Press, 1960. 74 p., maps.

The Intellectual Background to Charles Lyell's *Principles of Geology*, 1830–1833

Leonard G. Wilson

The *Principles of Geology* changed the thought of its time about the history of the earth, or, to be more precise, it radically altered the outlook of a small number of young men who, with newly opened eyes, proceeded over the next few decades to wreak a revolution in science. Charles Darwin and Charles Bunbury read its three successive volumes almost as they emerged; Herbert Spencer and Alfred Russel Wallace read it some years later, probably in a later edition.

What Charles Lyell offered in the *Principles* was a point of view, a philosophical outlook for the consideration of geological questions. This outlook required that the geologist assume two things: first, that the information needed for interpreting the past history of the earth was available in the rocks and sediments distributed over its surface; second, that conditions during the past were comparable with those of the present, that there was a uniformity of physical laws and geological processes between the past and the present. The extraordinary quality of Lyell's point of view is suggested by the fact that his book was received with shock and indignation by his contemporaries; also, except for a small group of younger men, it was steadily resisted by the geological profession over succeeding decades. In 1865 it appeared to receive its death blow from a frontal assault by Lord Kelvin who insisted that the history of the earth must have been short, violent, and catastrophic.[1] Kelvin deemed mere geological evidence of little value and to a surprising degree geologists were prepared to acquiesce in his judgment. The geologists might encounter in the heart of North

[1] Sir William Thomson (later Lord Kelvin), "The 'doctrine of uniformity' in geology briefly refuted," *Proc. Roy. Soc. Edinburgh* 5 (1866), 512–513.

America thousands of feet of sediments laid down in tranquil seas of vast extent, and this, the debris of whole continents long since vanished to fill the beds of ancient oceans, which themselves were now high and wind-blown plains marked by the winding wrinkles of river valley systems outspread over thousands of miles, this geological evidence of things past, so suggestive of very gradual change occurring slowly and imperceptibly through immense periods of time, was peremptorily dismissed by the physicists.[2]

The change began around 1903, when Ernest Rutherford showed that the discovery of radioactivity provided a slow, steady, and unceasing source of heat within the earth.[3] Kelvin's calculations of the age of the earth had been based on the assumption that it had originally been in an intensely hot, molten state and had gradually cooled by black body radiation. He estimated the rate and time of cooling from the present internal temperature of the earth, calculated from the average increase of temperature measured at descending levels in mines. Since he knew of no source of heat within the earth except that which was stored there, a diminishing residue from the past, he thought he could look both backward to a time when the earth had been an incandescent molten body and forward to a time when, as a consequence of relentless cooling, its surface would be utterly dead and frozen. Lord Kelvin had acknowledged that the basis for his calculations would be removed if a continuing source of heat should be found within the earth, and Rutherford showed in 1903 that such a source had now been found.

The damage done to geology during the period from 1865 to 1903 was not to be so quickly remedied. A whole generation of geologists had grown up uncertain of their principles in theoretical geology. Lyell had been discredited, at least so far as the rate and intensity of geological causes were concerned, and geologists felt free to invoke catastrophic explanations of past events, particularly to account for certain phenomena of mountain building. In many respects the situation in geology at the beginning of the twentieth century was similar to that in biology. In biology the fact of the occurrence of evolution had been accepted, but Darwin's theory of natural selection to explain the process of evolution had been cast into serious doubt. This was due in part to Lord Kelvin, in part to the unremitting antagonism to Dar-

[2] At the end of the century an effective protest was made by one distinguished American geologist. See Thomas C. Chamberlin, "Lord Kelvin's address on the age of the earth as an abode fitted for life," *Science*, *9* (1899), 889–901, *10* (1900), 11–18.
[3] Ernest Rutherford, "Radioactive Change," *Phil. Mag.*, *4* (1903), 370–396.

Plate IV

RODERICK IMPEY MURCHISON (1792–1871) & ARCHIBALD GEIKIE (1832–1924)

"*Deux géologues rapportant à Mrs. Ellice, un petit echantillon des Montagnes moutonnées découvertes en 1860, entre Glenquoich et Lochbourne.*" In August 1860, the young Archibald Geikie was assisting Murchison in mapping parts of the Highlands. They stopped for a few days at Glen Quoich (Invernessshire), the Highland summer home of the magnate Edward Ellice. Among the guests was Prosper Merimée, on his first (and last) visit to the Highlands. Geikie probably never saw this caricature by Merimée [A. L. Packe, "A Highland Visitor's Book, *Country Life*, *107* (1950), p. 370, Fig. 2] and refers to him in his autobiography with unconscious humor:

> One of the company was Prosper Merimée, whose chief occupation out-of-doors seemed to be sketching the scenery in water colour on course brown paper. He showed me some of his drawings, which filled me with surprise that so accomplished a man of letters should spend his time in the production of such crude and inartistic efforts to represent the noble scenery around him. [Geikie, *A Long Life's Work* (London, 1924), p. 79.]

Deux géologues rapportant à Mrs Ellice, un petit échantillon des Montagnes moutonnées
découvertes en 1860, entre, Glenquoich et Lochhourne
[en gaëlique : Sbhhgdh]

win's theory on religious grounds, and in part to the fact that the newly discovered laws of Mendelian genetics did not have an immediately obvious relationship to the working of natural selection. By 1924, however, knowledge of population genetics had developed to a point that allowed J. B. S. Haldane to explore the effects of natural selection in populations and, by 1930, for R. A. Fisher to present his *Genetical Theory of Natural Selection*, fully articulated in mathematical terms.[4] Since 1930 the cumulative work of geneticists, ecologists, and evolutionists has brought about a steadily increasing acceptance of Darwin's theory of natural selection, and, if anything, Darwin is reproached for having had insufficient confidence in the power and validity of his own concept in the years after 1859.

In geology, the twentieth century has seen the frequent extension of the time span of the earth's history to the current estimates of 3,000 to 4,000 million years. Methods of radioactive dating have enabled geologists to make more exact estimates of the ages of mountain ranges and have shown that the processes of mountain building have been distributed over long periods of time and therefore have occurred gradually. In general, the enormously increased age of the earth has made available the time needed for the slow and gradual working of geological processes and for the evolution of complex and highly adapted living forms by natural selection. Yet the influence of Lord Kelvin on geological thought over the past century remains profound and will not easily be erased. One of its consequences has been to obscure the historical significance of Charles Lyell and his concept of the uniformity of nature.

In 1830, when Lyell published his *Principles*, the difference between his position and that of his critics was very clear. Lyell presented a panorama of geological processes, similar to those active today on the surface of the earth, which had also been active through an indefinitely long period of past time. He suggested that conditions on the earth's surface were stable and enduring. Changes might occur, but they did so in ways comparable to the changes and variations now to be seen on the earth's surface. The history of the earth was long and, in effect, it was eternal. By contrast, Lyell's critics saw the history of the earth as relatively brief. They were influenced both by the account of creation given in the book of Genesis and by the knowledge of past forms of life which had already, by 1830, accumulated in significant amount. They considered the living world as having been created by

[4] R. A. Fisher, *The Genetical Theory of Natural Selection* (Oxford: Clarendon Press, 1930), 272 p.

God, and the discovery of a succession of fossil faunas suggested a succession of earlier creations that preceded the present one. In the English strata geologists had discovered the remains of a number of gigantic fossil reptiles, together with the remains of other plants and animals, facts which required the existence of a tropical climate in England during the geological past. Thus, they saw fossil remains as the remains of former living worlds that had passed away and that bore little resemblance to the conditions of the modern world. Since these former worlds had come to an end and all their species had been destroyed, it was natural to assume that such former worlds had been wrecked by some great catastrophe that had shaken the earth to its foundation.

There were, however, additional reasons for believing in the occurrence of large and immeasurable catastrophes during the earth's history. At the end of the seventeenth century, naturalists had generally admitted that the fossils found in stratified rocks were the actual remains of previously living animals, and the admission of this fact had required them to acknowledge that the strata had been laid down under the sea. The fundamental geological question, then, was how strata of marine origin had come to form elevated land. This was the great question that concerned geologists during the eighteenth century when they attempted to formulate an account or an explanation of the history of the earth. Basically, there were two possible interpretations: first, that the sea had formerly stood at a higher level than it now did, and had since receded, or, second, that the land had been upraised. Abraham Gottlob Werner had attempted to give an account of the history of the earth on the assumption that the sea had formerly stood at a higher level and had since receded. James Hutton, on the other hand, who insisted with Werner that stratified rocks had been laid down under the sea, believed that the strata had been uplifted rather than that the sea had receded.

For those geologists who followed Hutton in postulating that the strata had been uplifted, it was almost necessary that they accept the occurrence of catastrophes during geological history, for they found the strata not only elevated to form hills and mountains but also in inclined, even vertical, positions, and frequently bent and distorted in an extravagant way. For instance, the English geologist Thomas Webster, who went in 1811 on a geological excursion to the Isle of Wight, found that the central ridge of chalk hills that runs in an east-west direction through the Isle of Wight was made up of vertical, or nearly vertical, chalk strata that were magnificently exposed in Alum Bay at the western end of the island.[5] On the southern margin of the Isle of Wight, Webster found chalk strata dipping more gently into the sea. From these facts and from the general arrangement of the strata lying beneath the chalk, Webster decided that the Isle of Wight had been formed by a great fold of the strata (what we should now call an anticlinal fold) and that the vertical chalk strata of the central hills had once formed a continuous arc with the gently inclined strata of the southern shore. An immense mass of sediment must since have been eroded away. In the elevation of the chalk strata of Alum Bay to a vertical position, Webster saw the action of an immense force that had here raised and violently folded the crust of the earth in a manner that certainly appeared to have been catastrophic. In 1814 Webster also showed that the ridges of the North and South Downs of southeastern England were also the remnants of an anticlinal fold, and he suggested that the chalk strata of the Isle of Wight were continuous with those of the South Downs, sinking beneath the Solent and rising again in their vertical position in the central hills of the Isle of Wight.[6] This great undulation of the chalk thus formed a basin that Webster named the Hampshire Basin, which he considered comparable with the Paris Basin. Similarly, he found that the chalk of the North Downs dipped northward and sank beneath the valley of the Thames to appear again to the north in the Chiltern hills, thereby forming the London Basin.

In 1822 Conybeare and Phillips, in their *Outlines of the Geology of England and Wales*, similarly attributed the displacement of strata and their elevation into inclined, vertical, or bent positions to "violent convulsions."[7] The acceptance of the occurrence of violent convulsions during the past history of the earth appears, therefore, to have been a consequence of the acceptance of the Huttonian theory of the earth. The Reverend W. D. Conybeare considered himself to be more an advocate of the Huttonian position than of the Wernerian. He did not accept, for instance, that the ocean had formerly stood at a higher level and had subsequently, in various stages, receded. Similarly, when

[5] Webster's description of the Isle of Wight is found in his letters to Sir Henry Englefield, published in Sir Henry Englefield, *A Description of the Principal Picturesque Beauties, Antiquities and Geological Phenomena of the Isle of Wight. With additional observations on the Strata of the Island and their continuation in the adjacent parts of Dorsetshire*, by Thomas Webster, Esq. (London: Payne and Foss, 1816). xxvii + 238 p., 49 plates.

[6] Thomas Webster, "On the Fresh-water Formations in the Isle of Wight, with some observations on the Strata over the Chalk in the South-east of England," *Trans. Geol. Soc. Lond.*, *2* (1814), 161–254.

[7] Rev. W. D. Conybeare and William Phillips, *Outlines of the Geology of England and Wales, with an introductory compendium of the general principles of that science, and comparative views of the structure of foreign countries* (London: William Phillips, 1822). lxi + 470 p.; p. xv.

Charles Lyell began to interest himself in geology about 1816, the geology that he learned tended to be both Huttonian and catastrophic.

The acceptance of a catastrophic view of the history of the earth was greatly strengthened by the French anatomist Georges Cuvier. In his studies of the vertebrate fossils of the neighborhood of Paris, Cuvier had distinguished essentially two fossil faunas, the most recent of which occurred in the gravel and alluvium of river valleys and was represented by the fossil rhinoceros, hippopotamus, tapir, elephant, and mastodon. The fossil species were distinct from living ones, but they all belonged to genera with living representatives and all were mammals of the family Pachydermata, although the living pachyderms occurred only in tropical countries. In the gypsum quarries of Montmartre, Cuvier found a quite different set of animals, including a number of fossil pachyderms, but these fossils were so different from any living forms that he was obliged to create two new genera. Anoplotherium and Palaeotherium, to contain them. The significant point seemed to be that the fauna represented by the fossils of Montmartre was quite distinct from the fauna of the more recent gravel and alluvium of the river valleys and that there was a sharp gap between the two successive groups of animals rather than a continuity. This fact suggested that the life of the first fauna had been terminated by some general convulsion of the earth's surface. Similarly, when Cuvier and his collaborator Alexandre Brongniart studied the succession of strata of the Paris Basin, they found an alternation of fresh water and marine sediments. This alternation suggested to Cuvier the occurrence of repeated incursions of the sea into the Paris Basin, incursions that he interpreted to mean a series of catastrophic elevations of the general sea level.

Charles Lyell's systematic study of geology began when he attended William Buckland's course of lectures at Oxford in 1817 and again in 1818. Just two years earlier, in 1815, William Smith's geological map of England and Wales had been published at London, and in 1817 his *Stratigraphical System of Organized Fossils* also appeared. Buckland knew Smith personally and had learned from him the fossils characteristic of the different strata of the west of England. In 1812, the year that also saw the publication of Cuvier's *Recherches sur les ossemens fossiles*, the remains of an enormous fossil reptile were discovered in the blue Lias formation at Lyme Regis on the coast of Dorsetshire, and this discovery demonstrated that the rocks of England also contained the remains of enormous and bizarre extinct animals. These recent discoveries, combined with Buckland's warmth and enthusiasm as a lecturer, made geology an exciting pursuit for Oxford undergraduates and, quite understandably, for Charles Lyell.

During the summer of 1818, Lyell traveled with his family on the continent and visited the Alps and northern Italy. He wandered over glaciers and observed the effects of avalanches and mountain torrents. After his graduation from Oxford in 1819, Lyell began to study law at Lincoln's Inn, but he had recurrent trouble with his eyesight and his studies were frequently broken to take long holidays. In the summer of 1822, during a horseback ride through Sussex, he visited Dr. Gideon Mantell, a country surgeon at Lewes. Just a few months earlier, Mantell had found a remarkable group of fossils in a quarry in the Tilgate forest. They included bones of tortoises, crocodiles, and birds, pieces of carbonized wood and other remains of plants, including tree ferns and large leaves, and casts of shells. The beds consisted of clays and sands such as might have been deposited in the delta of a river. The feature of Mantell's discovery that struck Lyell very forcibly was that these beds had been laid down in fresh water. They represented a river valley or river delta; that is, they represented a portion of an ancient land surface. Yet the formation in which they occurred lay many layers of sediment beneath the chalk formation, which itself had been laid down in the deep sea. Thus, since the time when the Tilgate forest beds had been deposited, the land surface had sunk through many thousands of feet and there had been deposited over it a great depth of sediment, including the full depth of the chalk formation. At a later time, this immense depth of sediments had been elevated from the bottom of a deep sea to form the present land of England, and the overlying chalk and greensand strata had been eroded away to lay bare this former, most ancient land surface. To Lyell, it was a convincing demonstration of the magnitude of the earth's movements that had occurred in the southeast of England. It was also proof that the Wernerian theory could not be true because no conceivable changes in level of the sea could account for the succession of sediments existing between the Tilgate forest beds and the chalk formation. Also of deep influence of Lyell's mind was the fact that in Mantell's collection of fossils he was able to visualize a community of animals and plants, in fact, a complete physical environment comparable with what might exist in a modern river delta or estuary such as that of the Ganges in India.

In the summer of 1823, Lyell went to Paris to spend several months. There he met all the leading naturalists and scientists, including Cuvier, Brongniart, Humboldt, and many others. Perhaps the French scientist whose influence upon him was most significant was the geologist Constant Prevost. In their study of the Tertiary sediments of the Paris Basin, Cuvier and Brongniart had stressed the alternation of marine and freshwater sediments. Prevost showed Lyell that this alternation was not always exact. Sometimes a bed that was marine in one part of the Paris Basin proved to be freshwater in another portion. Prevost suggested that the Paris Basin had been a great inlet of the sea that may have been more or less salt according to the degree to which it was connected with the open ocean. Its water might be more salt toward the sea and more or less brackish, or fresh, where rivers flowed into the basin. Relatively small land changes might convert the Paris Basin, therefore, from an arm of the sea into a freshwater lake, and vice versa. Prevost's study of the Paris Basin thus showed that these sediments need not have the catastrophic implications given to them by Cuvier and Brongniart. His influence tended to strengthen Lyell's desire to interpret each sediment as deposited in a particular kind of physical environment.

During the following summer of 1824, Lyell visited his family's estate of Kinnordy in Scotland and while there studied the freshwater marl deposits occurring in some small lakes in the vicinity. Lyell found that these marl deposits sometimes graded into a hard, compact limestone having embedded in it the seed pods and stems of the green alga Chara. This limestone was exactly similar to some of the ancient freshwater limestones occurring in the Paris Basin formations for which Cuvier and Brongniart had thought there existed no modern counterparts. In this discovery, therefore, Lyell found a series of deposits and a physical environment that represented a counterpart of the conditions under which the freshwater strata of the Paris Basin had been formed.

In 1819 Lyell had been elected a Fellow of both the Geological Society of London and the Linnean Society. In 1823 he also accepted the post of one of the secretaries of the Geological Society and in that capacity helped to prepare papers for publication in the *Transactions of the Geological Society*. The *Transactions* had been allowed to lapse since 1820, but in 1824 the first volume of the second series appeared. By serving, in effect, as one of its editors, Lyell was made thoroughly acquainted with its contents and, therefore, with the geological discoveries included in it. These discoveries included William Buckland's discovery of the megalosaurus at Stonesfield and W. D. Conybeare's discovery of the plesiosaurus. Of the plesiosaurus, Lyell wrote to Mantell:

> The new animal is a very perfect skeleton, and a prodigy, for it has forty cervical vertebrae, whereas existing quadrupeds range from seven to nine, reptiles from three to nine. Aves reach no higher than twenty, the swan being the maximum. What a leap have we here, and how many links in the chain will geology have to supply.[8]

[8] C. Lyell to G. Mantell—February 17, 1824, in Katherine M. Lyell, *Life, Letters and Journals of Sir Charles Lyell Bart* (2 vols; London; John Murray 1881), Vol. I, p. 151.

Thus, Lyell saw in the discovery of these remarkable fossils a deeper significance for geology.

In 1826 Lyell wrote for the *Quarterly Review* a long article that was a review of the volume of the *Transactions* that he had helped to edit, which included some of his own papers. In this article he also included mention of other recently published geological books and a discussion of the state of geology at the time. Of the then recent additions to geological knowledge Lyell considered the numerous discoveries of fossil remains of greatest interest. These included the skeleton of the extinct Irish elk, the horns of a Scandinavian elk found in Scotland and the remains of elephant, rhinoceros, and hippopotamus found in England. There were also William Buckland's remarkable discoveries of bones in caves. Perhaps the large number of bones of the hyena had been the most striking discovery. This wealth of fossil animals found in the British Isles contrasted sharply with the small number of large mammals living there now. Lyell referred also to Conybeare's discoveries of the ichthyosaurus and plesiosaurus. Buckland's of the megalosaurus, Mantell's of the iguanodon, and Cuvier's of the pterodactyls. Taken together, these discoveries presented a startling array of ancient animals. Lyell remarked:

> None of these fossil plants or animals appear referable to species now in being, with the exception of a few imbedded in the most recent strata; yet they all belong to genera, families or orders established for the classification of living organic productions. They even supply links in the chain, without which our knowledge of the existing systems would be comparatively imperfect. It is therefore clear to demonstration, that all, at whatever distance of time created, are parts of one connected plan.[9]

Thus, for Lyell the emerging fossil record provided a previously missing dimension in time of the plan of the natural universe. He referred to Bishop Butler's *Analogy of Religion,* quoting Butler's statement, "We are placed in the middle of a scheme, *not a fixed but a progressive one.*"[10] Lyell went on:

> Relations between different portions of the system, however distant, are proved sometimes to subsist, and to extend even from extinct to living races of plants and animals. Sources of apparent derangement in the system appear, when their operation throughout a series of ages is brought into one view, to have produced a great preponderance of good; and to be governed by fixed general laws, conducive, perhaps essential, to the preservation of the habitable state of the globe.[11]

[9] Charles Lyell, "Art. IX [review of] *Transactions of the Geological Society of London,* Vol. I, 2d Series. London 1824," *Quarterly Review, 34* (1826), 507–540, p. 538.

[10] *Ibid.*, p. 539.

[11] *Ibid.*

In this passage we see revealed Lyell's emerging sense of a comprehensive natural order existing through time and working out its various manifestations as the result of enduring natural laws and orderly processes. In his geological thinking, he was still in some measure catastrophic for he said of the "strata composing the crust of the globe," "that they have been subject at different and often distant, epochs to violent convulsions."[12] Yet he also added later:

> But in the present state of our knowledge, it appears premature to assume that existing agents could not in the lapse of ages, produce such effects as fall principally under the examination of the geologist. It is an assumption, moreover, directly calculated to repress the ardor of inquiry by destroying all hope of interpreting what is obscure in the past by an accurate investigation of the present phenomena of nature.[13]

In 1827 George Poulett Scrope's *Geology of Central France* was published, and Lyell reviewed it for the *Quarterly Review*.[14] This book, with its magnificent illustrations of the mountains and valleys of the Auvergne and the Ardêche districts of France, with their dramatic formations of basalt, aroused in Lyell a strong desire to visit the area. The attraction of Auvergne was not merely the dramatic quality of its geological formations, but also the fact that through a remarkably long period of geological time it had been a recognizable geographical entity. Its volcanoes represented a wide range of different ages, and one could trace in the area a long succession of natural operations. Thus, in May 1828, Lyell set out on a geological expedition to Auvergne in company with Roderick Murchison. The geological formation that Lyell and Murchison first encountered as they rode southward into Auvergne were the freshwater limestones and marls of the valley-plain of the Limagne, through which ran the Allier River. In the Limagne, they found a very thick series of freshwater sediments that had been laid down in an ancient lake formerly occupying the present valley. There were strata of marl as much as 700 feet thick, and these were usually thinly laminated, a characteristic given to them, wrote Lyell, by "the innumerable thin plates or scales of that small animal called *cypris*, a genus which comprises several species, of which some are recent and may be seen swimming rapidly through the waters of our stagnant pools and ditches."[15] Cypris has two small valves, like the halves of a

[12] *Ibid.*, p. 507.
[13] *Ibid.*, p. 518.
[14] Charles Lyell, "Art. IV. [review of] *Memoir on the Geology of Central France; including the volcanic Formations of Auvergne, the Velay, and the Vivarrais; with a Volume of Maps and Plates.* By G. P. Scrope, F. R. S., F. G. S., London 1827." *Quarterly Review*, *36* (1827), 437–483.
[15] Charles Lyell, *Principles of Geology, being an attempt to explain the former changes of the Earth's surface by reference to causes now in operation* (3 vols.; London: John Murray, 1830–1833), Vol. III, p. 230.

clam shell, which it casts off and renews each year, and the myriads of these valves, cast off each year, appeared to give rise to the paper-thin subdivisions of the marl. The delicate laminations of the marl thus suggested the conditions that had existed in the ancient lake. "A more convincing proof," Lyell wrote, "of the tranquility and clearness of the waters and of the slow and gradual process by which the lake was filled up with fine mud cannot be desired."[16] If each paper-thin layer represented one year's deposition and the old deposit was 700 feet thick, the vast vista of years required to lay down this formation could easily be imagined. The formation also suggested that the conditions in the ancient lake were very similar to those existing in modern lakes. This impression was further strengthened by the thick strata of indusial limestone that they found in Auvergne. This limestone was characterized by indusia or the larval cases of the caddis fly, which is abundant in modern streams and lakes. The ancient fossil caddis fly larval cases, like the modern ones, had many small freshwater shells attached to them, a fact that suggested to Lyell a further analogy between past and present conditions.

Near Aurillac, Lyell and Murchison found a range of marl hills containing an especially thinly laminated marl, the layers being as little as a thirtieth of an inch in thickness. When the layers were separated, Lyell found preserved in each the flattened stems of Chara and other plants, sometimes multitudes of small paludinae and other freshwater shells. "These minute foliations of the marl," he later wrote, "resemble precisely some of the recent laminated beds of the Scotch marl lakes, and when divided may be compared to the pages of a book, each containing a history of a certain period of the past."[17] The freshwater formations of Auvergne, therefore, provided Lyell with detailed analogies to the conditions in his Scottish lakes and suggested calm and uniform conditions extending into the dim distances of geological time.

In Auvergne, Lyell and Murchison also studied the volcanoes and traced the manner in which successive lava flows had filled up river valleys and how the rivers, thus displaced, had proceeded over long periods of time to cut new channels through the layers of lava and basalt occupying their former valleys. They thus saw, demonstrated at first hand, the power of rivers to cut deep gorges through solid layers of rock, not once but several times in the long history of Auvergne.

From central France, Lyell and Murchison went south to Marseilles and thence to Nice, where Murchison fell ill, apparently of a malarial

16 *Ibid.*, p. 231.
17 *Ibid.*, p. 239.

fever, and where they therefore remained for several weeks while he recovered. During this time, Lyell wrote a paper on the excavation of valleys based on their observations in Auvergne, and his enforced leisure gave him an opportunity to reflect on the great wealth of observations that they had made in central France. The analogies he had found between conditions of the past and those of the present so far exceeded his expectations that he was eager to go on and pursue his geological studies further so as to develop a full scheme of such analogies. In this spirit he talked over his plans with Murchison, who encouraged him to develop his ideas to his utmost and suggested he go to Sicily, where, said Murchison, "there are the most modern analogies, volcanic, marine, elevatory, subsiding, etc. I know the island as a soldier, and if you make straight for Etna, will just time it right for work, for the season will be exactly suitable."[18]

When Murchison recovered, they went along the coast into Italy, where, at Turin, Signor Bonelli showed them his large collection of fossil shells from the Subapennine beds of northern Italy and told them that the fossils of the Subapennine beds were distinctly different from those of the Tertiary beds of the hill of Superga near Turin, which resembled more the fossils from the Tertiary beds of Bordeaux in France. Bonelli's information tended to confirm what Lyell had already suspected; namely, that the Tertiary formations were not all of one roughly contemporaneous group, but represented a series of formations formed in successive ages. He believed that marine strata were uplifted from the bed of the sea by earth movements associated with earthquakes and volcanoes. Therefore, he thought that he should find younger and more recently uplifted strata the closer he approached volcanoes that were currently active. At Nice he had studied a marine Tertiary formation in which about 18 percent of the fossil shells belonged to species living in the Mediterranean. The fossils of the Subapennine beds also contained approximately 20 percent of living species, a proportion significantly greater than that of the beds of the hill of Superga or those of Bordeaux, or of the Paris Basin. If he went to the south of Italy and to Sicily, where there were the active volcanoes of Vesuvius and Etna, he hoped to find younger strata with still higher proportions of living species as fossils in them.

At Padua, in September 1828, Lyell took his leave of Murchison and Mrs. Murchison, and traveled alone south through Italy to Naples, examining collections of shells at Parma and Florence on the way. On

[18] Charles Lyell to Gideon Mantell—August 22, 1828. Mantell MSS., Alexander Turnbull Library, Wellington, New Zealand.

his arrival at Naples he went immediately out to the Island of Ischia, where he found, high on the side of the volcanic mountain that forms the center of the island, a stratum of clay containing some 30 species of marine shells, all identical with species living in the surrounding Mediterranean. This discovery astonished and delighted Lyell, for it confirmed beyond his expectations what he had hoped to find. Here were strata so young as to have been laid down within the period of the fauna now living in the Mediterranean, yet elevated through many hundreds of feet in a region of very recent volcanic activity. These conclusions were strengthened even further when he crossed from Naples to Messina and rode southward along the eastern coast of Sicily. After a visit to the Val del Bove and after ascending Etna, he crossed the plain of Catania to the hills of the Val di Noto on his way to Syracuse. In the plateau of the Val di Noto Lyell found a hard, white limestone formation some 800 feet thick. This rock, which was intersected by many volcanic intrusions, was solid, was regularly bedded, and contained only the casts and impressions of fossils. It reminded Lyell of certain hard, white Secondary limestones he had seen in England. Beneath this white limestone was a white calcareous sandstone and a series of blue marls. Lyell did not at first have an opportunity to study the blue marl, but on the morning of his departure from Syracuse he was told of an outcrop of the marl on the south side of the harbor. When he went there, he was able to collect some 30 species of fossil shells and corals, so perfectly preserved as to retain their original colors and all belonging to species living in the Mediterranean. Lyell was astonished to the point of bewilderment. These fossils meant that the assemblage of corals and shells living in the Mediterranean was very old. They had been living there when this marl was deposited and through all that immense period of time during which the sandstone was laid down over the marl and the 800 feet of hard, white limestone had been deposited and consolidated, and in addition, through the further period of time that had been required to raise up this thousand feet or more of rock from the bottom of the sea to form the highlands of the Val di Noto. His observations of the white limestone also meant that there was nothing in the characteristics of the limestone—its hardness or the fact that it contained mere casts or impressions of shells—that required it to be of great age. Previously, he would have associated these characteristics with limestones of the Secondary period and would have considered that their hardness and their containing only the impressions of shells were the consequences of age. Now he saw that these features were, rather, the results of the particular set of conditions under which the sediment had been formed. Lyell wrote later that although he had frequently been reminded of Descartes' precept "that a philosopher should once in his life, doubt everything he had been taught," he was still startled to find:

> a limestone of enormous thickness filled with recent shells, or sometimes with mere casts of shells, resting on marl in which shells of Mediterranean species were imbedded in a high state of preservation. All idea of attaching a high antiquity to a regularly stratified limestone, in which the casts and impressions of shells alone were discernible, vanished at once from my mind.[19]

Lyell also observed both that the mountain of Etna had been built up gradually by many repeated eruptions and that volcanic activity elsewhere in Sicily had been intermittent and spread over a very long period of time. Near Vizzini he saw "a bed of oysters, perfectly identifiable with our common eatable species, no less than 20 feet in thickness . . . resting on a current of basaltic lava; upon the oyster bed again is superimposed a second mass of lava together with tuff or peperino."[20]

From Syracuse Lyell went south to Cape Passaro and then west along the southern coast of Sicily. Near Girgenti he found that the formation that Dr. Daubeny had termed "calcareous breccia" at Girgenti was identical with the sandy limestone of Floridia near Syracuse, the sandy limestone including the series of strata lying just above the blue marl in which Lyell had found such a wealth of perfectly preserved shells and coral. The rock at Girgenti was a soft, calcareous sandstone or conglomerate resting upon the blue clay, but unlike the sandstone at Floridia it was rich in shells. Furthermore, whereas the strata at Syracuse were horizontal, those at Girgenti were upraised and inclined so that from beneath the blue marl appeared strata of white laminated marl. The discovery of these additional strata beneath the blue marl at Girgenti, and the full section of the blue marl there exposed, so much increased the depth of sediments that had been laid down in Sicily as to make Lyell doubt the correctness of his findings. If his correlation of the hill at Girgenti with the horizontal strata at Syracuse were right, sedimentary strata and volcanic formations must have accumulated to a thickness of 2,000 feet since the Mediterranean had become populated with most of the shells and corals now living in it. Alarmed lest he had made some mistake in interpreting the order of superposition of the strata, Lyell turned inland at Girgenti and rode

19 Lyell, *Principles,* Vol. III (1833), pp. x–xi.
20 *Ibid.*, p. 73.

across country to Caltagirone. His purpose was to obtain a new section of the strata of the Val di Noto with their interspersed volcanic rock. From his observations, especially those at Caltagirone, he decided that the blue clay formation must be divided into two quite distinct deposits, the upper being the blue marl so rich in shells at Syracuse; the lower was the white, laminated marl that frequently contained beds of gypsum, sand, sulphur, and sometimes rock salt. He found no fossils in these lower beds until near Raddusa, where he found some fish scales and fish bones. At Castrogiovanni the same fossils occurred in these beds, even more plentifully, but there Lyell found the section that he wanted. Revealed in one immense escarpment rising to a height of 3,800 feet above sea level were all of the strata of the great Val di Noto formation, from the white limestone down through the white laminated marl to a blue clay and gypsum without shells. Since all members of the series were represented at Castrogiovanni, their order was decisively revealed. They also demonstrated the continuity of the strata from east to west in Sicily.

It is difficult to recapture the sense of complete novelty with which Lyell confronted these facts. He saw stretching back to the remote distances of geological time a continuity of living species and a continuity of process. Sediments had accumulated, volcanic eruptions had occurred at long intervals; everything, in fact, had gone on as it was now going on. On his return from Sicily to Italy he wrote to Murchison from Rome of his plans for a new book to deal with the principles of reasoning in geology. These principles, he wrote, "are neither more nor less than that no causes whatever have from the earliest time to which we can look back, to the present, ever acted but those now acting and that they never acted with different degrees of energy from that which they now exert."[21] From Rome he returned through Switzerland into France.

At Paris Lyell learned from Jean Desnoyers that he, too, had come to the conclusion that the Tertiary formations were not all contemporaneous but represented in different areas formations of successive ages. Lyell also conferred with Dr. Gerard-Paul Deshayes, a physician whose hobby was conchology, who had a very large collection of Tertiary fossil shells. When Lyell sought Deshayes' help in identifying and naming his Italian and Sicilian shells, he found that Deshayes, like himself, had formed the idea of classifying the different Tertiary forma-

[21] Charles Lyell to Roderick Murchison — January 15, 1829, Murchison MSS., Geological Society of London. Published in Katherine M. Lyell, *Life and Letters*, Vol. I, pp. 234–235.

tions according to the proportion of living species among their fossils. On this basis, Lyell and Deshayes found that the Tertiary formations fell into four great groups of successive ages. The Tertiary beds of the Paris and London Basins contained less than 5 percent of living species; those of the beds of Touraine and Bordeaux, about 16 percent of living species; those of the Subappenine hills of northern Italy, about a third to a half of living species; and those of southern Italy and Sicily contained an overwhelming majority of living species. Lyell was later to name the geological periods represented by these different formations *Eocene*, *Miocene*, and *Older* and *Newer Pliocene*, respectively. The beds of the Paris and London Basins, which had seemed to be the most recent geological formations known, now appeared at the bottom of a numerous series of formations representing sediments many thousands of feet thick; and the Tertiary period, which had seemed both recent and brief as a geological epoch, now stretched far back in time and assumed an immensity of age almost proportionate with the time represented by the Secondary rocks.

From Paris, Lyell went on to London, where he arrived in February 1829. He immediately set about writing the book that he ultimately completed at the end of June 1830, which appeared as the first volume of the *Principles of Geology*.

Biographical Notes

HANS BAUMGARTEL is the director of the Metallurgical Engineering School, Unterwellenborn, D.D.R. He studied ore dressing at the Bergakademie Freiberg and holds a Ph.D. degree from Karl-Marx University, Leipzig. He is the author of more than 40 published works dealing principally with the history of mining and geology at Freiberg and the work of Lomonosov and von Humboldt.

FRANCK BOURDIER is the Director of the Laboratory of Quaternary Geology and Prehistory (Ecole Pratique des Hautes Etudes) in Paris. He is the author of numerous articles on the history of natural sciences, and he wrote a handbook, *Préhistoire de France* (1967) that includes a history of French prehistoric archaeology.

JOHN G. BURKE is Associate Professor in the Department of History at the University of California, Los Angeles. He has a B.S. in physical metallurgy from the Massachusetts Institute of Technology and an M.A. and Ph.D. in history from Stanford University. Dr. Burke is the author of *Origins of the Science of Crystals* (1966) and several articles on crystallography and mineralogy.

WALTER F. CANNON is Curator of the History of Physics and Geoscience at the Smithsonian Institution. His primary field of research has been the early Victorian period in Britain.

ALBERT V. CAROZZI, a native of Switzerland, was graduated from the University of Geneva with a Dr.Sc. in geology and mineralogy in 1948. He taught sedimentary petrology at the University of Geneva until 1957, when he joined the University of Illinois, where he has been Professor of Geology since 1959. His fields of research include sedimentation and the petrography of sedimentary rocks, on which he has contributed two textbooks, and the history of geology with emphasis on German and French protogeologists. In this latter field, he has published annotated translations of A. G. Werner, J. B. Lamarck, Louis Agassiz, B. de Maillet and R. E. Raspe.

ALLEN G. DEBUS is Professor of the History of Science at the University of Chicago. Trained as a chemist, historian, and historian of science, Dr. Debus has concentrated his research on sixteenth- and seventeenth-century science. He is the author of chemical patents and a frequent contributor to journals specializing in the history of science. Dr. Debus has recently prepared a new edition of a seventeenth-century collection of English alchemical works — Elias Ashmole's *Theatrum Chemicum Britannicum* — and he has edited a biographical dictionary, the *World Who's Who in Science from Antiquity to the Present.* He is also the author of *The English Paracelsians,* a book-length study of the English followers of the Swiss physician-chemist, Paracelsus. Dr. Debus is currently engaged in research for a new book on the development of science in the Renaissance.

R. H. DOTT, JR., is Professor of Geology at the University of Wisconsin. He worked on the Pacific Coast in the petroleum industry and served on active duty at the Air Force Cambridge Research Center in Massachusetts before joining the Wisconsin faculty in 1958. His chief interests are earth history, sedimentology, and the history of geology. Research in orogenic belts stimulated his interest in the history of ideas of earth dynamism.

JOAN M. EYLES (Mrs. V. A. Eyles) is an honors graduate in geology of the Universities of London and Wales. She has nearly completed a biography of William Smith.

V. A. EYLES is District Geologist (retired) of the Geological Survey of Great Britain. Dr. Eyles' study of geology at the University of Bristol was interrupted by war service in France from 1915 to 1919 with the Royal Engineers and the Royal Air Force. As Geologist, 1920 to 1955, he served mainly in Scotland. He is a D.Sc. (Bristol) and was elected a Fellow of the Royal Society of Edinburgh in 1939. He lectured on the history of geology at University College, London, from 1951 to 1956 as part of a course for the M.Sc. in the history of science. He visited the United States in 1956, lecturing by invitation at the University of Illinois.

Dr. and Mrs. Eyles, individually or as coauthors, have published more than 30 papers and articles on the history of geology.

JOHN C. GREENE is Professor of History at the University of Connecticut. His publications have concerned early American science, especially natural history, and the development of evolutionary concepts. He is currently working on a book entitled *American Science in the Age of Jefferson* and has written numerous articles on the early history of science in the United States.

HOLLIS D. HEDBERG is a professor of geology at Princeton University, where he gives a course on the stratigraphic systems that includes the early history of stratigraphy and the development of stratigraphic principles. He has been president of the International Subcommission on Stratigraphic Classification since its inception in 1952 and is the author of numerous papers on stratigraphic classification and terminology.

MELVIN E. JAHN is Macy Fellow in the History of Medicine and the Biological Sciences at the University of California, San Francisco Medical Center. His published writings include an annotated translation of J. B. A. Beringer's *Lithographia Wirceburgensis* and articles on Beringer and Edward Lhwyd, From 1963 through 1965 he held a National Science Foundation grant to conduct research on early paleontology in the Royal Society of London. Current research continues this project as well as a projected series of articles on the history of museums.

SISTER SUZANNE KELLY is Assistant Professor in the General Program of Liberal Studies at the University of Notre Dame. Her A.B. was from Benedictine Heights College, with a major in mathematics; M.A. and Ph.D., University of Oklahoma, in history of science. Her field of concentration was sixteenth- and seventeenth-century cosmology. Sister Kelly has several years' teaching experience as a science and mathematics teacher in high schools. She is the author of *Gilbert's Influence on Bacon: A Revaluation* (1963) and *The De mundo of William Gilbert* (1965).

AURÈLE LA ROCQUE is Professor of Geology at Ohio State University, where he has taught paleontology, historical geology, paleoecology, and history of geology since 1948. He has written some seven books and 70 papers, of which two books and several papers deal with the history of geology, especially the French sixteenth- and seventeenth-century protogeologists. He is a member of several geological, paleontological, and biological societies, as well as the History of Science Society.

ALEXANDER M. OSPOVAT is Associate Professor of History at Oklahoma State University. His special area of research is the life and work of Abraham Gottlob Werner and the influence of the Bergakademie Freiberg on the development of geology. He has visited East Germany three times to do research at the Bergakademie Freiberg and to give papers at meetings sponsored by that school. He is the author of numerous papers on Werner and Werner's influence on the development of geology in America and Europe.

LEROY E. PAGE is Associate Professor in the Division of Natural Science at Monteith College of Wayne State University. His research field is the history of geology in the early nineteenth century, particularly in Great Britain. Dr. Page is the author of several articles on this subject.

RHODA RAPPAPORT is Assistant Professor of History at Vassar College. She studied history of science at Cornell University and received her Ph.D. in 1964. She is the author of articles on the history of geology and chemistry in the eighteenth century.

M. J. S. RUDWICK is Lecturer in the History of Science at Cambridge University, and was previously Lecturer in Geology there. He teaches the history and philosophy of biological and geological sciences, and at present his special area of research is on the structure of controversy in early nineteenth-century geology. He is the author of several papers on the history of geology and paleontology.

CECIL J. SCHNEER, Professor of Geology at the University of New Hampshire, began planning this Conference in 1964. His publications are in physical-chemical mineralogy and the history of science. He currently is working on the connection between the form and the structure of crystals and is also studying the neo-Platonic background to the science of the Renaissance. Dr. Schneer is an Associate Editor of the *Dictionary of Scientific Biography*.

MORGAN B. SHERWOOD is Associate Professor of History at the University of California at Davis and associate editor of *Agricultural History*. He is the author of *Exploration of Alaska* and coeditor of *The Politics of American Science*.

CYRIL STANLEY SMITH, Institute Professor, Massachusetts Institute of Technology, is a metallurgist whose main professional work has been in the microstructure of polycrystalline materials. He is author of *A History of Metallography* (1960) and editor of *The Sorby Symposium on the History of Metallurgy* (1965) and *Sources for the History of the Science of Steel* (1968). He has produced English translations of Biringuccio's *Pirotechnia* and other metallurgical classics.

KENNETH L. TAYLOR is Assistant Professor of the History of Science at the University of Oklahoma. His doctorate is from Harvard University, where he completed his dissertation on the career of Nicholas Desmarest in 1968. In 1965–1966 he was in France doing research for this project.

VLADIMIR VLADIMIROVICH TIKHOMIROV is Doctor of Geological-Mineralogical Sciences, Professor, Membre effectif l'Academie International d'Histoire des Sciences, and head of the Department of the History of Geology in the Geological Institute of the U.S.S.R. Academy of Sciences (Moscow). He carried out investigations in the Caucasus and published works on geological structure, paleogeography, and the history of geological problems (development of the earth's crust and upper mantle). Most of his works are devoted to the history of geological sciences, primarily in Russia. He is the author of more than 150 books and articles and the founder and editor of *Contributions on the History of Geological Sciences*. He is the organizer and President of the Commission on the History of Geological Sciences of the International Union of Geological Sciences.

EUGÈNE WEGMANN *studied in Neuchâtel, Grenoble, Paris, and Trondhjem.* He did research in the Alps under the direction of Emile Argand; went to Norway (1924–1926) and Finland (1927–1931), where he worked with J. J. Sederholm and others, northeast Greenland with Dr. Lauge Koch's expeditions (1932–1935) and southern Greenland (1936–1938). He succeeded Professor Argand in Neuchâtel from 1941 to 1964. He holds degrees from the Universities of Alger, Besançon, Clermont-Ferrand, Grenoble, and Neuchâtel. Professor Wegmann is a member of the Academies of Science of Norway and Göttingen and of the Royal Academy of Arts and Sciences of Belgium. He has written on structural geology, migmatites and granites, tectonic levels, living tectonics, and the history of some geological notions and principles.

GEORGE W. WHITE is Research Professor of Geology at the University of Illinois. Dr. White was formerly Professor of Geology at the University of New Hampshire and Ohio State University and State Geologist of Ohio. He is vice-president for North America of the International Committee for History of Geological Sciences. He is the author of more than 100 papers and bulletins on glacial geology and the history of early American geology. He is editor of *Contributions to the History of Geology*, a series of facsimile reprints of early geological books, published by Hafner Publishing Company. Dr. White is a member of numerous professional societies, including the American and British History of Science Societies.

LEONARD G. WILSON is Professor of the History of Medicine at the University of Minnesota. He has written a number of papers on the history of biology and geology and has completed the second edition of the late J. F. Fulton's *Selected Readings in the History of Physiology*. He is presently working on a biography of Sir Charles Lyell.

JOHN W. WELLS is Professor of Geological Sciences, Cornell University, where he teaches paleontology and stratigraphy. He is interested in the history of geology in New York State. Dr. Wells is the author of *Early Investigations of the Devonian System in New York, 1656–1836* (Geological Society of America, Special Paper 74, 1963), in addition to numerous papers on stratigraphy and the history of geology.

Index